Lecture Notes - Joint EPS-SIF International School on Energy - Course 6 Energy Innovation and Integration for a Clean Environment

EPJ Web of Conferences Volume 268 (2022)

Online
19 – 23 July 2021

Editors:

L. Cifarelli
F. Romanelli

ISBN: 978-1-7138-6617-6

Printed from e-media with permission by:

Curran Associates, Inc.
57 Morehouse Lane
Red Hook, NY 12571

Some format issues inherent in the e-media version may also appear in this print version.

No changes have been made to the content of these proceedings. There may be changes to pagination and minor adjustments for aesthetics.

Printed with permission by Curran Associates, Inc. (2023)

For additional information, please contact EDP Sciences – Web of Conferences
at the address below.

EDP Sciences – Web of Conferences
17, Avenue du Hoggar
Parc d'Activité de Courtabœuf
BP 112
F-91944 Les Ulis Cedex A
France

Phone: +33 (0) 1 69 18 75 75
Fax: +33 (0) 1 69 28 84 91

contact-edps@webofconferences.org

Additional copies of this publication are available from:

Curran Associates, Inc.
57 Morehouse Lane
Red Hook, NY 12571 USA
Phone: 845-758-0400
Fax: 845-758-2633
Email: curran@proceedings.com
Web: www.proceedings.com

TABLE OF CONTENTS

Preface

In order to contribute to the training of young scientists working in the energy sector or intending to do so, the European (EPS) and the Italian (SIF) Physical Societies started the Joint EPS-SIF International School on Energy as a collaborative initiative. The Courses are foreseen to take place on a biennial basis in the beautiful venue of Villa Monastero in Varenna, Lake Como, Italy.

The 1st Course was held in summer of 2012 on "*New strategies for energy generation, conversion and storage*". The 2nd Course in 2014 was devoted to an overview of "*Basic concepts and forefront ideas on energy*", covering the major scientific areas. The 3rd Course was exceptionally held in 2016 at the Ettore Majorana Foundation and Centre for Scientific Culture (EMFCSC), in Erice, Italy, in collaboration with the Materials Research Societies (MRS and EMRS), on "*Materials for energy and sustainability*". The 4th Course was organized in 2017 back to Varenna, with focus on "*Advances in basic energy issues*". In 2019 the 5th Course, titled "*Energy: Where we stand and where we go*", was successfully held again in Varenna.

In 2021, due to the COVID-19 pandemic, the 6th Course of the Joint EPS-SIF International School on Energy, titled "*Energy Innovation and Integration for a Clean Environment*", took place as a virtual course on 19-23 July 2021.

This was a challenge for a School which is typically meant to favour and foster interactions among students and lecturers. Nevertheless, a special online platform was devised with a "get together hallway" and parallel meeting rooms in addition to the main lecture room to allow virtual discussions and encounters among the participants. Overall, this system worked well despite the obvious limitations due to the impossibility of meeting in person and managed to keep all parties connected for the whole week.

The 2021 School's programme matched the primary goal of the School to present all research and development fields with relevance for the technologies of energy production, conversion, transmission and savings. The unique feature of the School actually lies in its multidisciplinarity and interdisciplinarity including basic and applied topics but also climate and economic aspects. This wide scope is essential in order to provide the students with a global insight into the complex nature of energy supply and consumption.

For this purpose, this School effectively brings together a large number of scientists working in different disciplines, but all related to energy technologies with young scientists from all over the world.

The 6th Course of the School gathered nearly 80 participants —lecturers, observers and students— of 10 different nations. The lectures were delivered by about 20 experts in various fields. The concepts and ideas involved in modern energy systems and technologies were clearly addressed by the lecturers who invested every effort to be as didactic as possible. In this way knowledge gaps were filled, and answers were provided to long-standing questions in the mind of the students during the lively discussion-times that followed each lecture. In this way, an online community formed and the basic idea of such a summer School was preserved.

The proceedings of the School, published as Lecture Notes, conserve the teaching material presented and make it available to those who did not attend the School. They serve as a reference book for both specialists working in one of the energy fields but with interests in the status of other energy-related areas, and non-technical readers who want to get a general overview on the involved concepts and techniques and their prospects. Papers are ordered according to topics: global overview of world energy resources, renewable energies, energy storage, nuclear energy (fission and fusion), energy and grids. These proceedings are published both on paper, in a volume of the "*Lecture Notes of the Joint EPS-SIF International School on Energy*" series, and on-line open-access in *The European Physical Journal Web of Conferences* (*EPJ WoC*).

We are especially grateful to our distinguished colleagues and to our "best students" who have all accepted to write and have timely provided their contributions for these Lecture Notes. Our heartfelt thanks go as well to our many sponsors. Finally, we are grateful to the members of the Scientific Committee of the School and to our Scientific Secretary for their dedicated work concerning the programme of this Course and these proceedings, and to the staff of the Italian Physical Society for having made this School possible despite the difficult times of the pandemic.

By the time these Lecture Notes were published, Professor Hermann-Josef Wagner (Ruhr Universität Bochum, Germany), our distinguished colleague and dear friend, unexpectedly passed away. Since the establishment of the School in 2012, his lectures on wind energy have been an extremely valuable tradition and a very strong point of the School. His last lectures in 2021 were recorded and are visible on the web site of the School. He will be sorely missed in the years to come. These Lecture Notes are published in his memory.

L. CIFARELLI and F. ROMANELLI

edp sciences

Web of Conferences

Statement of Peer review

In submitting conference proceedings to *Web of Conferences*, the editors of the proceedings certify to the Publisher that

1. They adhere to its **Policy on Publishing Integrity** in order to safeguard good scientific practice in publishing.
2. All articles have been subjected to peer review administered by the proceedings editors.
3. Reviews have been conducted by expert referees, who have been requested to provide unbiased and constructive comments aimed, whenever possible, at improving the work.
4. Proceedings editors have taken all reasonable steps to ensure the quality of the materials they publish and their decision to accept or reject a paper for publication has been based only on the merits of the work and the relevance to the journal.

Title, date and place of the conference

Lecture Notes of the Joint EPS-SIF International School on Energy
Course 6
Energy Innovation and Integration for a Clean Environment
19 - 23 July 2021, online

Proceedings editor(s):

L. Cifarelli, F. Romanelli

Date and editor's signature

3.Oct.2022

17, avenue du Hoggar - PA de Courtabœuf – BP 112 - 91944 Les Ulis Cedex A (France)
Tél. : 33 (0)1 69 18 75 75 - Fax : 33(0)1 69 07 45 17 – www.edpsciences.org

EPJ Web of Conferences **268**, 00002 (2022) https://doi.org/10.1051/epjconf/202226800002
LNES 2021

Climate change: what we know and what not

HAUKE SCHMIDT

Max Planck Institute for Meteorology - Hamburg, Germany

Summary. — There is ample scientific knowledge about climate change which is assessed at a frequency of about once every seven years by large, international groups of scientists in the comprehensive reports of the Intergovernmental Panel on Climate Change (IPCC). This presentation attempts a short and digestible summary of the certainties and uncertainties existing about the physics of climate change. For my summary, I strongly rely on the contribution of IPCC's working group 1 to the most recent IPCC assessment report that was published in 2021. To set the stage, I first present some basics about climate and climate modeling before summarizing the current state of knowledge and then providing a few personal conclusions.

1. – Introduction

"*It is unequivocal that human influence has warmed the atmosphere, ocean and land.*" This is the first sentence of the Summary for Policy Makers of the Working Group 1 contribution to the Sixth Assessment Report published in September 2021 by the Intergovernmental Panel on Climate Change (IPCC). The IPCC is the United Nations' body responsible for assessing the status of climate science. Working group 1 (WG1) covers the physical science basis which is also the topic of this text. The assessment reports are major documents issued by the IPCC at a frequency of about every 7 years. The latest WG1 report [1] has more than 2000 pages, more than 230 scientists were involved as lead authors, and about 5 times as many as contributing authors. They had to respond to almost 10^5 reviewer comments. Anyone who wants to get an exhaustive overview of "what we know and what not" about the physical basis of climate change science is referred to this report. A less comprehensive version of the assessment is provided by the "Technical Summary" [2], and an even shorter summary which is more accessible for the general public by the "Summary for Policy Makers" [3].

This presentation provides my personal, very incomprehensive view on the matter which is, nevertheless, strongly influenced by the IPCC reports. The title and the content are prompted by reactions to a number of presentations on climate science I have given over the years to a variety of audiences. Often, I sensed surprise about how many details on our changing climate are still unknown. Possibly, I sometimes fall into the trap of talking too much about things we do not know, which are, of course, to a scientist much more attractive than things which are known. Therefore, I think it is important to emphasize from the start that there is exhaustive knowledge on climate change, but at the same time not to brush over the unknowns.

The first sentence of the "Summary for Policy Makers" cited above characterizes knowledge about anthropogenic global warming as "unequivocal". To my knowledge, such a strong statement is made for the first time in an assessment report. The recent IPCC reports use a calibrated language to qualify the certainty or uncertainty of findings (see Box 1.1, [4]). Confidence in evidence is expressed using five levels between "very low" and "very high", the assessed likelihood of an outcome or result is characterized by terms as "very likely", for a probability larger than 90%, or "more likely than not" for a probability larger than 50%. The title of this presentation could be challenged epistemologically. Is knowledge, is certainty about understanding possible at all? As much as possible, I will use IPCC terminology in this presentation to qualify the degree of certainty instead of using absolute statements like "We know ...". I interpret the use of the term "unequivocal" by the IPCC in the sense that there is as much confidence in the existence of anthropogenic global warming as one can have. Uncertainties, however, still exist in particular concerning the magnitude of warming that can be expected in the future, and what global warming means on the regional scale, or, to put it in the words of Marotzke *et al.* [5], "how weather will change with climate".

As confidence in findings is difficult to justify without sound understanding of underlying processes I introduce some climate basics in sect. **2**. Section **3** provides an overview of the current state of climate research in terms of observations, attribution of observed changes, and climate model projections. Some concluding remarks are made in sect. **4**.

2. – Climate basics

2·1. *What is climate? – "Climate is what you expect; weather is what you get."* The American science fiction author Robert Heinlein is often quoted with this aphorism, but variants of it have been discovered in other, earlier works. Funny as it may sound, it contains the essential characteristic of climate being the statistics of weather "*over a long period of time*" [6]. How complex the interactions in the Earth system are that produce the climate has been noted already by Alexander von Humboldt in the 19th century: "*The*

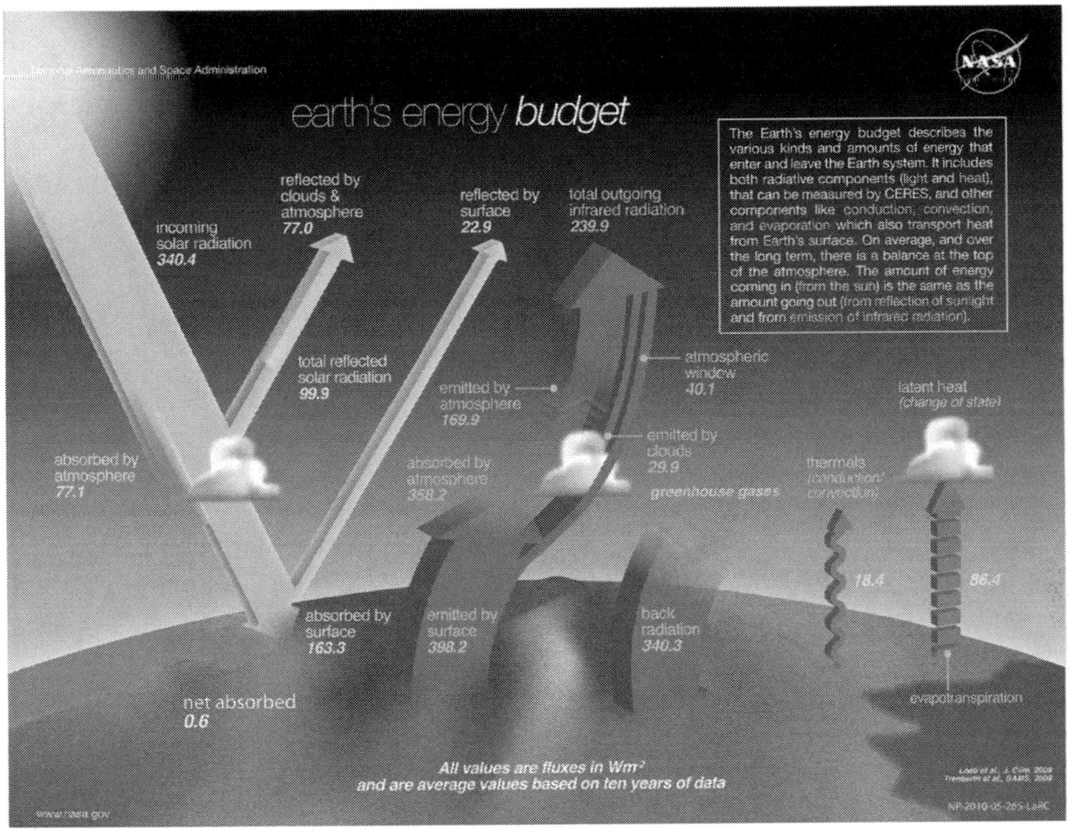

Fig. 1. – Earth's energy budget. (Figure from NASA [9].)

word climate, however, denotes first and foremost a specific property of the atmosphere, but this property depends on the perpetual interactions of a fully and deeply moving, of currents of contrasting temperatures furrowed sea surface with the heat-radiating dry Earth, which is manifoldly structured, elevated, coloured, naked, or covered with woods and herbs." [7] (personal translation from German original). This complexity is what makes understanding and predicting Earth's climate complicated.

2·2. *Earth's energy budget*. – However, central elements of the behaviour of the climate of planet Earth (and more generally of planets with an atmosphere) can be understood from a one-dimensional view on the energy budget of the Earth's atmosphere as presented in fig. 1. On average, the Earth receives about 340 W/m^2 solar radiation. About 30% of it (Earth's albedo) is reflected back to space directly (mostly by the atmosphere and in particular clouds, but partly also by the surface), a bit more than 20% is absorbed by the atmosphere, and almost half of it absorbed at the surface. The surface emits energy in the form of infrared radiation following the temperature dependence of the Stefan-Boltzmann law, but also transfers energy to the atmosphere in the form of latent and sensible heat, *i.e.* basically by evapotranspiration and heat conduction, respectively. Only a small fraction of the emitted infrared radiation reaches space directly. The largest part is absorbed by greenhouse gases (H_2O, CO_2, and others) and clouds in the atmosphere, reemitted in all directions, and partly absorbed at the surface, again. This greenhouse effect is why the Earth is much warmer than one could expect from a planet without an atmosphere absorbing in the infrared. It is also intuitive why changes in greenhouse gas concentrations would change the energy balance, and that clouds are important for it. They do both, cool the surface by preventing solar radiation to reach it, and warm it by preventing terrestrial, infrared radiation to exit the atmosphere. The net effect is a cooling (of about $20\,\mathrm{W/m^2}$), but the effect of individual clouds depends on their environmental conditions defined in particular by their altitude, geographical position, and time of occurrence. For a more comprehensive presentation of the energy budget, see, *e.g.*, Trenberth *et al.* [8].

2˙3. *Climate modeling.* – Numerical climate models are used by scientists to study the complex interactions of climate processes and are the only tool that enable detailed projections of future climates. Today's models are the result of more than half a decade of development efforts which are, of course, ongoing. In the 1960s, attempts were made to represent Earth's energy budget as presented above in numerical models known as one-dimensional radiative convective equilibrium (RCE) models. Earth's troposphere is considered to be in RCE as the energy transfer happens either via radiation or, the mostly upward transfer of latent and sensible heat, via convection. The latter does not play a role in the stratosphere which is to first order in pure radiative equilibrium. A ground-breaking study was published by Manabe and Wetherald [10] who calculated with such a 1D-RCE model that the surface temperature would increase by 2.7 K in a new equilibrium established as a response to the doubling of the atmospheric CO_2 concentration. This quantity is known today as equilibrium climate sensitivity (ECS) and its prediction by Manabe and Wetherald has turned out remarkably robust. Kluft *et al.* [11] repeated their calculation with much higher vertical resolution and a state-of-the-art radiative transfer model and arrived at a warming of 2.8 K. Of course, such models can only be informative about global average climate and neglect a lot of details of the climate system. Manabe and Wetherald, for instance, ignored clouds. However, the ECS predicted by them is in the range of today's observation-based estimates for present-day Earth and also values predicted by today's state-of-the-art climate models (see below). In 2021, Syukuro Manabe was co-awarded the Nobel prize in physics for his contributions to understanding climate change. 1D-RCE models are still used today in particular to better understand dependencies of ECS and aid our analysis of more complex models.

One may argue that 3D climate modeling started in the 1950s with a study by Philips [12] who built on ongoing work for short-term numerical weather predictions. He studied the flow patterns evolving in a global 2-level quasi-geostrophic model with 17×16 horizontal grid points that was started with an atmosphere at rest and simulated a period of about 30 days. Since then, numerical climate models have basically evolved in two directions. Scientists have, on the one hand, added complexity and on the other, increased resolution [13]. Over time, the atmospheric general circulation models have been enhanced by coupling them to land surface processes and to models representing the circulation of the oceans. Since 1995, these coupled atmosphere ocean general circulation models from a variety of climate research centers around the world are being compared in the Coupled Model Intercomparison Project (CMIP). With all models, a set of well-defined numerical simulation experiments is performed that now include, *e.g.*, a long equilibrium run for preindustrial conditions, a simulation of the recent past since 1850, projections of the future following specified emission scenarios, and more idealized experiments, like an instantaneous doubling of the CO_2 concentration. Results of the current phase (CMIP6; [14]) have provided important input for the Sixth Assessment Report (AR6) as the earlier project phases have for earlier IPCC reports. In the fifth phase of the project (CMIP5), for the first time, several models included an interactive representation of the carbon cycle, and since then also the number of models with complex representations of atmospheric chemistry is increasing. Typically, the models of CMIP6 represent the global atmosphere with horizontal mesh sizes of about 100 km. This means that a large number of processes that occur on sub-grid scales cannot be resolved adequately in the simulation and thus need to be accounted for by additional assumptions made in so-called parameterizations. However, with the advent of exascale computing looming, climate modeling is attempting the step change to so-called storm-resolving simulations with mesh sizes of a few kilometers horizontally [15]. A rationale is that at such resolution we can get rid of some of the parameterizations, for instance of atmospheric convection, to base crucial processes like vertical atmospheric energy transport (see above) on laws of physics instead, and thereby increase the fidelity of the simulations further.

3. – The current status of climate research

3·1. *Observations.* – Earth's atmospheric volume fraction of carbon dioxide has increased from a pre-industrial value of about 280 ppm ("parts per million", *i.e.* 280 of one million air molecules were CO_2) to an average of about 416 ppm in 2021 (measured at Mauna Loa, Hawaii). The concentrations of other, largely anthropogenic greenhouse gases like methane have increased, too.

Earth's globally averaged surface temperature has already warmed by about 1.1 °C when comparing the average over the years 2011–2020 with a preindustrial reference (1850–1900) [3]. Of course, there is no direct measurement of this quantity but it is derived from the large number of available surface observations which are, however, inhomogeneously distributed. Nevertheless, the different institutions providing independent analyses arrive at very similar trends. Figure 2 shows the time evolution of global annual mean surface temperatures analyzed by the US National Oceanic and Atmospheric Administration (NOAA). Averaged over all land areas the temperature has increased by approximately 1.6 °C, *i.e.* about twice as much as averaged over all ocean areas [3]. A larger warming over continents is not only due to the thermal inertia of the oceans, but would also be expected in a hypothetical warmer equilibrium, which can be explained to

Fig. 2. – Annual and global mean surface temperature relative to the 20th century average as calculated by the US National Oceanic and Atmospheric Administration (NOAA). (Figure from NOAA [18].)

first order by energy budget considerations over land and ocean [16]. Due to the limited water availability over land, evaporation can contribute less to the cooling of the surface and a new equilibrium would require higher temperatures than over oceans. It is also relatively well understood from largely thermodynamical considerations of feedbacks to warming that high (and in particular Arctic) latitudes warm more than lower latitudes. Pithan and Mauritsen [17] argue that dominant contributions to these feedback differences are related to a larger increase of black-body emissions for unit warming at higher temperatures and the different vertical structures of the warming (causing differences in the so-called lapse-rate feedback) at high and low latitudes. Of course, also the frequently mentioned ice-albedo feedback at high latitudes, *i.e.* less reflection of sunlight by a surface with less ice and snow cover, contributes to the effect.

Besides surface temperature, changes in a large number of other climate parameters are reported in AR6, *e.g.*: Global mean sea level increased by 0.20 m ("very likely" range: 0.15 to 0.25 m) between 1901 and 2018. The average rate of sea level rise has increased over time and reached, with "high confidence", 3.7 (3.2 to 4.2) mm/year between 2006 and 2018. Precipitation averaged of all global land masses has "likely" increased since 1950, with a faster rate of increase since the 1980s. The storm tracks of the mid-latitudes have "likely" shifted poleward in both hemispheres since the 1980s. The Arctic sea ice area in September, *i.e.* when the seasonal minimum occurs, has decreased by about 40% between 1979–1988 and 2010–2019. It is not surprising that with rising mean temperatures it is "virtually certain" that since the 1950s, hot extremes have become more frequent and more intense over most land regions, while cold extremes have become less frequent and less severe. There is "high confidence" that, also since the 1950s, and at least for most land regions with sufficient data availability, the frequency and intensity of heavy precipitation events have increased. The limitation of these statements to sometimes relatively short periods does not necessarily mean that there were no changes earlier, but it could also be that the availability of observations is insufficient to detect trends.

3·2. *Causes of climate change.* – The previous section provides examples for changes in a variety of climate parameters that have been clearly detected according to the IPCC AR6. Many of these changes can be attributed to the anthropogenic increase of greenhouse gases [3]. The attribution of climate change was arguably pioneered by another co-recipient of the 2021 Nobel prize in physics, Klaus Hasselmann, who used the so-called fingerprint method to identify the anthropogenic signal in global mean temperature on the background of the noise introduced by natural variability in the 1990s [19-21]. As many other attribution techniques that have been developed and used since then, this method involves the numerical simulation of factual and counter-factual climate states, *i.e.* states with and without a specific forcing, in this case with and without the anthropogenic increase of atmospheric greenhouse gases. Attribution techniques are today not only used to identify causes for observed trends, but also to estimate how much more or less likely specific extreme events may have become due to anthropogenic forcing. A good starting point for more information on attribution is the AR6 "Cross-Working Group Box: Attribution" [4] which states that "attribution often builds on the understanding of the mechanisms behind the observed changes". Personally, I would emphasise the importance of understanding even more strongly. Today's complex global models, for instance, are used in attribution studies, and, of course, do simulate warming from increasing greenhouse gas concentrations, but they are often questioned due to their opaqueness. For the understanding of global warming, I hence consider the above-mentioned view in terms of the global energy budget as crucial. It may be argued that the 1D-RCE models that basically represent this view are too simple to cover the complexity of the system. My trust in the attribution of global warming to anthropogenic emissions comes from the understanding provided by simple tools and the fact that adding more complexity (three dimensions, many more processes) to models cannot stop them to warm for higher greenhouse concentrations.

However, the magnitude of the warming is still somewhat uncertain. A key parameter frequently used to characterize it is the so-called equilibrium climate sensitivity (ECS) which measures how global surface temperature would change in the long term if the atmospheric CO_2 concentration doubled. Arguably, the Swedish physicist and chemist Svante August Arrhenius was the first to investigate the change in surface temperature as a function of CO_2 concentration [22]. He estimated a warming of about 6 K for CO_2 doubling, but speculated that it might be overestimated. Based on only two climate models, the U.S. National Academy of Sciences estimated in 1979 an ECS of between 1.5 and 4.5 K [23]. Until recently, estimates of ECS and its uncertainty range had changed only slightly, illustrating that this is an aspect of climate science where there was little progress over the last decades. The same temperature range is also found in the IPCC's Fifth Assessment Report. A review by Sherwood *et al.* [24] was able to narrow this range based on historical and paleoclimate data and process understanding. In the current AR6, the "likely" range of climate sensitivity has narrowed to 2.5 to 4 K, with a best estimate of 3 K [3].

Thanks to ongoing model development and longer observation time series, AR6 states even more clearly than previous reports that humans are influencing many aspects of

Earth's climate. As mentioned initially, anthropogenic global warming is, for the first time, called "unequivocal". The "likely" range for the human contribution to the current warming of the Earth's surface since the preindustrial reference period (1850–1900) is given to be 0.8 to 1.3 °C, with a best estimate of 1.07 °C, *i.e.* virtually all of the observed warming is anthropogenic. Furthermore, it is now considered "likely" that humans have contributed to the observed change in precipitation patterns since the mid-20th century, and very likely that they are primarily responsible for the retreat of glaciers since the 1990s and for the decline in Arctic sea ice since 1979 [13]. Concerning extremes, it is now considered "virtually certain" that heat extremes have increased since 1950, and there is "high confidence" in a major human contribution. Some of the extreme heat events of the past decade would have been extremely unlikely without human influence. It is further assessed that the increase in extreme precipitation events over most land areas since the 1950s is "likely" primarily anthropogenic [25]. However, dividing Earth's land masses into 45 regions, 19 show an increase in heavy precipitation events, and none a decrease. But existing attribution studies allow only for two of the 19 regions to have at least "medium confidence" in the increase being anthropogenic [3]. Of course, this does not necessarily mean that the increase is not anthropogenic, but that evidence is limited, yet.

Of course, climate change can also occur due to natural, non-anthropogenic forcings. Large volcanic eruptions are known to cool the planet by injecting sulfur-rich gases into the stratosphere where they build aerosols that scatter sunlight away from the surface, another effect that can be understood to first order in a one-dimensional energy-budget view. The aerosols have a stratospheric lifetime of about a year. The most recent relatively large eruption, the one of Mt Pinatubo, Philippines, 1991, caused a peak global cooling of about 0.4 K. Variability of solar irradiance also has an impact. However, the well-known 11-year solar cycle causes irradiance variations of only about 0.1% and a variability of the global mean surface temperature of less than 0.1 K. On much longer, palaeo-timescales, in particular, orbital variability and atmospheric composition changes have influenced climate fundamentally.

3˙3. *Projections of the future.* – For the CMIP6 simulations that informed AR6, a set of shared socioeconomic pathways, or "SSPs" for short, has been developed that describe different potential future trajectories of developments in global society [26]. In combination with different emission reduction targets, they map how societal decisions affect greenhouse gas emissions. Depending on the model and experiment, the concentrations or emissions of these scenarios enter simulations with complex global climate models which provide the climate projections.

Traditionally, in IPCC reports, the multi-model mean results of the simulations were communicated as most likely climate change resulting from a given scenario. However, as a number of CMIP6 models show very high ECS in comparison to the estimate from observations (see above), the estimate of future warming was determined in AR6 for the first time by combining the model simulations with some bias correction [27]. The spread of model results provides an indication for the uncertainty of the projections. For a very high-emission scenario SSP5-8.5, the increase in near-surface air temperature by the end of the century (2081–2100) relative to the pre-industrial reference period (1850–1900) is given in AR6 as 4.4 K. The "very likely" (5% to 95%) range is between 3.3 and 5.7 K. For the low-emission scenario SSP1-1.9, the corresponding range is 1.0–1.8 K. In the near term (2021–2040), it is "very likely" to exceed the 1.5 °C target from the Paris Climate Agreement in scenario SSP5-8.5, "likely" in scenarios assuming somewhat lower emissions (SSP2-4.5 and SSP3-7.0), and still "more likely than not" ($> 50\%$) in the low-emission scenarios SSP1-2.6 and SSP1-1.9. It should not be forgotten that, as discussed above for the observed warming, a stronger increase than the global mean is also expected in the future over the continental regions (see ref. [27] for all data in this paragraph).

The numbers presented above show that the global mean temperature projected for a given time in the future depends strongly on the emission scenario. However, due to the long lifetime of carbon dioxide and the inertia of the climate system, the warming of the Earth's surface on the time scales of decades to at least a century is determined largely by the total amount of carbon emitted, *i.e.* by cumulative historical emissions, such that very similar temperatures will be reached in different scenarios at the respective times of similar cumulative emissions [3]. Because of this relationship, remaining emission budgets for achieving specific warming targets can be determined. For example, it is estimated in AR6 that there is a 50% chance of reaching the 1.5 °C target or 2 °C target if 500 or 1350 billion tons of CO_2 ($GtCO_2$), respectively, will be emitted globally since the beginning of 2020. Currently, global emissions amount to about 40 $GtCO_2$ per year, meaning that emissions sufficient to cross the target of 1.5 °C will be reached in the early 2030s if global emissions stayed constant.

Hawkins and Sutton [28] argue that the projections of future climate are influenced by three sources of uncertainty: the assumptions made in the scenario, imperfections in the models, and the influence of internal variability. It is clear from the numbers presented above that the uncertainty of global mean temperature at, say, the end of this century, is strongly influenced by the scenario uncertainty. Hawkins and Sutton estimate its contribution on the time scale of a century to about 50%, while it contributes only about 10% for decadal projections. Internal variability is, by far, the largest source of uncertainty on this short time scale, but is less important on longer timescales. Model uncertainty, or one may say model imperfectness, contributes about 1/3 over all time scales, enough to think about how to improve the fidelity of current climate models.

The uncertainty is often larger for other climate parameters than temperature, and in general much larger for regional than global climate change, *i.e.* for changes on spatial scales that matter for people. One example for the inadequacy of present-day climate models and, more fundamentally, of our understanding of some global circulation features incorporated in these models is presented by Shepherd [29]. He presents results from four state-of-the-art climate models for the response of lower tropospheric zonal winds in the North Atlantic region in winter in a high-emission scenario for the end of this century. The results differ widely. Westerlies over the city of Hamburg, for instance, strongly increase in one model, moderately in another, remain almost unchanged in a third, and decrease in the fourth.

Stevens *et al.* [30] present simulations with also four different state-of-the-art climate models in a very idealized setting usually called an aquaplanet, *i.e.* an Earth-like planet where the surface of the planet is covered fully by an ocean. Two simulations are performed which only differ by sea surface temperatures being prescribed globally 4 K warmer in one experiment than in the other. The four models produce responses of tropical precipitation to this global warming which can hardly be imagined more different. Of course, the "true" response to such an experiment is unknown, but the large difference of the simulated responses in such a simple setting creates doubts on the fidelity of projected changes in tropical precipitation simulated for more realistic, and more complicated settings. It is plausible that convection parameterizations, needed to represent vertical energy transport at scales smaller than those resolved by traditional climate models, are at the core of the problem, and that km-scale modeling, as mentioned in sect. **2**˙3 can alleviate it.

Sea level rise is a partly well understood phenomenon, in particular as almost half of the current increase comes from the thermal expansion of water with increasing temperature. For the high-emission scenario SSP5-8.5, models on average predict a global mean sea level rise until the end of this century of a bit less than 1 m compared to the preindustrial reference. However, according to a low-likelihood but high-impact storyline, the increase could also reach almost 2 m, and until 2300, even "sea level rise greater than 15 m cannot be ruled out with high emissions" [3]. The uncertainty in this quantity is to a large degree due to uncertainties related to the evolution of ice sheets and its simulation.

Relatively large uncertainties are also associated with future projections of some extreme weather events. This is partly due to the poor statistics existing for very extreme events which occur at very low frequency, but the modeling of some extremes, as tropical cyclones or floodings in specific valleys may also require very high model resolutions. Again, physical understanding may help to increase confidence in projections. This understanding creates, for instance, high confidence in the projection of a 7% intensification of extreme daily precipitation events on the global scale. Regionally this is often much less clear due to relatively large uncertainties in potential circulation changes as mentioned above for the example of Europe. It is, however, clear that every additional warming increases the intensity and frequency of hot extremes. For some regions, there is high confidence, that, *e.g.*, the temperature of the hottest days will increase 1.5 to 2-fold compared to global mean warming. AR6 also states high confidence in an increase of peak wind speeds of the most intense tropical cyclones (for all changes reported in this paragraph and other extremes, see [3]).

4. – Some concluding remarks

I've started this text citing the IPCC AR6 that anthropogenic global warming is "unequivocal", one may also say that this is something that we know, or at least, something of which the physical climate science is as certain as it can possibly be. As reported above, there are many other aspects of global warming of which scientists are more or less certain, and others with a higher degree of uncertainty. In general, one may say that features related to global thermodynamics are much better understood than possible changes of global circulation patterns. Personally, I think that none of the reported uncertainties justifies to postpone efforts to massively reduce the emission of greenhouse gases. Changes can turn out both less or more severe than best estimates. In the end, decisions cannot and should not be taken by scientists, but societies. The IPCC reports provide a very valuable basis for this decision making. Here, I've only tried to summarize parts of the physical basis of climate change, which is also the subject of the reports of IPCC's working group 1. This is certainly an insufficient input for decision making, as I have not said anything about the impacts of climate change and how to possibly react specifically. These topics are covered by the respective IPCC reports by the working groups 2 (Impacts, Adaptation and Vulnerability) and 3 (Mitigation of Climate Change).

While I do not think that uncertainties in projections provide a justification for postponing emission reduction efforts, they do limit our capacities to adapt to climate change. I find it disconcerting, that the "likely" range of global equilibrium climate sensitivity is still as large as reported (2.5 to 4 K), that we have fairly limited knowledge on how circulation systems in the North Atlantic region or tropical precipitation patterns may change under global warming, and that we cannot rule out extreme sea level rise. The job of scientists is to better understand these phenomena and reduce the uncertainty to enable societies to better prepare. In addition, while we know areas for which the uncertainties are large (the known unknows) it is entirely possible that there are limits to our understanding we are not yet aware of (the unknown unknowns). I think it is very likely that surprises will happen with continuing global warming. Although much warmer and colder periods have occurred on palaeo time scales, we simply have no empirical evidence of how a present-day world that is warmer by 2, 3, or 4 °C may look like. Improved scientific understanding can help to be better prepared for such surprises. As an example, the discovery of the springtime ozone hole over Antarctica in 1985 [31] came as a surprise, but humanity could react to it quickly thanks to global political consensus and a rapidly developed scientific explanation of the phenomenon that would have been unthinkable without the fundamental scientific knowledge about atmospheric ozone accumulated earlier.

* * *

I'd like to thank Paul Keil and Friedrich Wagner for very useful comments on an earlier version of this text, and the organizers of the summer school for giving me the opportunity to participate.

REFERENCES

[1] Masson-Delmotte V., Zhai P., Pirani A., Connors S. L., Péan C., Berger S., Caud N., Chen Y., Goldfarb L., Gomis M. I., Huang M., Leitzell K., Lonnoy E., Matthews J. B. R., Maycock T. K., Waterfield T., Yelekçi O., Yu R. and Zhou B. (Editors), *IPCC, 2021: Climate Change 2021: The Physical Science Basis*, contribution

of Working Group I to the Sixth Assessment Report of the Intergovernmental Panel on Climate Change (Cambridge University Press) 2021.

[2] Arias P. A., Bellouin N., Coppola E. *et al.*, "Technical Summary", in *Climate Change 2021: The Physical Science Basis*, contribution of Working Group I to the Sixth Assessment Report of the Intergovernmental Panel on Climate Change, edited by Masson-Delmotte V. *et al.* (Cambridge University Press, Cambridge, United Kingdom and New York, NY, USA) 2021, pp. 33–144.

[3] Masson-Delmotte V. *et al.* (Editors), "IPCC, 2021: Summary for Policymakers", in *Climate Change 2021: The Physical Science Basis*, contribution of Working Group I to the Sixth Assessment Report of the Intergovernmental Panel on Climate Change (Cambridge University Press, Cambridge, United Kingdom and New York, NY, USA) 2021, pp. 3–32.

[4] Chen D., Rojas M., Samset B. H. *et al.*, "Framing, Context, and Methods", in *Climate Change 2021: The Physical Science Basis*, contribution of Working Group I to the Sixth Assessment Report of the Intergovernmental Panel on Climate Change, edited by Masson-Delmotte V. *et al.* (Cambridge University Press, Cambridge, United Kingdom and New York, NY, USA) 2021, pp. 147–286.

[5] Marotzke J., Jakob C., Bony S., Dirmeyer P. A., O'Gorman P. A., Hawkins E., Perkins-Kirkpatrick S., Le Quéré C., Nowicki S., Paulavets K., Seneviratne S. I., Stevens B. and Tuma M., "Climate research must sharpen its view", *Nat. Clim. Change*, **7** (2017) 89.

[6] WMO, `https://public.wmo.int/en/about-us/frequently-asked-questions/climate` (accessed on April, 21, 2022).

[7] von Humboldt A., *Kosmos – Entwurf einer physischen Weltbeschreibung* (1845-1862).

[8] Trenberth K. E., Fasullo J. T. and Kiehl J., "Earth's global energy budget", *Bull. Am. Meteorol. Soc.*, **90** (2009) 311.

[9] NASA, `http://science-edu.larc.nasa.gov/energy_budget/pdf/Energy_Budget_Litho_10year.pdf`, capture from October, 1, 2019.

[10] Manabe S. and Wetherald R. T., "Thermal equilibrium of the atmosphere with a given distribution of relative humidity", *J. Atmos. Sci.*, **24** (1967) 241.

[11] Kluft L., Dacie S., Buehler S. A., Schmidt H. and Stevens B., "Re-examining the first climate models: Climate sensitivity of a modern radiative–convective equilibrium model", *J. Clim.*, **32** (2019) 8111.

[12] Phillips N. A., "The General Circulation of the Atmosphere: A Numerical Experiment", *Q. J. R. Meteorol. Soc.*, **82** (1956) 123.

[13] Verronen P. T. and Schmidt H., "Numerical models of atmosphere and ocean", in *Earth's Climate Response to a Changing Sun* (EDP Sciences) 2015, pp. 179–185.

[14] Eyring V., Gillett N. P., Achuta Rao K. M. *et al.*, "Human Influence on the Climate System", in *Climate Change 2021: The Physical Science Basis*, contribution of Working Group I to the Sixth Assessment Report of the Intergovernmental Panel on Climate Change, edited by Masson-Delmotte V. *et al.* (Cambridge University Press, Cambridge, United Kingdom and New York, NY, USA) 2021, pp. 423–552.

[15] Satoh M., Stevens B., Judt F., Khairoutdinov M., Lin S. J., Putman W. M. and Düben P., "Global cloud-resolving models", *Curr. Clim. Change Rep.*, **5** (2019) 172.

[16] Byrne M. P. and O'Gorman P. A., "Land–ocean warming contrast over a wide range of climates: Convective quasi-equilibrium theory and idealized simulations", *J. Clim.*, **26** (2013) 4000.

[17] Pithan F. and Mauritsen T., "Arctic amplification dominated by temperature feedbacks in contemporary climate models", *Nat. Geosci.*, **7** (2014) 181.

[18] NOAA National Centers for Environmental Information, *State of the Climate: Monthly Global Climate Report for 2021*, published online January 2022, retrieved on June 10, 2022 from `https://www.ncei.noaa.gov/access/monitoring/monthly-report/global/202113/supplemental/page-1`.

[19] Hasselmann K., "On the signal-to-noise problem in atmospheric response studies", in *Meteorology of Tropical Oceans*, edited by Shaw D. B. (Royal Meteorological Society, Bracknell, UK) 1979, pp. 251–259.

[20] Hasselmann K., "Multi-pattern fingerprint method for detection and attribution of climate change", *Clim. Dyn.*, **13** (1997) 601.

[21] Hegerl G. C., v. Storch H., Hasselmann K., Santer B. D., Cubasch U. and Jones P. D., "Detecting greenhouse gas induced Climate Change with an optimal fingerprint method", *J. Clim.*, **9** (1996) 2281.

[22] Arrhenius S., "On the influence of carbonic acid in the air upon the temperature of the ground", *London, Edinburgh Dublin Philos. Mag. J. Sci.*, **5** (1896) 237.

[23] Charney J. G., Arakawa A., Baker D. J., Bolin B., Dickinson R. E., Goody R. M., Leith C. E., Stommel H. M. and Wunsch C. I., *Carbon Dioxide and Climate: a Scientific Assessment* (National Academy of Sciences Press, Washington) 1979.

[24] Sherwood S. C., Webb M. J., Annan J. D. *et al.*, "An assessment of Earth's climate sensitivity using multiple lines of evidence", *Rev. Geophys.*, **58** (2020) e2019RG000678.

[25] Seneviratne S. I., Zhang X., Adnan M. *et al.*, "Weather and Climate Extreme Events in a Changing Climate", in *Climate Change 2021: The Physical Science Basis*, contribution of Working Group I to the Sixth Assessment Report of the Intergovernmental Panel on Climate Change, edited by Masson-Delmotte V. *et al.* (Cambridge University Press, Cambridge, United Kingdom and New York, NY, USA) 2021, pp. 1513–1766.

[26] Riahi K., van Vuuren D. P., Kriegler E. *et al.*, "The shared socioeconomic pathways and their energy, land use, and greenhouse gas emissions implications: An overview", *Global Environ. Change*, **42** (2017) 153.

[27] Lee J.-Y., Marotzke J., Bala G. *et al.*, "Future Global Climate: Scenario-Based Projections and Near-Term Information", in *Climate Change 2021: The Physical Science Basis*, contribution of Working Group I to the Sixth Assessment Report of the Intergovernmental Panel on Climate Change, edited by Masson-Delmotte V. *et al.* (Cambridge University Press, Cambridge, United Kingdom and New York, NY, USA) 2021, pp. 553–672.

[28] Hawkins E. and Sutton R., "The potential to narrow uncertainty in regional climate predictions", *Bull. Am. Meteorol. Soc.*, **90** (2009) 1095.

[29] Shepherd T. G., "Atmospheric circulation as a source of uncertainty in climate change projections", *Nat. Geosci.*, **7** (2014) 703.

[30] Stevens B. and Bony S., "What are climate models missing?", *Science*, **340** (2013) 1053.

[31] Farman J. C., Gardiner B. G. and Shanklin J. D., "Large losses of total ozone in Antarctica reveal seasonal ClOx/NOx interaction", *Nature*, **315** (1985) 207.

Energy: from yesterday to tomorrow

CHRISTOPH BUCHAL(*)

Forschungszentrum Jülich - Jülich, Germany

Summary. — The extensive global changes and achievements of the Industrial Revolutions combined with the ever growing utilization of plentiful fossil energies are considered as a challenge for today —and even more so for tomorrow. After a short historical review of the technological developments and some remarks on climate, interesting changes of the paradigm of energy supply and usage are explained, including possible future developments. Some of the ideas and the consequences of the new analysis ("Net Zero 2050") of the International Energy Agency are presented and discussed.

1. – Introduction

The value of energy has gone through very different stages of appreciation:

For centuries, energy was rather scarce and of significant value. For example, coal frequently was mined underground in deep pits under terrible conditions, which caused health problems and deadly accidents.

The discovery of plentiful oil changed the picture. The prices dropped and the consumption of oil products soared. Short political crises with reduced supply did not

(*) E-mail: c.buchal@fz-juelich.de

influence the long-term trend of a permanent growth of production. The sufficient availability of oil and gasoline was taken for granted by the public and the positive feeling of ample and safe supplies was reinforced by reliable electrical power from nuclear plants.

Within the last decades of this abundance, public awareness and discussions on climate change due to CO_2 emissions spread and caused a change of acceptance. Negative news on energy use became more frequent and the blessings of energy for the daily life and industry were more or less ignored. While energy was still sufficiently available, oil, coal and natural gas acquired a prevailingly negative connotation of a malicious drug for mankind. All of a sudden, the war in Ukraine and the perspective of radical energy embargoes changed this picture. Facing the consequences of national energy starvation, political leaders and even deeply committed environmentalists had to admit that energy is not a pathogenic drug, but the essential "life sustaining blood" of all modern industrial societies. The prime importance of a reliable, affordable and safe energy supply for life, farming, industry, mobility and information was taught "the hard way". Under the present conditions of imminent shortage and rising prices a reliable energy supply returned into the centre of attention.

Looking back in history, the development of different energy sources and energy systems form the basis of the Industrial Revolutions. Two centuries of fantastic achievements by scientists, engineers and technicians, by tinkerers and nerds have changed our world irreversibly and drastically, and mostly to the very much better.

In the preceding volume of this conference a compact review of the benefits of the Industrial Revolutions and the up-to-date developments has been presented [1]. For more than two centuries Science and Technology had invented new machines, factories, electricity, communication and transport systems. They were fuelled by fossil energies, first coal, thereafter oil and later natural gas. These new fuels greatly empowered the general advancement of mankind in historically unique dimensions.

The most obvious results are

- a strong growth of the world population, supported by improved nutrition, hygiene and medicine,
- in parallel much better health conditions and even a doubling of the average lifespan,
- a rapidly advancing standard of living in most nations,
- a widespread reduction of difficult, hard and dangerous working conditions,
- and, last but not least, much more personal freedom with more time and resources for education, travel and recreation.

The above-mentioned review also covers the basic energy facts of our daily needs [1]. There is little doubt that the energy related emissions into the atmosphere should be reduced as much as possible. Unfortunately this is not easy to accomplish, especially in time of military conflicts and limited international cooperation. Certainly the search for all kinds of improvements of energy systems with lower emissions will dominate the next decades and offer an abundance of new jobs in these fields.

We may add that the classical branches of energy technologies presently are expanded into the new research fields of "Bioenergy". Their aim is a better understanding of the details of photosynthesis, plant growth and the energetic use of biomass. Of high priority remain the improvements of global food supply, animal feed and to some extend also technical biofuels. The associated conflicts are obvious: animal feed, human food and meat consumption on the one hand and on the other hand, the use of bio-energies for cooking, heating and increasingly also as biofuels.

2. – The changing General Narrative of Life, Climate and Energy

What has been on people's mind in earlier times and what is their present mind-set? What are people discussing as important guidelines for life? What motivates families and especially the youth? The common mind-set on important issues within the population of a nation is of strong influence for personal plans and decisions.

For thousands of years mankind had been struggling against the strong forces of nature. The constant fight for survival was simply "natural" and this judgement frequently was confirmed by personal experience. Nature per se does not know any mercy: "Mother Nature" as well as "Mother Earth" sometimes are bountiful, but sometimes dangerous, cruel and deadly. History describes the pre-industrial era as a constant battle against hunger, diseases and premature death [2].

The pre-industrial concept of energy was fairly simple. Energy was needed for heat: daily firewood for cooking, for surviving the cold winters and charcoal for working with metals. Work was performed primarily by human muscles and by some animals as horses or oxen. In addition, water and wind power could be used for mills and ships. Rich nations and wealthy people could enjoy the help of many servants, and frequently even slaves as victims of war.

The invention of the steam engine and the science of thermodynamics slowly, but irreversibly changed the picture. Coal became the most precious source of energy. Coal mines, growing factories and steam trains emerged in the first phase of the industrial revolution. Later followed by electrical power, then by petroleum, cars and airplanes. The modern world became dependent on sufficient supply of energy in the form of coal and oil as the basis of all industrial societies and their every days life.

Looking back, modern industrialization and especially the general availability of energy and electrical power changed the historic narrative of a dangerous life due to the perils of nature towards a growing admiration of our technical achievements. With very good reasons, scientific and technical progress was regarded as a wonderful blessing. Progress opened possibilities of greater personal liberty. The modern and advanced forms of industrialization terminated the wide spread child work, stopped overly hard or slave-like working conditions and permitted general schooling and even university education.

"Electricity", "Smoking chimneys", cars and "Roaring airplane engines" stood for progress and better life. Within the developed rich countries the awareness of the constant battle for survival completely faded out, while the narrative of a safe and better life by superior technology became generally accepted. This optimistic and positive picture of technology and future emerged more strongly than ever in the second half of the 20th century. Cars became very common. Cities were reconstructed for better conditions for automobiles. New highways permitted easy long-distance travel and vacation trips for most people.

In addition, the overwhelming power and impact of the newly discovered nuclear energy became visible for everyone. In the following years, nuclear energy found an important civil application at many nuclear electrical power stations. Therefore, the verdict on energy was generally clear and undisputed:

> *Technological progress is important and desirable. Oil must provide the lion's share of the modern energy supply. The future will be even brighter with friendly, clean and safe nuclear power stations everywhere. Coal and oil should not be burned in the long run, but are needed for the production of the wonders of the chemical industry, for instance the new world of plastics, colours, chemicals and medicines. Electricity will be generated mostly by nuclear energy "too cheap to meter" and should provide ample power for machinery, light, radio, TV, telephone and possibly even the newly emerging art of electronic computing.*

As a highly motivated young student of physics, I was sharing this optimism. Coal in Germany was still wide spread, but had always been rather dangerous to mine, dirty to use, generating very unhealthy smoke, fumes and particles. There was hope to phase out this ugly and cumbersome energy carrier smoothly, at first from all the households, where it was used for heating and cooking.

At the same time, the very hard human labour of farming was replaced by Diesel engines everywhere in the Western world. In the developed nations, farming was increasingly a matter of using huge machines, converting "*Diesel to bread*". The population became used to constantly reliable and successful harvests. This was due to scientific plant breeding and the optimized use of fertilizers and pesticides, provided by the chemical industries.

> *In rich and advanced nations, only very few people have to worry about the availability of their food. Instead, a reliable energy supply at affordable prices for living, housing and mobility has become a matter of general concern and controversies.*

Around 1970 this positive narrative about nuclear energy became somewhat clouded and disputed by the "Anti-Nuclear"-movement. The protesters mistakenly emphasized the dangers of nuclear energy by confusing the obvious dangers of nuclear war with the carefully controlled operation of civil nuclear power plants.

The two major oil crises of 1973 and 1979 fortunately did not result in the expected dreadful effects, but rather boosted new coal-fired and nuclear power stations. Unimpaired nearly everybody continued to enjoy the mobility with new cars. Living quarters spread out into the country, as cars provide convenient means for individual commuting. Combined with a growing tourism industry, a new European network of highways invited to vacations in foreign countries, using the family car. Within the entire Western World and the OECD countries this modern lifestyle coexisted in happy harmony with the advancement of technology and the ample supply of the needed fuels.

2·1. *Some intermediate remarks on the ever changing climate.* – Before discussing the present-day Narrative of Energy, it should be pointed out, that the 20th century was characterized by a mild climate with very few weather catastrophes, especially very few severe river floodings [3]. While such good centuries had happened before, they are in contrast to many previous time periods, which were either too wet, or too dry, or too cold. In fact, the scientific "Climate Narrative" around the Sixties must sound strange to young people of today. Climate discussions were dominated by the fact that the climate of the Earth had strictly followed a periodic pattern during the last million years [4, 5]. Long glacial ("ice covered") and rather short interglacial ("warmer times") periods characterize the "present geological ice age" of our planet. After the last 10000 years of relatively mild climate, science expected the new glaciation arriving soon. Everybody hoped, that this natural cycle would be somewhat delayed and not soon picked up again, because the consequences for all living conditions, especially farming and the food supply of the extremely grown world population would be unimaginably catastrophic. The challenges and dangers of the ever changing weather were well known, and the cold time of the Little Ice Age (1450–1850), the cold winters of the war times or, for instance, the complete freezing of Lake Constance and others in 1963 represented the obvious dangers of a possible climate change —the threat of another change towards lower temperatures. There was no fear of too much warming. Warmer times even would be welcomed, exactly as before during the Roman temperature optimum or the medieval optimum. The general Narrative on Energy remained optimistic, because a sufficient supply of cheap and reliable energy was available and would be most helpful to battle the challenges of a coming colder climate.

As far-fetched as it may sound, one might argue that humans changed the climate already much earlier than generally discussed, because the CO_2 and methane emissions from farming in earliest human history until the Industrial Revolutions of the 20th century probably retarded the natural geological trend towards the expected next glaciation [6]. This observation is completely compatible with the present understanding of the excessively rising emissions and a climate change towards unfavourably high temperatures. Nevertheless, one should keep in mind, that climate science is extremely complex, even considering only the past and the present. The changes of the global climate zones depend on many coupled nonlinear global interactions between them, involving all of the Earth, the Sun and human impacts. Physical models and mathematical simulations may reveal significant trends, but do not allow exact solutions ("predictions") for many decades into the future [4, 5, 7].

3. – A time of more and new fears

Today, the Narrative has changed again. A new fear was arising concerning the possible dangers of global warming due to the observed CO_2-enrichment in the atmosphere. Unfortunately, the previous public expectations for nuclear energy had been shattered by the accidents of Chernobyl (1986) and later Fukushima (2011). The German reaction was especially strange and hasty. In an unexpected political move during the weeks after the Fukushima tsunami, the German government decided to phase out all older nuclear power plants in Germany immediately and to terminate the newest and best of our nuclear plants latest by 2022. By sudden political ruling, more than 35 GW of reliable and CO_2-emission free nuclear electrical power became officially unwanted. Although in 2011 many arguments about the unpleasant consequence of rising CO_2-emissions were already raised, they were not strong enough or they were simply ignored. Instead, *ad hoc* ethical arguments against nuclear power were thought up and the expectation was spread, that a national program for installing renewable power from photovoltaics and wind ("Energiewende") would be able to replace the nuclear energy swiftly —and even satisfy the entire energy demand of Germany.

It seems, that the public easily and strongly reacts to fear and has a hard time understanding numbers and probabilities, balancing risks and benefits in a clear and rational way. Now, in addition to the nuclear fear, the climate fear had entered the public discussion. The IPCC-reports on climate change plus the discussions of possible tipping points of the future climate together with the Paris Agreement (2015) urged termination of the use of fossil fuels rapidly. Otherwise, an uncontrollable temperature rise is feared. Therefore, the global goal of Zero Emissions in 2050 has become a widely used slogan. Until 2022, "Climate Protection and Zero Emissions" dominated the German news and the public showed a great amount of good will towards all related projects, however expensive and inefficient they may turn out to be. The complex and sobering details of physics and technology on the global scale remain of limited interest. After the military aggression of Russia against Ukraine in February 2022, the details of these ambitious renewable energy projects have become even more uncertain and controversial, because the intended embargoes, the different national interests and the limited national potentials are facing enormous global problems.

4. – The analysis of the International Energy Agency (IEA)

The IEA is a subdivision of the OECD. It was founded in 1974 by 16 nations as a reaction to the oil crisis of 1973. The IEA has acquired the reputation of a competent and respected institution. They publish the famous *World Energy Outlook* and always have been concerned about a reliable data base, analysing production and demand as well as providing a realistic forecast. For example, in 2015 the IEA new policies scenario made a prediction of the global energy needs in 2040. Fully aware of the emission problem, they expected a reduction of the fraction of fossil fuels from 85% to 70% until 2040. To achieve this goal, the installation of renewable energies (RE) had to be strongly supported and increased. Nevertheless, the yearly production of fossil energies would keep on growing until 2040. In other words: in spite of the strong worldwide growth of RE production, the IEA predicted that RE would not even satisfy the steady increase of the global demand. This IEA outlook presented little hope for a significant reduction of the coal, oil or gas production until 2040. Of course, this view was in disaccord to the Paris emission goals and the IEA contracted the image of a protector of "yesterday's fossil industries".

In a carefully prepared approach, the IEA was able to bridge the gap between their well-founded scenarios and other possibly overoptimistic climate goals. In May 2021, the IEA "took the bull by the horns" and responded directly to the goals of IPCC. The IEA published a comprehensive new analysis, which looks very different at first glance. Nevertheless, it turns out to be a realistic continuation of their previous work. It is called "NET ZERO by 2050" and presents the emission budgets of three examples of global pathways into the future. This study does not provide any new climate model. Instead, it relies completely on IPCC data and climate studies. The IEA uses the calculated emission data from their own database and scenarios and combines these "energy related emission data" with the climate projections, which have been published in the IPCC reports. The study covers 220 pages, is clearly organized and well worth reading [8].

4·1. *Pathway No. 1: Continue along the presently stated policies (STEPS).* – The STEPS scenario takes into account all policies, which are backed up by existing legislation and ruling. Future "possibilities", hopes and wishes are not considered. Under these circumstances most probably the global emissions will not decrease. Instead, they will remain at about 36 Gt CO_2/year and keep this level until 2050. It is noted, that the strong recession of global travel due to COVID-19 pandemic resulted in a small dip of 2 Gt CO_2/y or approximately 6% only. A recovery after the pandemic is expected. Under these conditions, a global temperature rise of 2.7 °C until 2100 is estimated (50% probability). This scenario ends up far away from the Paris goals and comes close to previously published expectations.

4·2. *Pathway No. 2: All open promises become true (Announced Pledges Case, APC).* – The APC pathway is much more optimistic. It is assumed, that all promises and already-mentioned possibilities are fulfilled completely and in a timely matter. With other words: It is the lowest emission path, which we can hope for under the present political conditions. This path demands significant efforts from all nations and shows that the CO_2-emissions would drop to 22 Gt/y until 2050. Certainly this would be progress —but only half the way to NET ZERO in 2050. The largest contributions to the emission reductions are provided by very high investments into sustainable electrical power generation. The global demand in electrical power doubles. New power grids for underdeveloped countries and areas are installed. The investments in photovoltaics and wind turbines are growing especially fast and in 2050 their production of electrical energy is ten times higher than in 2020. Nuclear power and hydro are growing moderately. A fraction of the remaining coal power plants is fitted with carbon capture and storage (CCS) technology. The total end-energy needs for industry, buildings and mobility will stop growing and remains essentially stable. Also, the mineral oil demand and supply for mobility remains stable, not increasing and not yet decreasing. Although many new cars are running on electricity, the need of mineral oil for cars, trucks, airplanes and ships drops slightly. Hydrogen very slowly becomes a fuel for transportation. It is the authors (ChB) opinion, that this APC case will imply rigorous global changes and regulations, but indeed may describe a realistic option, which one might use for future planning.

4.3. *Pathway No. 3: The NET ZERO EMISSIONS by 2050 Scenario (NZE).* – What would you do, if you were directly confronted with a possibly over-optimistic scenario, which describes the complete abandonment of fossil energies with the claim: "Yes, we can do it!"? The best answer may be to take the optimists by their words and check all the needed requirements and conditions. The IEA has chosen this approach as well and carefully executed the calculations on the basis of the available data and information.

It takes a couple of hours or days to study ref. [7] and its colored graphics in detail. In the following, a few facts are summarized. They provide an excellent guideline for scientific and political discussions and planning, although the actual conditions and goals will be difficult to meet.

The entire scenario is based on very strong international cooperation and needs worldwide implementation. The IEA is very optimistic about the general economy and assumes further economic growth with a doubling of the global GDP and a world population of 10 billion until 2050. In view of the total restructuring of the energy world, the spending on energy investments has to double at minimum —but no investments whatsoever are acceptable for new coal, oil or gas projects. The annual investments in photovoltaics (PV) and wind turbines have to be increased by a factor of 10(!). In parallel, the role of nuclear power becomes more significant with the doubling of the number of nuclear power plants. Thorium technology and fusion are not mentioned. All the remaining coal power plants must be retrofitted with carbon capture and storage (CCS) —or they will be shut down, with serious consequences for the availability of electrical power.

The known oil and gas fields are sufficient to provide the entire remaining demand and the oil companies have to change into different projects of geological exploration, as CO_2 storage as well as the further exploration and search for metals like lithium and other minerals. All nations with large fossil resources have to expect strongly falling sales and revenues. Their cooperation and the reconstruction of their economies will be essential, but remain a challenge. In contrast to the APC scenario, oil will become much less important even for mobility and will be reserved for the chemical industries. The oil price probably will drop by 50% due to the shrinking demand. Of course, this opens frightening prospects for many nations living on oil revenues.

For coal, there is little room left either. Clean coal technologies together with CCS provide the only chance for survival of coal mines and coal power plants. To some degree, even bio-energy as wood shall be burned in power plants, if these are equipped with CCS. This permits an additional direct removal of carbon from the biosphere. The further lowering of the total emitted CO_2 becomes an important element of the overall net zero emission concept.

The NZE scenario demands several immediate decisions and many checkpoints for 2030 are provided as benchmarks for success towards 2050. The final scenario comprises more than 400 milestones until 2050. In summary, it is extremely drastic and overly demanding. The NZE scenario requires the immediate start and implementation of all possible energy *efficiency* programs. In addition, the IEA guidelines demand to forcefully expand all technologies, which are already sufficiently developed and available. For example, the yearly investments and installations of market-ready technologies as PV and wind turbines must quadruple and the sales of electric cars have to grow by more than an order of magnitude. The IEA considers battery-electric cars as the most important change in mobility and ranks fuel cells and other less developed technologies as synthetic fuels as less relevant. Nevertheless in parallel, the research for future concepts as carbon capture and storage (CCS), better batteries, electrolysers and the entire range of hydrogen technologies need to speed up. The further development of immature technologies needs strong support to permit their implementation as soon as possible.

In the IEA scenarios, hydrogen becomes very important. Soon after 2030, huge quantities of hydrogen, produced in electrolysers by renewable electricity, will be needed for the chemical industry, for fuel cell vehicles and for the fabrication of synthetic fuels for aviation and ships. A world market for sustainable fuels (methanol, hydrogen, synfuels, ammonia, *e.g.*) has to be established. New chances are opening up for nations within the sunbelt for large-area PV installations. In many coastal and windy regions, a forest of wind turbines must be erected. The generated renewable electrical energy must be transported by high-voltage dc lines or converted onsite to synfuels by large plants. A world market for sustainable fuels is expected. Again, international cooperation becomes essential in order to bring these investments and technologies into the less developed areas. This supports the demand of the IEA that the developed nations have to strongly support the rest of the world in order to achieve the global climate goals.

Electrical power from environmental friendly sources will become of prime importance everywhere and for every use. In contrast to the German decision, the number and output of nuclear plants is expected to double. Hydropower is greatly favoured, but not available everywhere. Wherever coal plants remain indispensable for supply security of electrical power, they must be coupled with CCS. Will clean coal technology be market ready in time?

What about the human aspect? Will this long list of demands, regulations and drastic changes be acceptable by the general public? Will people change their lifestyle? The IEA suggests intense public involvement, increased awareness and better education on all aspects of this agenda, but remains very cautious about deep cuts in behavioural patterns. In summary, they do not expect more than 4% of the total reductions from direct changes in behaviour. For instance, air travel is not banned, but should not grow further. Rail transport, especially high-speed trains for passengers and freight, should gain importance.

Aside from technology and behaviour, there is a third sector with significant impact. Farming, forestry and land use contribute 25–30% to the greenhouse gas (GHG) emissions. For example, the IEA estimates that a moderate reduction of the meat consumption in countries with the highest levels of meat consumption down to the global average level would reduce the GHG emissions by 1 Gt-equiv. per year.

5. – Conclusion

In the past, the extraction and use of fossil fuels had offered fantastic benefits for mankind. We all became dependent on lots of easily available fossil energy. Today, we are used to the best possible fuels for combustion engines, because Diesel and gasoline are perfect for a planet with oxygen available everywhere in the atmosphere. These fuels have a high-energy density and they are available as long as oil is plentiful. The expected end of the fossil reserves is far in the future, possibly many decades or in the case of coal even two centuries away.

The perspectives and promises of science are reassuring: In the future, nuclear energy plus the radiation from the sun will be sufficient to provide us with all the energy needed. Of course, it will take many devoted scientists and engineers to develop and deploy all the future new technologies and installations needed.

This scientific promise is not exaggerated. Unfortunately, there are problems arising: Do we have sufficient time for a swift and gentle change? Do we have peace and international cooperation for the development of huge wind parks and solar plants in those foreign areas with the optimum harvesting conditions? The goal itself, the complete change of our energy system to zero emissions, including renewable energies and nuclear technologies, remains unavoidable in the long run.

Presently, the ongoing Russian invasion of Ukraine has shattered the plans for the German Energy Turn-Around. One basic "bridge" technology towards renewable energies was the intermediate use of natural gas predominantely imported from Russia to replace coal and nuclear power. Presently, all trade with Russia is reconsidered and a full embargo for natural gas, oil and coal from Russia is discussed. Outcries about the doubling of oil and fuel prices are dominating the public discussions. They shifted the focus away from the emission problems towards the secure availability of energy at acceptable prices. Domestic coal and imports from different suppliers are attractive again and nuclear energy is reconsidered.

The IEA had suggested a peaceful and very tough, but under peace conditions possibly feasible pathway into a Net ZERO Emission Future by 2050. It even expected continuous economic growth —but not automatically for everyone. Instead, it predicts significant hardship for nations, companies and individuals, who work with fossil fuels and depend on them. The most difficult situation is expected for nations with large fossil resources, because they have to stop extracting and selling them. Is it possible to convince these nations to give up their valuable assets? Or will they continue to sell their resources in order to maintain their income, even at much lower prices? [9] Without intense international cooperation and without mutual technical and financial support the NET ZERO proposal finally might turn out to be unrealistic. Considering the fact, that humans unwillingly have changed the climate for millennia by deforestation and farming, even studies of climate engineering should not be excluded and ignored anymore.

* * *

I thank all my colleagues of the Italian and European Physical Societies for their encouragement and efforts to lead this international conference to success as a lively ZOOM-based meeting in times of strict Covid-19 restrictions.

REFERENCES

[1] BUCHAL CH., *The challenge of discussing energy facts and climate change*, in *Energy: Where we stand and where we go*, edited by CIFARELLI L. and WAGNER F., *Course 5, EPS-SIF International School on Energy, Varenna 2019*; All lecture notes are available as pdf in open access in the internet, see `https://epjwoc.epj.org/articles/epjconf/abs/2020/22/contents/contents.html`.
[2] FOGEL R. W., *The Escape from Hunger and Premature Death 1700-2100* (Cambridge University Press) 2004.
[3] PFISTER CH., *Wetternachhersage – 500 Jahre Klimavariationen und Naturkatastrophen* (Verlag Paul Haupt, Bern) 1999.
[4] SCHÖNWIESE CH.-D., *Klimatologie*, 5th edition (Ulmer, Stuttgart) 2020.
[5] BUCHAL CH. and SCHÖNWIESE CH.-D., *KLIMA*, 3rd edition (MIC Verlag, Köln) 2016.
[6] RUDDIMAN W. F., *Plows, Plagues and Petroleum – how humans took control of climate* (Princeton University Press) 2005.
[7] KOONIN S. E., *Unsettled?* (BenBella, Dallas) 2021.
[8] `www.iea.org/reports/net-zero-by-2050`.
[9] SINN H. W., *The Green Paradox – A Supply-Side Approach to Global Warming* (MIT Press) 2012.

EPJ Web of Conferences **268**, 00004 (2022) https://doi.org/10.1051/epjconf/202226800004
LNES 2021

The "Energy Transition": myth or reality?

JONATHAN CRAIG

Eni Natural Resources - Milan, Italy

Summary. — Rapid technological advances across a variety of industries are enabling society's quest for sustainability. The resulting "Energy Transition" is causing an unprecedented shift in how the world's population consumes energy and natural resources, driven by a range of factors including decarbonisation, economics, access to energy, societal expectations around environment and climate, energy efficiency, political and regulatory forces and emerging technologies. The "Energy Transition" will occur at different speeds and shapes in different geographies, depending on local economic and societal pressures. World energy demand is continuing to rise, mainly in the emerging economies of India, China and Africa, driven by population growth and economic development. Many countries are trying to do the "right thing" for their population and their economy and are not necessarily driven by a desire to "save the planet". Maximising income to fund important social and education programmes is often higher in the priorities of some governments than reducing carbon emissions. The energy transition is not as simple as doing what is right for the planet. Governments, businesses and people in fossil fuel producing regions have shown that they will protect their economies despite the threat of environmental harm. The fossil fuel era will not be over soon. Despite the rapid growth in the use of renewable energy resources, in all reasonable scenarios there will still be a need for oil, gas and coal for the foreseeable future —if we are to meet the social and economic expectations of the world's growing population. The energy mix will certainly change as we try to reduce carbon emissions, but the speed and magnitude of that change will depend heavily on national economic priorities, political will and public opinion. How fast can we realistically move towards a low carbon future and still meet the world's growing demand for energy? What are the key constraints on the rate of change? —and what role will geoscientist play in driving and delivering the new energy solutions? The "Energy Transition" is a huge responsibility, but also a huge opportunity, one in which we all have our part to play.

1. – Introduction

Many people see the "Energy Transition" in rather simplistic terms as the "*switch from carbon-based to non-carbon-based fuels*", but there is much more debate about the time scale over which this switch can and, indeed, should occur. Expectations of the general public in this regard vary significantly, ranging from "today", to a few years to a few decades.

One widely accepted definition of "The Energy Transition" is *"a pathway towards transformation of the global energy sector from fossil-fuel–based to zero-carbon by the second half of this century"* (IRENA —International Renewable Energy Agency).

Transition implies change, radical change in this case, and radical change cannot happen overnight or it will leave economic disparities and insecurity. Transitioning from our current energy system, with more than 80% of total energy supply coming from fossil fuels, to a more diverse mix of resources dominated by renewables in just a few decades will require unprecedented effort. Many people do not understand just how difficult it will be, and how it will impact them personally [1]. Our fossil-fuel–based energy system is much more fragile than many realize and any significant disruption of this delicately balanced system will inevitably lead to energy and commodity demand/supply imbalances, market volatility and, unless demand is moderated faster than supply, will result in associated price spikes and, ultimately, increased cost to consumers. The global energy market is currently fragmented, vulnerable and precarious.

The global energy scene is already in a state of flux due to both economic and societal drivers. Large-scale shifts in the global energy system are occurring thanks to rapid deployment of, and rapid reduction in the cost of, major renewable energy technologies and a growing shift towards electricity in energy use across the globe. Many countries are trying to do the "right thing" for their economy and populations, but they are not necessarily driven by a desire "to save the planet". Maximising income to fund important social and education programmes is often higher in the priorities of some governments than reducing carbon emissions. At least 25 countries around the world are dependent on their national oil companies for at least 20% of their total government income, with several relying on oil revenues for as much as 90% [2].

Energy transition is happening —and will continue to happen— but it will take different speeds and shapes in different geographies, given various local economic and societal pressures. In all scenarios, there is still a need for oil and gas for the foreseeable future, and the demand for certain earth metals will increase materially —estimates vary, but a 200% increase in demand for aluminium, iron, lead and nickel and a 1000% increase in demand for cobalt, lithium and manganese have been suggested. It is important to remember that renewable energy and other "cleaner" sources of energy also have environmental impacts. It is not the fuel that is the issue, but the environmental impacts.

2. – Historical energy transitions

The current energy transition is not the first to affect humanity. Until the 18th century, the energy supply for industry was dominated by water-powered mills and the use of wood charcoal. Wood charcoal was used for both iron smelting and glassmaking. In the 16th and 17th centuries water-powered hammers began to be used in the production of iron and brass and water supplied the power for the manufacture of gunpowder and paper, and for saw-milling and knife grinding. In the 18th century, the use of charcoal for iron smelting was gradually replaced by coke, first introduced by Abraham Darby at Coalbrookdale in Shropshire, England, in the 1720s. Patents for the production of coke by heating coal in the absence of air, were granted as early as 1590 and, as a fuel, it was critical to the Industrial Revolution in Britain. There simply was not enough wood to power Britain's demand for iron and steel —coke made British blast furnaces more competitive, reducing the price of steel.

By the late 16th century and early 17th century so much woodland in the Weald area of southern England was being cut down to make charcoal for glass and iron production that it pushed up the price of wood for building and domestic heating, leading to social unrest. Queen Elizabeth I introduced laws in 1559 and 1584 restricting the felling of trees in the Weald to make charcoal for iron production and in 1615 King James I prohibited the use of charcoal in glass making. The switch from charcoal and water to coke and coal represents the first great "Energy Transition" in the industrialised world.

The industrial use of coal began in Scotland in 1575 when a pit was sunk under the Firth of Forth by Sir George Bruce of Carnock of Culross using an artificial island with ships drawing up directly alongside to be loaded with coal. Coal, in turn, led to the development of steam power, first introduced by Thomas Savery and Thomas Newcomen nearly simultaneously in around 1712. It provided the technology for Britain to switch to coal power and for 200 years, steam became the power source of industry and transport worldwide.

The first oil wells were dug in China as early as 347 B.C., but the modern oil industry began in 1847 in Derbyshire when a Scottish Chemist called James Young distilled oil seeping into a local coal mine into a thin lamp oil and a thicker oil for lubricating engines. Twelve years later, in 1859, the world's first commercial oil well was drilled in Titusville, Pennsylvania, ushering in the second great global Energy Transition and the beginning of the "Oil Age", leading, in turn to the development of the internal combustion engine and all the energy consuming technologies of modern times. Drake's discovery came at an opportune time as it coincided with the decline in Whale Oil supplies, due to over-hunting. Whale Oil had previously both stimulated and supplied a growing market for lubricants and as a fuel for lighting, heating and cooking.

What can we learn from these historical energy transitions?

1) Transition certainly is not "instantaneous" —charcoal continued to be used for iron making for at least another 100 years and coal continued to be used for power also for at least a century.
2) Transition was generally not the result of running out of a particular source of energy —it just became economically and/or socially more attractive to use something else because of the environmental impact and/or concerns about sustainability.
3) Often government intervention was required to drive the transition.

4) Ultimately, the transition resulted in huge —and often unexpected— social and commercial changes including the growth of new industrial centres and the collapse of industries based on the older energy source.

3. – The benefits of abundant and affordable energy

Since the early 1900s abundant and affordable energy has fuelled huge improvements in the quality of life of billions of people around the world (fig. 1). Abundant and affordable energy helps speed and simplify basic tasks allowing people (usually women) to have more time for other productive work while lighting homes helps improve education by allowing people to read in their homes after dark. Other benefits include better medical care, refrigeration and the use of air conditioning to reduce heat-related deaths in the "heat islands" of over-crowded inner-city areas. Sustainable Development Goal 7 (SDG7) —one of the 17 goals set out by the United Nations General Assembly in 2015— set the ambition of achieving universal access to affordable, reliable, sustainable and modern energy by 2030.

Today 2.5 billion people still have little or no access to affordable energy and rely on cooking and heating homes by burning wood, coal or dung, fuels that fill their homes and lungs with particulates. In 2012, nearly 25% of the population of 105 developing countries (including 57% of the population of Africa) had no access to electricity, leaving a total of some 930 million people in the world "living in the dark". Access to electricity in sub-Saharan Africa is the worst in the world. According to "Tracking SDG7: The Energy Progress Report", produced by a consortium of organisations including the World Bank, the International Energy Agency and the World Health Organisation, 759 million

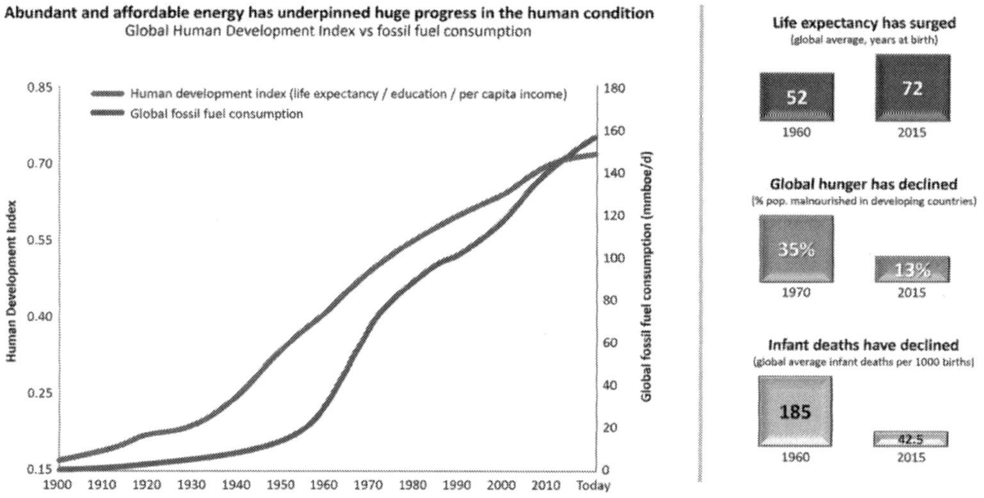

Fig. 1. – Left: energy consumption and quality of life. Right: abundant and affordable energy from fossil fuels has underpinned huge improvements in the quality of life for billions of people around the world for more than a century.

people were without electricity in 2019 and three-quarters of them were in sub-Saharan Africa [3]. More than 70% of health facilities in sub-Saharan Africa have no access to reliable electricity. The limitations that result from a shortage of reliable electrical power range from teachers struggling to conduct lessons in the dark to midwives forced to work with only the weak light from a mobile phone —providing they can find someone with a generator to recharge the phone [3]. Progress in other areas of the world has been much better. By 2019, 94% and 95% of the population in Western Asia and North Africa, respectively, had access to electricity.

Access to energy and energy consumption varies dramatically between different countries (fig. 2). About 3 billion people in the world subsist on the energy equivalent of 2 litres of gasoline a day. People in Europe consume between 7 and 12 litres of gasoline per person per day while, in the U.S. the per capital consumption is 20 litres per day. 90% of households in the United States had air conditioning in 2022 [4]. In India, the annual per capita income (2022) is about 91 000 rupees and the average cost of an air conditioner is about 22 000 rupees or about 25% of total annual income.

Annual per-capita electricity consumption varies from between 4000 and 16000 KWh in much of Europe, and in Australia, the U.S. and Canada to a few 10s of KWh in much of sub-Saharan Africa. A normal domestic fridge consumes 9 times more energy in a year than someone living in Ethiopia!

For the last 200 years, half of humanity has been burning fossil fuels to generate more abundant and affordable energy, and their appetite for energy is still increasing. Meanwhile, the other 3.5 billion people struggle to survive because they do not have access to sufficient or reliable energy, and they aspire to one day enjoy the additional benefits and comforts that energy abundance provides.

Energy poverty is not just a problem in the "developing world". Even in Europe the majority of EU countries have "moderately high" to "extreme" levels of energy poverty

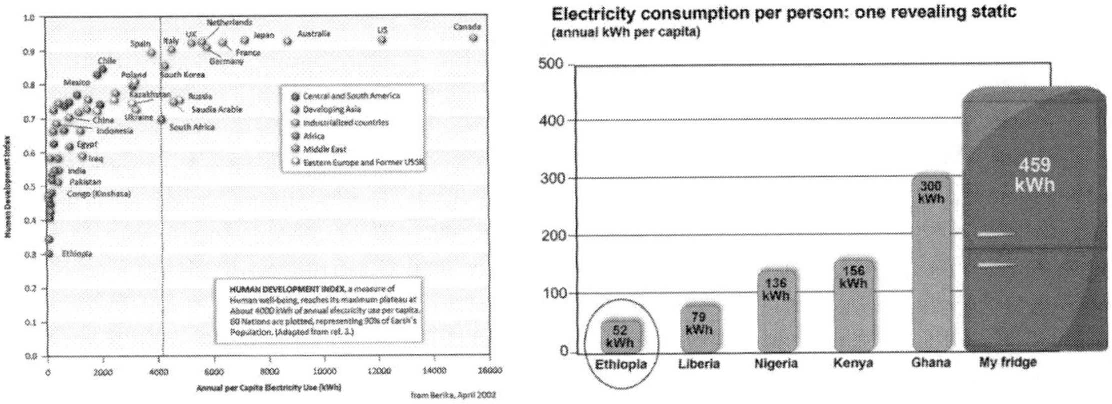

Fig. 2. – Left: human development index *vs.* annual per capita electricity use for 60 nations, representing 90% of the world's population. Right: energy consumption per person (annual kWh per capita) in selected African countries compared with the annual energy consumption of a domestic fridge.

among low-income households (OpenExp, 2019). In some European countries in 2019 up to 22% of households were returning to heating systems based on firewood (20.4%) or coal (1.6%) because of the high cost of other forms of energy, while in one European country 400000 households, primarily in rural areas, were not yet connected to the electricity grid.

Renewable energy sources (particularly off-grid solar power) can make a considerably contribution to alleviating energy poverty in many areas of the world but, in the meantime, the number of coal-fired power stations in Southeast Asia and China continues to grow at an alarming rate as countries in these regions try to keep pace with surging energy demand from their growing populations. Those who live in extreme energy poverty have little capacity to embrace renewable energy sources and for them, at least, decarbonisation is not a priority.

It is estimated that some 2 billion people will still live in energy poverty in 2040 —perhaps 25% of the world's population at that time. How do we lift 2 billion people out of energy poverty while, at the same time minimizing the environmental impacts?

4. – Climate change and greenhouse gas emissions

Humans have emitted about 2500 tons of CO_2 ($GtCO_2$) into the atmosphere since 1850, leaving less than 500 $GtCO_2$ of the remaining carbon budget if we are to keep global warming below 1.5 °C. In other words, the world has collectively burned through 86% of its carbon budget [5]. The U.S. has released more than 509 $GtCO_2$ since 1850 and is responsible for the largest share of historical emissions, about 20% of the global total. China is second with 11%, followed by Russia (7%), Brazil (5%) and a tie between Indonesia, Germany and the U.K. (4% each).

Despite having known about the detrimental effects of climate change since at least the 1980s, nothing much has changed in the last 40 years. Most of us still use fossil fuels to heat our homes and power our cars, we take regular cheap flights and are addicted to energy-consuming technology of mobile phones, wifi and endless TV and radio streaming. Internet use currently accounts for 10% of the world's electricity use and is projected to consume most of the equivalent of the world's entire renewable wind and solar energy production capacity for the next few decades. Air conditioning accounts for another 10% of the world's total electricity demand and is expected to triple by 2050, requiring more new electricity than that already used by the US, EU and Japan combined [6]. Greenhouse gas concentrations in the atmosphere, global sea levels and ocean heat all set record highs in 2021. A pronounced dip in carbon emissions during 2020 during to Covid pandemic was only temporary. Carbon-equivalent emissions from energy, including methane, industrial processes and flaring increased by 5.7% in 2021 [7]. A transition to a cleaner energy is imperative, but global energy demand has quadrupled in just 50 years (fig. 3) and today fossil fuels still supply about 83% of global primary energy demand.

It is important to understand the painful truth that climate change cannot be addressed just with renewable energy —we simply cannot scale solar and wind fast enough to impact emissions at the scale and in the time frame that climate models suggest is necessary without significant technological advances. There are no easy solutions that bal-

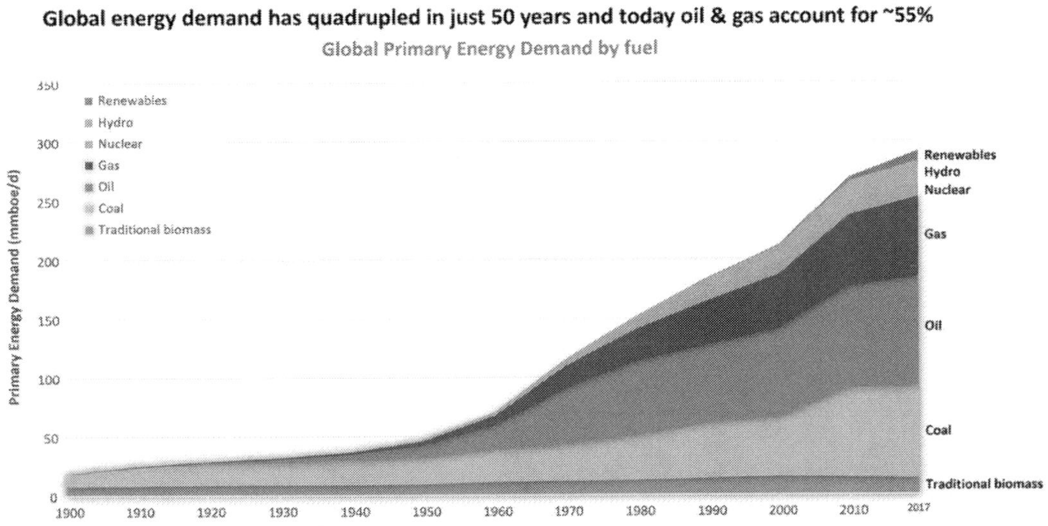

Fig. 3. – Global Primary Energy Demand by fuel type in mmboe/d between 1900 and 2017 (Source: Eni S.p.A.).

ance the growing need for energy with the need to dramatically reduce carbon emissions.

It is clear from almost all analysis that the Accelerated Energy Transition now needed to maintain the global temperature rise to below 1.5 °C will require multiple decarbonisation solutions to be deployed across all sectors of the global economy —power, buildings, industry, transport, plus carbon capture and, potentially, direct air capture of CO_2 (fig. 4). It is also clear that with the currently stated policies the world will not meet

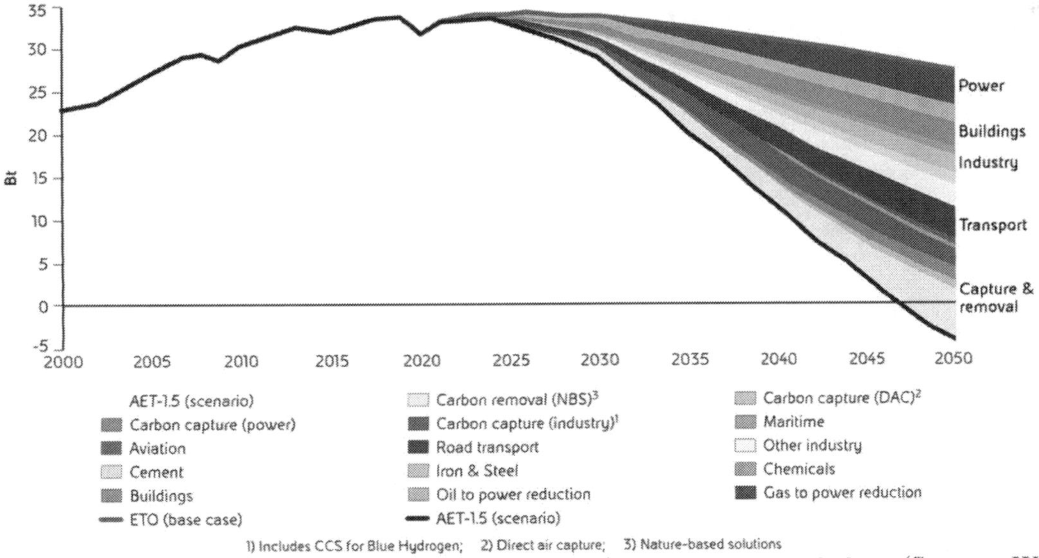

Fig. 4. – We need to deploy multiple solutions to achieve net zero emissions (Source: Wood Mackenzie, 2022 [8].)

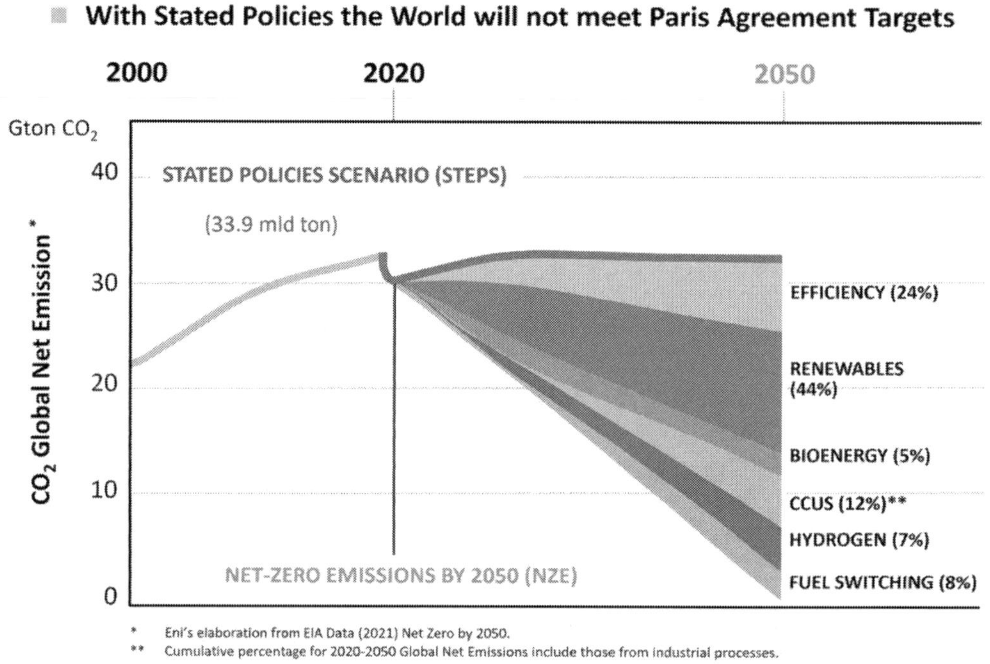

Fig. 5. – Greenhouse Gas Emissions —The Climate Change Challenge (Source: Eni/EIA, 2021).

the Paris Agreement targets. Reaching net zero by 2050 will require further massive reductions through increased efficiency, the use of renewables, bioenergy, CCS, hydrogen and fuel switching (fig. 5).

Different types of oil and gas have different levels of "greenhouse" gas (GHG) emissions (figs. 6, 7). Total lifecycle GHG emissions from gas are ca. 25% lower than GHG emissions

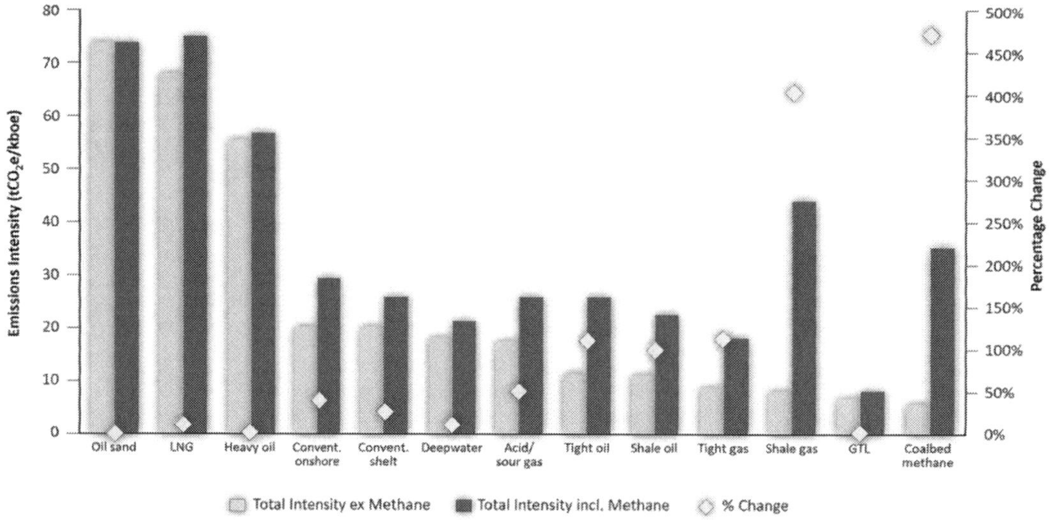

Fig. 6. – Average upstream emissions intensity for different types of oil and gas resource (Source: Wood Mackenzie, 2019).

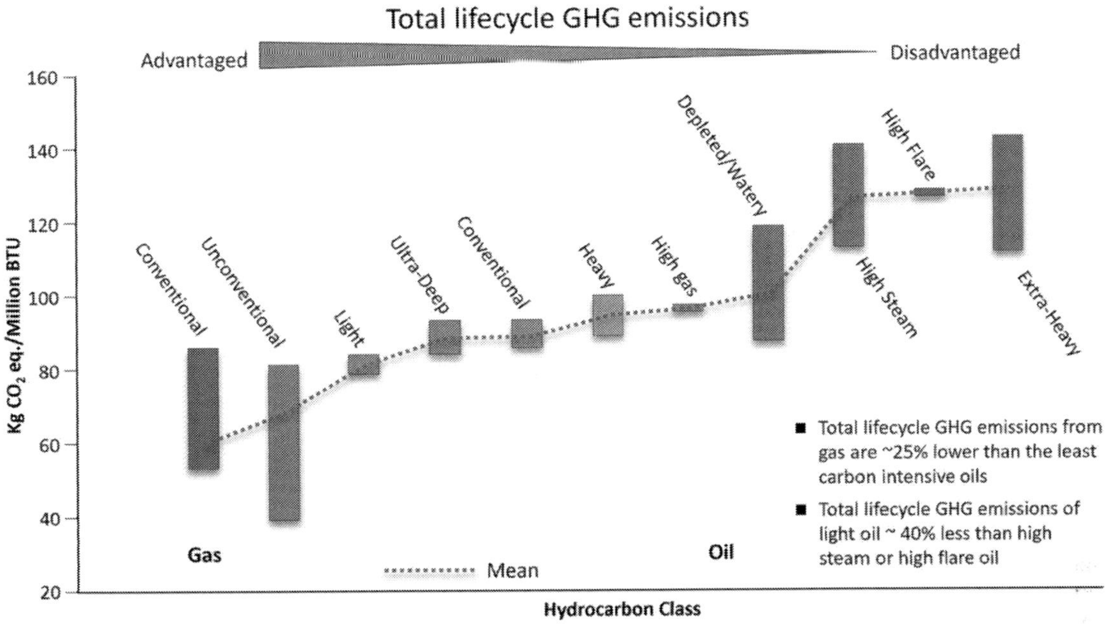

Fig. 7. – Total lifetime GHG emissions for different types of oil and gas resource (Source: Davies and Simmons, 2021 [9]).

from the least carbon intensive oils, while total lifecycle GHG emissions of light oil are at least 40% less than those from oil sands and heavy oils. Liquified Natural Gas (LNG) has high emissions intensity (fig. 6) because of the heavy power requirement for liquefaction, transportation and regasification. In the future we will have to use renewable energy for the liquefaction and degasification and some of the LNG cargo to power the tankers which transport LNG around the world.

Low-cost, low carbon-intensive oil and gas resources (so-called "advantaged" hydrocarbons; fig. 7) will be needed to replace high-cost, high carbon-emission production.

5. – Global energy demand: history and prediction

World energy consumption is continuing to rise, mainly in new emerging economies, where accelerating energy demand is driven by population growth (fig. 8) and economic development. The latest estimates from the United Nations, released in July 2022, are that world population will continue to grow until 2086, peaking at just over 10.4 billion people, somewhat lower than earlier estimates of 13 billion by 2100. The growth is partly the result of declining levels of mortality, particularly in children, and an increase in life expectancy. With rising population, but moderating energy intensities, overall global energy demand will slow. It is expected to increase from 13 Btoe in 2018 to 16 Btoe in 2040 —a growth rate of just 1%, half the rate of the past decade. The global energy mix is changing only gradually and, if this does not change, the world is likely to rely heavily on fossil fuels for decades to come. Most current forecasts suggest that coal, gas and

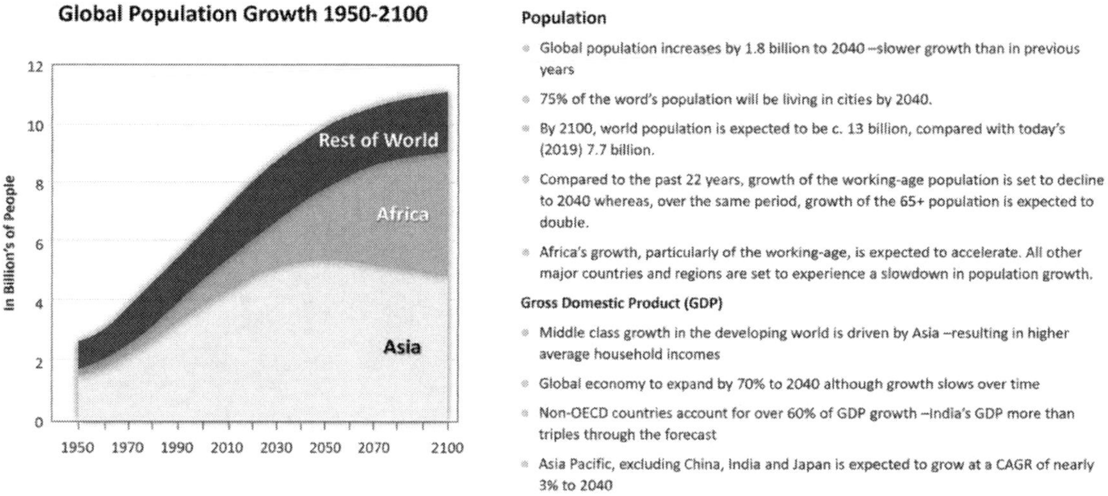

Fig. 8. – Global population growth, 1950–2100 and population and Gross Domestic Product (GDP) trends to 2040.

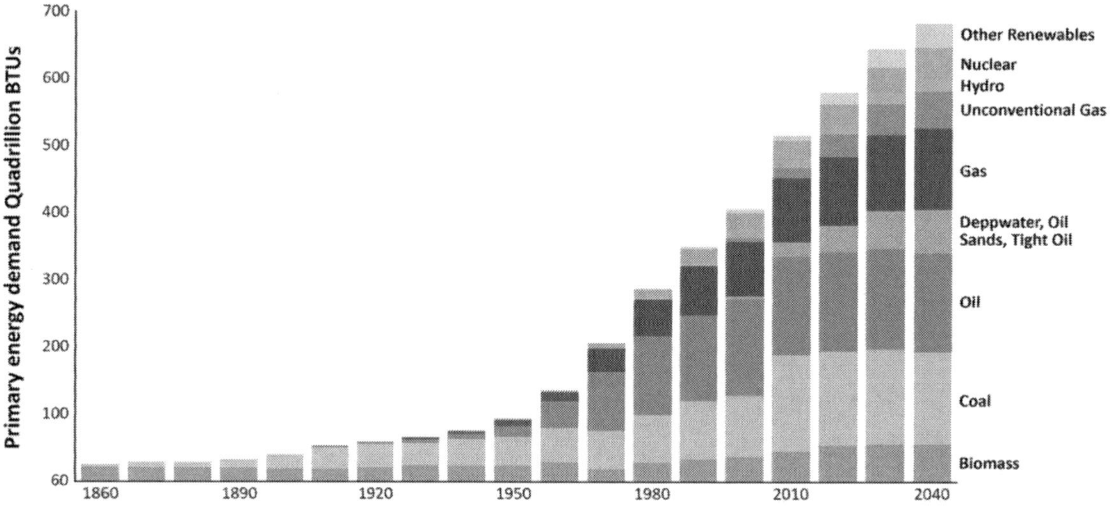

Fig. 9. – Evolution of global energy demand, 1860–2040. Primary energy demand in quadrillion British Thermal Units (BTUs).

oil will still contribute around 85% of primary energy supply by 2040, much the same as today (fig. 9). Availability of resource, infrastructure and cost competitiveness (lack of a realistic carbon price) keeps fossil fuels resilient. Overall, global energy demand is expected to increase by about 25% between 2010 and 2040, due primarily to population growth in emerging economies. Cutting-edge technologies must be employed to meet that demand while simultaneously decreasing greenhouse gas emissions to slow climate change.

Wind and solar will contribute about 25% of global power supply by 2040 compared with about 7% today, but zero-carbon energy (nuclear, hydro, wind and solar) will remain a relatively small part of the overall energy mix. In rural areas, small-scale or off-grid renewable energy systems (mostly solar) are expected to play an increasingly important role, but the infrastructure requires a lot of maintenance. Renewables, excluding hydro-electric, continue to increase their share of global power generation with strong expansion in solar and wind energy. In 2020, approximately 10% of global electricity generation was from renewable sources, but this was at least in part due to lower demand from large coal, oil and gas-fired power stations during "lockdowns" because of the Covid pandemic. Renewables share in global power generation reached almost 13% in 2021, considerably higher than nuclear energy at 9.8% [7].

In 2021, wind energy already formed a significant proportion of total electricity generation in some European countries – Denmark (48%), Ireland (33%), Portugal (27%), Luxembourg (25%), Spain (23%), United Kingdom (21%), Germany (20%) and Greece (20%) [10]. In the United States, wind energy provided 9% of total electricity supply in 2021, with an average levelised cost of $32/MWh, about the same cost as to produce energy from any other source [4]. In the United Kingdom, the cost of electricity generated by offshore wind farms reached an all-time low of £ 37.35/MWh (c. $44.25/MWh) in 2021.

About 22% of electricity in the United States was generated from renewable sources (hydropower, wind, solar) in 2022 and this is expected to reach 24% in 2023. In 2022, the United Kingdom (29%), China (28%) and Germany (22%) accounted for more than 75% of global installed offshore wind power capacity. The 1.2 GW Hornsea Project One in the UK was the world's largest offshore wind farm in 2022. One full rotation of a wind turbine will power a typical home in the west for one full day.

In major OECD markets, wind and solar's share of power output will range from 35% to 50% by 2040, but globally, wind, solar, geothermal and hydro will still only contribute about 10% to global power demand by 2040, largely because we will continually trying to keep up with the rising global energy demand. To achieve net zero by 2050 probably requires that wind and solar energy scale rapidly to account for 61% of a significantly larger power market in which electricity dominates, providing 48% of the world's energy consumption [8]. At present there are significant technological limitations —particularly around long-duration storage— in reaching a fuel mix comprised of 50% or greater share for solar and wind. Longer duration and cheaper storage are the key to high renewables penetration.

How do we meet our future energy needs while, at the same time, reducing carbon emissions from fossil fuels?

6. – World energy growth by region 2015–2035

India and China will lead energy demand growth over the period 2015–50, while China peaks and OECD (Organisation for Economic Cooperation & Development, 36 member states) markets decline. Since 2010, the growth rate in energy demand in the Asia Pacific region has been ca. 35%, while it has been nearly flat in North America and Europe. By 2040, energy demand in the Asia Pacific region will be 3× larger than North America's

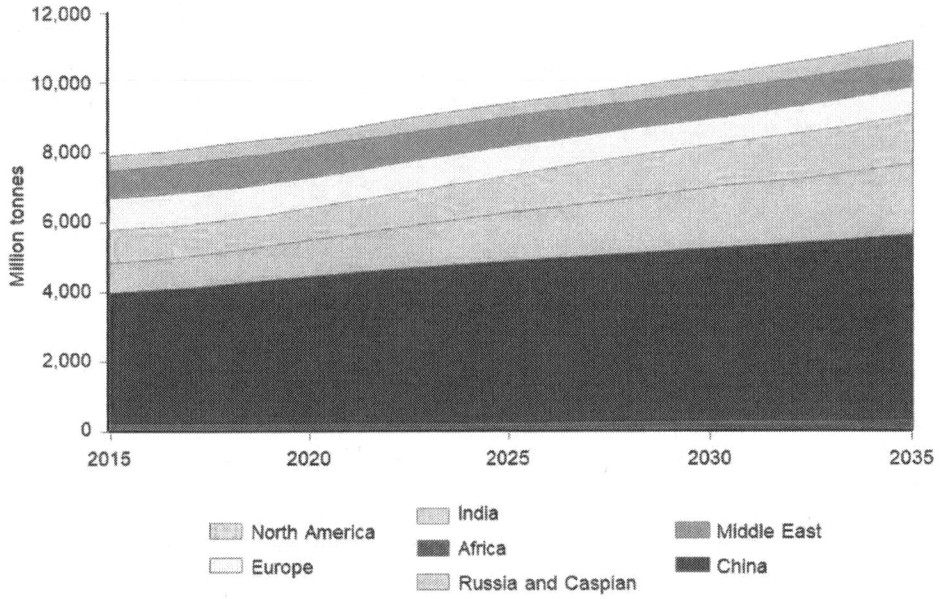

Fig. 10. – World primary energy demand by region, 2015–2035 (Source: Wood Mackenzie, October 2015).

total energy market. About 70% of world energy demand growth to 2040 will be in Asia, driven by China and India (fig. 10).

Four regions of the world —United States, Middle East/North Africa, Russia and China— account for the majority of global hydrocarbon production. Monthly crude oil production in the United States exceeded 10 mbpd in 2019, the highest since 1970. Currently, crude oil, natural gas, and other petroleum products represent 92% of primary energy resources in the United States (EIA, August 2020). The same sources produce 39% of the electricity used in the United States and 91% of the transportation fuel used in the United States is petroleum-derived (EIA, 2019).

In Wood Mackenzie's 1.5 °C "accelerated energy transition scenario" (AET-1.5), oil demand falls from ca. 100 million b/d today to 70 million b/d in 2035 and 35 million b/d in 2050 [8]. Even in this "1.5-degree scenario" world, investment in oil and gas is required as supply from onstream fields declines by about 40 mb/d between now and 2040 —ca. 2 million barrels per year— and low-cost, low carbon-intensive oil and gas resources (so-called "advantaged" hydrocarbons) will be needed to replace high-cost, high carbon-emission production.

The hydrocarbon era will not be over soon. According to a recent McKinsey study, fossil fuel use might flatten from 2035, with oil and coal in decline, but gas use will continue to increase. The great benefit of oil is its portability and its high energy intensity per unit volume. If we could use crude oil without the emissions we certainly would because it is one of the most energy dense forms of energy we have. One barrel of oil contains 5.7 Million BTU of energy —the equivalent of a human working 8 hours a day, 5 days a week for 10 years! (fig. 11).

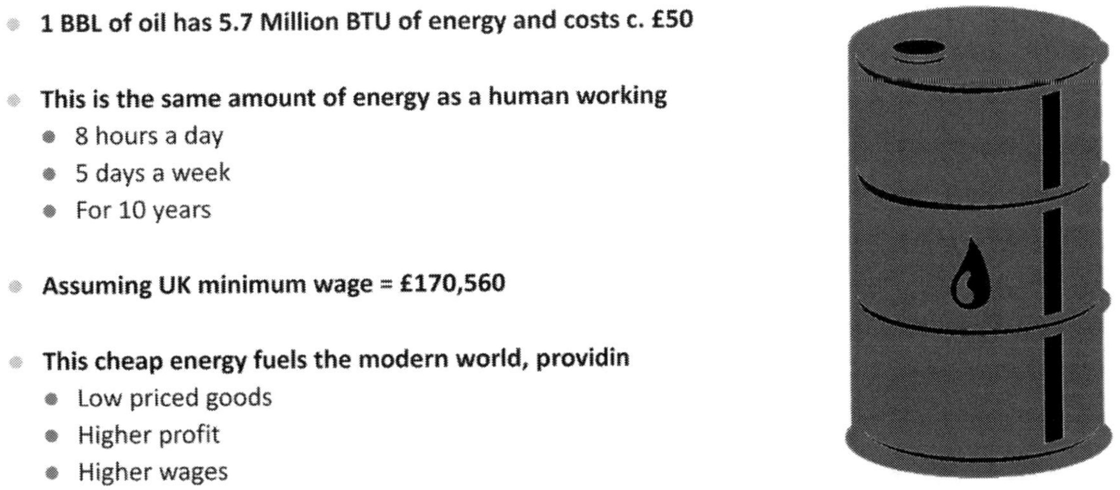

Fig. 11. – The real value of a barrel of oil (Courtesy of Mike Simmons, Haliburton Landmark).

In the next decade, liquids demand growth will be fuelled by chemicals and transport. Today, the world consumes about 100 million barrels of oil per day —the equivalent of about 2 litres of oil for every man, woman and child on the planet every day. We have a serious addiction to oil!

The accelerated energy transition requires more low-carbon energy, sooner and faster. In the early stages we will have to rely heavily on commercially proven technologies — mainly wind, solar and electric vehicles, but hydrogen/ammonia, CCUS, long-duration battery storage, nuclear (Small Modular Reactors), geothermal and bioenergy will all have a role to play. The world cannot just switch off fossil fuels —we have to wait until the low-carbon alternatives and innovation can deliver at scale. Modern society must create a new energy system in which both generation and sustainability are compatible.

Global coal demand is expected to peak in the next decade (some analysts consider it peaked in 2014), largely due to the rapid and intentional displacement of coal with gas as the power sector goes low carbon. In the United States and the United Kingdom, for example, electricity generation is moving away, or has moved away, from coal-fired plants with natural gas and renewable energy sources predicted to meet future demand. However, governments elsewhere must be convinced that gas is a viable and profitable alternative to coal.

While coal use for power generation will fall to zero in some Western European markets, in some other economies, like India, large populations will be connected to the electricity grid using a coal-heavy fuel mix. In 2021 coal-fired power generation reached an all-time record (10 244 terawatt-hours). In early 2022 new coal-fired power plants were under construction in China, India, Indonesia, Turkey, Mongolia, Vietnam, Singapore, Zimbabwe, South Africa, Greece, Bosnia and Hertzegovina, Serbia, Poland, Kazakhstan, Colombia, Brazil and Mexico. Vietnam's energy policy calls for doubling its power production from coal. China, India and Indonesia were the top 3 countries in 2022 in terms of the percentage of their electricity generation that comes from coal-fired power stations [11]. China and India between them accounted for more than 70% of the increase in coal-fired power generation between 2021 and 2022. Currently, about 70% of India's power is coal-based. India's coal-fired power plants are so inefficient that it has been suggested that the government could feasibly shut down 20–30% of them without major disruptions to the country's energy supply. India has introduced a moratorium on all new coal projects from 2027, with solar currently being cheaper than coal in India. At COP26 in November 2021, Indian Prime Minister Narendra Modi pledged to cut India's carbon emissions by one billion tonnes and reduce the carbon intensity of the economy by 45% by 2030 by installing 500 GW of renewable-energy capacity. India remains decades behind the renewable capacity of other economic powerhouses, such as China. India, with a current population of 1.4 billion people, is set to surpass China as the world's most populous country in 2023.

China is investing heavily in renewables, but it will still be almost 50% dependent on coal for power in 2040. The energy mix in China is changing fast —China will invest 343 billion euros between 2017 and 2020 in wind, nuclear, hydro and solar to replace their coal-fired electricity generation. China's ambition is to create 13 million jobs in renewable energy and provide their population with cleaner air. China has also said that it will not build new coal-fired power stations abroad and that it will increase support for green energy projects in developing countries. China is a major producer and importer of coal, but in 2022 Russia discounted its coal prices so much that China began importing Russian coal and stockpiling its own. Before Russia discounted its price, the lowest price coal in the region was $210/ton shipped from the Port of Newcastle in Australia. Russia discounted its price to $150/ton. Indonesia responded to the price war by discounting its coal prices below Russia's [12].

Coal's share in world electricity output reached 41% in 2014 but will decline to 25% by 2040. However, currently strong growth in coal demand in India and other developing markets is partially offsetting declining demand in OECD countries and in China. In the United States CO_2 emissions have gone down due to a surge in shale gas production, which replaces coal. Cheap US coal exported to Asia and Europe is increasing emissions there. U.S. coal exports increased by 61% in 2017 as exports to Asia more than doubled. Meanwhile, the coal industry in the U.S. continues to grow, especially for metallurgical coal used by the steel industry.

Global primary energy production increased by almost 6% in 2021, reflecting the world's post-pandemic recovery. The increase in primary energy production between 2019 and 2021 was almost entirely driven by renewable energy sources. The level of fossil fuel consumption remained largely unchanged between 2019 and 2021, with lower oil demand offset by higher natural gas and coal consumption [7].

7. – Future global oil and gas demand

Global oil demand is expected to peak in the next two decades (ca. 2035) at around 112 million b/d of liquids with a slow decline by 2040 (fig. 12). Technological and

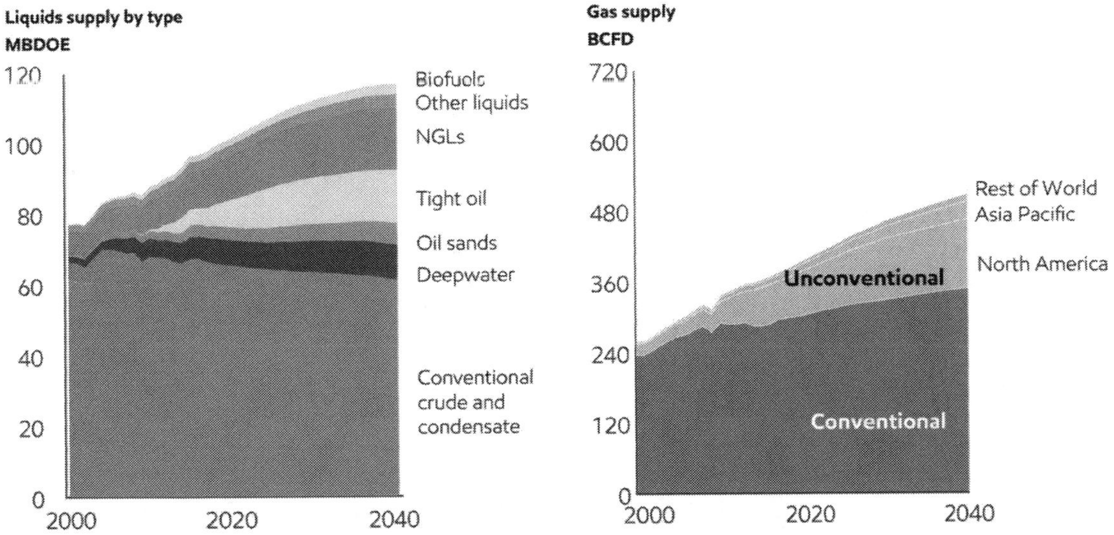

Fig. 12. – Global liquids supply by type in million barrels per day oil equivalent and gas supply by type and region in billion cubic feet per day, 2000–2040 (Source: ExxonMobil, 2017).

economic competition for oil is coming with a shift from oil-based transport to electricity-based transport. The question is "when" rather than "if". Oil's resilience depends on continued petrochemicals demand (the largest proportion of oil demand growth in next two decades) despite moves to address single plastics use. Oil demand is expected to fall in Europe and North America due to fuel efficiency and energy transition (fig. 13), but the International Energy Agency expects global demand for oil to increase from 96.5 MMbopd in 2021 to 104.1 MMbopd in 2026.

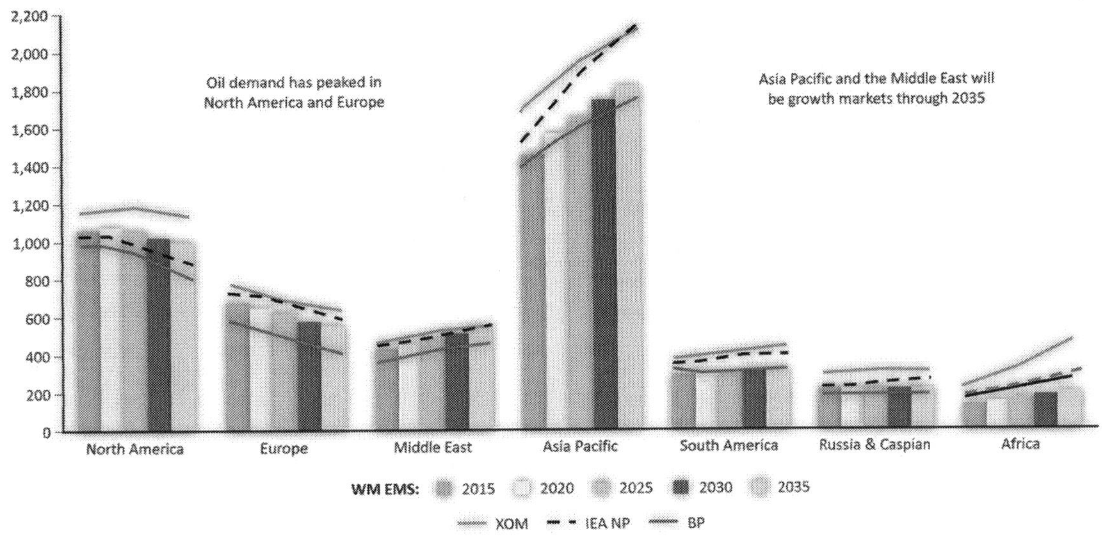

Fig. 13. – World oil demand by region, 2015–2035 (Source: Wood Mackenzie, 2017).

Global gas demand is expected to continue to grow until at least 2040 (fig. 12) with the pace determined by affordability of gas relative to other fuels and technologies — and the policies that governments put in place to tackle climate change. Demand for gas is expected to expand by 30% from 2020–2040. Gas is a foundation fuel for the energy transition —not a bridge— meeting low carbon goals in heating, and a flexible source of power generation. About 35% of the UK's electricity is currently produced by gas-fired power stations, but about 50% of all the gas consumed in the U.K. (including both electricity generation and domestic consumption) is imported. This heavy reliance on gas imports, primarily from Russia and Qatar, raises serious concerns about security of supply, concerns which are amplified by the fact that the UK has little or no gas storage capacity. Following the Russian invasion of Ukraine in 2022, Europe is looking to cease imports of Russian gas completely and is looking to alternative sources of piped gas and LNG. In May 2022, the price of natural gas averaged \$8.14 MMBtu (million British thermal units); exactly one year earlier, it was \$2.91/MMBtu. In June 2022, total European Union LNG imports from the US exceeded imports of gas by pipeline from Russia for the first time [13]. Global LNG demand will grow strongly as indigenous production in major consuming markets either declines or struggles to increase.

Why gas?
When burned cleanly, gas releases about half as much carbon as coal. Cleaner electricity generation means economies can grow and produce lower emissions if they replace coal-generation with gas-generation. In the U.S. gas is now so abundant that producers cannot find enough domestic demand, so they are rushing to export.

But. . . .
Excess gas from oil wells is sometimes burned off by "flaring", releasing carbon dioxide. Gas leaking from pipelines and processing plants adds methane to the atmosphere —methane is 84 times more efficient at trapping heat in the atmosphere than carbon dioxide over a 20-year period. This is currently the "Achilles heel" of the gas industry.

8. – Climate change and the future energy mix

The global energy system needs to move sharply towards a global warming pathway limited to 2 °C or lower, but despite great efforts to reduce costs in renewables, electricity, zero-carbon technologies and advanced transportation —not to mention burgeoning support in governmental polices worldwide— we are still way off track! The global backdrop has become more challenging and more polarized —with less collaboration. Polices being promoted by the EU and smaller economies are helpful, but larger, energy-dense countries and energy-rich segments generally lack any serious progress. The Russian invasion of Ukraine in 2022 has heightened concerns about energy security. Many countries that rely heavily on fossil fuel imports will want to diversify their supply of gas, coal and oil away from Russia to other sources and will place increased effort on developing low-carbon energy sources —more renewables, nuclear and hydrogen— to bolster energy security.

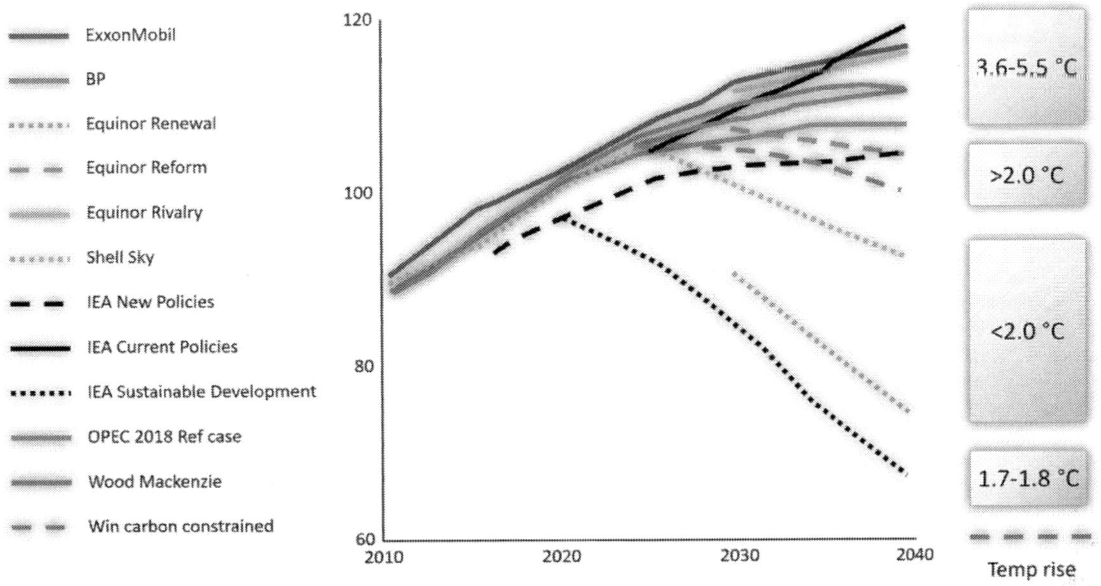

Fig. 14. – Global oil demand predictions in millions of barrels per day, 2010–2040 and the associated projected rise in global average temperature by 2100 (Source: Wood Mackenzie, April 2019).

Most predictions indicate that oil demand is set to rise to 110 million b/d or more by 2040 (fig. 14), but if these demand projections are realised, without equivalent carbon offsets elsewhere, climate science is telling us that the result will be a rise in global temperatures of somewhere between 3.5 and 5.5 °C —which would be catastrophic. Scenarios that work back from a 2 °C limit (IEA Sustainable Development, Shell SKY, Equinor renewable, for example) all infer a peak in oil demand in the 2020s followed by a drop to 70–90 million b/d by 2040 or 65–70 million b/d to achieve a 1.5 °C limit (fig. 14). To reach the International Energy Agency's zero-carbon target requires EV sales, efficiency gains and nuclear capacity at 2–3× most reasonable forecasts and carbon prices in excess of $100/t and probably reaching a global average of $175/t by 2050 [8]. At the peak of the Covid pandemic in 2020, with few planes flying and few vehicles on the roads, the world was still consuming about 80 million barrels of oil per day.

To accelerate the shift to a 2-degree world, governments need to incentivise development of low carbon technologies —decarbonisation of the industrial sector, the rise of hydrogen fuel and CCS are essential technologies to accelerate the energy transition.

Data from the International Energy Agency allows us to envisage two different scenarios for the future of world energy supply (fig. 15):

- A Stated Polices Scenario, based on existing commitments (but with an assumption that not all of these will be met) with a post-Covid recovery followed by increasing supply, with coal, oil and gas essentially unchanged through 2025 to 2040.

 https://doi.org/10.1051/epjconf/202226800004

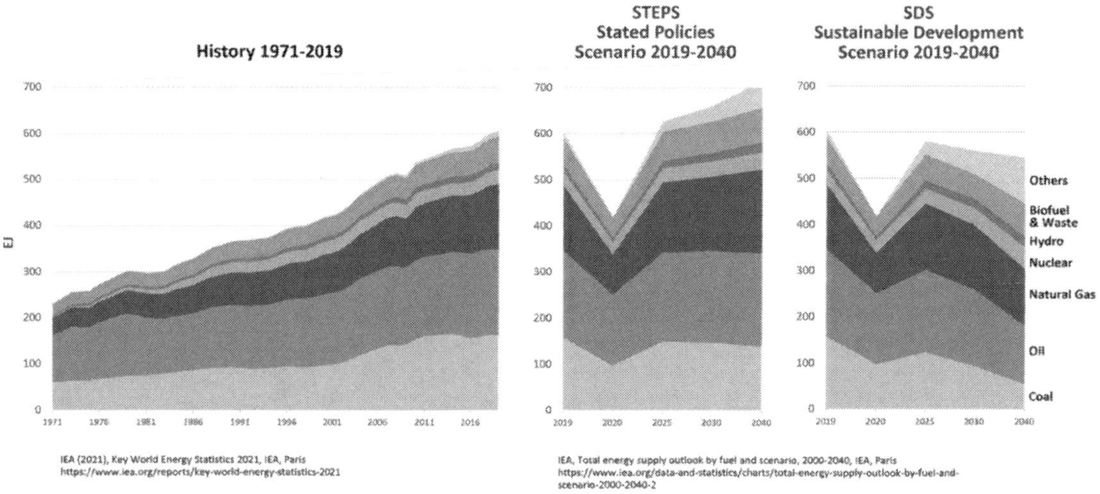

Fig. 15. – World energy supply by source, 1971–2019 and predicted 2019–2040 (Source: International Energy Agency, 2021).

- A Sustainable Development Scenario with a post-Covid recovery followed by a steady decline in overall supply, with coal, oil and natural gas all in decline, and coal "phased down" rapidly between 2025 and 2040.

Overall, the current level of investment in oil and gas is caught between two different visions of the future: it is too high for a pathway aligned with limited global warming to 1.5 °C, but not enough to satisfy rising demand [14].

The rapid increase in energy demand as the world emerged from the Covid-19 pandemic led to a severe demand-supply imbalance and a consequent rapid increase in global oil and gas prices. Between April 2020 and the end of 2021, natural gas prices rose from about \$3-\$4 per MMbtu to about \$40 per MMbtu, a 10-fold increase [15]. The demand-supply imbalance was then further exacerbated by the Russian invasion of Ukraine in February 2022, leading to concerns about energy security and potential gas shortages, particularly in Europe, resulting in further increases in the price of gas in Europe.

There are no easy solutions that balance the growing need for energy with the need to dramatically reduce carbon emissions.

So, how do we meet our future energy needs while, at the same time, reducing carbon emissions from fossil fuels? Will forecasts for sustained oil demand growth "prevail" over scientific warnings of the consequences of that growth?

9. – Why is it so hard to decarbonise?

Some parts of the global economy are more difficult to decarbonise than others. Most projections of global energy demand indicate significantly increasing demand for energy for electricity generation, industrial use and transportation (*e.g.* fig. 16). Only in the residential/commercial sector is the demand for energy expected to remain rel-

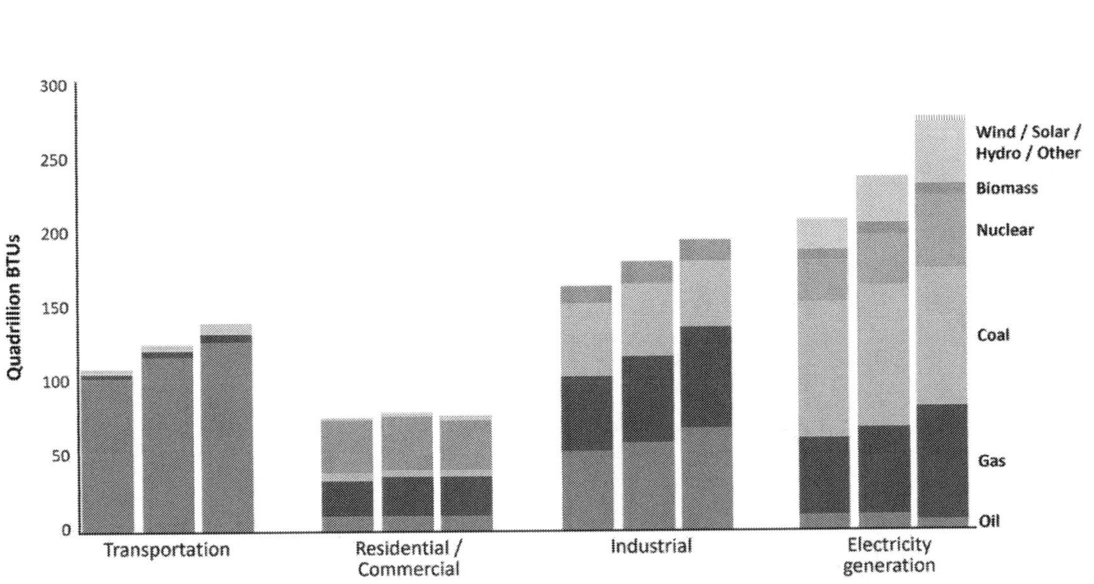

Fig. 16. – Global energy demand by sector, 2016–2040 (Source: ExxonMobil, 2017).

atively flat, partly due to the expectation of improved efficiency in the insulation of buildings. Heating remains the largest source of greenhouse gas emissions in the UK. Most of the heating in our buildings and industries is delivered by fossil fuels; natural gas remains the predominant source of heating for most customers connected to the grid.

Renewable energy sources are expected to make a significant contribution to energy demand in the future, but by 2040 oil and gas are still expected to account for about 30% with the proportion of oil declining and gas increasing.

There is predicted to be very little contribution of renewable energy to the industrial sector by 2040 as this is primarily feedstock for petrochemicals and industrial processes. Crude oil is a vital raw material used in the manufacture of everything from mobile phones to fabrics, lubricants, agricultural chemicals, pharmaceuticals, and many other items (fig. 17). Wind turbines will not work without durable hydrocarbon-based coverings for the blades and electric vehicles will not achieve driving ranges exceeding 300 miles without high strength and lightweight plastics. Each electric vehicle requires 772 pounds/c.350 kg of petrochemicals for its paint, vents, headlights, mirrors, tires, handles, seats, computers, dashboard, trim, body and steering wheel. About 1 billion barrels of oil will be required for the petrochemicals needed to make the 300–500 million new electric vehicles expected to be manufactured over the next 20 years [16].

The petrochemicals industry is growing extremely fast (+6% per annum in 2022), primarily due to increasing demand in the construction, textile, medical, pharmaceuticals, consumer goods, automotive and electronics industries. Asia is the fastest growing petrochemical market. The highest demand is for ethylene, propylene and benzene, used for packaging, electronics, plastics and rubber [11].

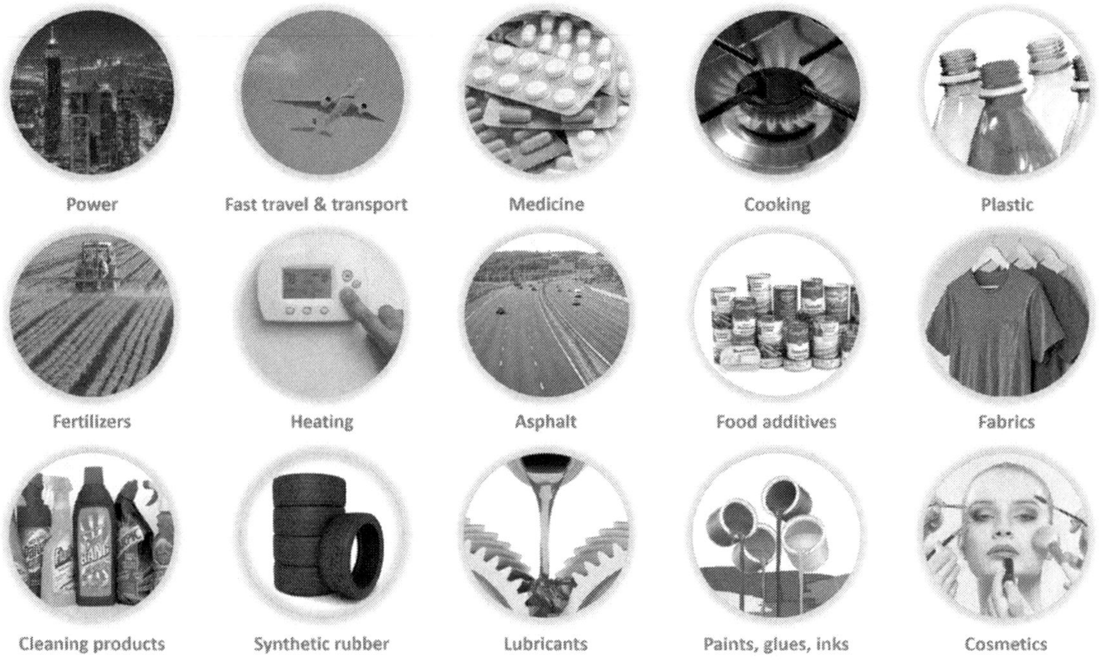

Fig. 17. – Benefits and uses of oil and gas. Natural hydrocarbons and their derivatives are central to modern life today. In many demand sectors there are no viable substitutes yet.

In 2019, 6% of global oil demand went into production of plastic. By 2050, as other parts of the economy decarbonise, plastics may well be responsible for 15% of all greenhouse gas emissions [17].

The transportation sector is expected to remain predominantly reliant on oil to 2040 and probably beyond. Of the ca. 100 million b/d of global oil demand today, fuel use for light vehicles and two-wheelers accounts for over 28 million b/d. This is expected to rise slowly to 29 million by 2025 (largely due to the impact of increases in engine efficiency) and then decline after 2030 due to an accelerating shift to electric vehicles. 97% of vehicles still run on fossil fuels and this proportion is only expected to reduce to 42% by 2040. Fuel demand for cars and two-wheelers in India is expected to increase by over 50% between now and 2025.

Increased use of electric cars could be an important "tipping point" in the Energy Transition and could have a significant impact on oil demand within the next few years. The passenger fleet will probably be the first sector to make the switch to electric vehicles. Electrifying the lorry and agricultural vehicle fleets faces many problems. Sales of new cars powered by internal combustion engines are already subject to planned bans in some countries. In the UK the uptake of electric vehicles is slow —only about 2%— largely because the cost of an electric car is still in the range of £30000 (Nissan Leaf and VW e-Golf) to £70000 (Audi e/Tron), too much for the majority (even with the Government's

£3500 grant), and the maximum range is still only ca. 300 miles. The current nationwide network of about 25000 charging points needs to be rapidly expanded.

Only about 2% of passenger car sales in 2018 worldwide involved electric car technology. This is expected to increase to nearly 35% of total passenger car sales in 2040. However, if the Chinese government decided to shift China to electric cars this could happen within 4/5 years and Chinese power systems are already designed to accommodate this. The use of electric vehicles is a good way to decarbonise the transportation sector except in countries —like China and India— where a significant proportion of electricity generation capacity is from coal-fired power stations. The limiting factor for transition to an electric vehicle fleet in many countries will probably be the charging infrastructure. Fuel efficiency gains in transport are expected to have a larger impact than electric vehicles on reducing oil demand to 2040.

Oil has, so far, stubbornly resisted giving way to batteries. This is largely because vehicles need to carry their own source of power with them —the lighter the better. A kilogram of petrol stores as much energy as 60 kg of batteries and has the convenient property of disappearing after use. Empty batteries, alas, are just as heavy as full ones!

While electric cars are finally starting to break through, electric Jumbo jets are a much tougher challenge! Improvements in energy efficiency and increasing the use of biojet fuels is key to decarbonising the aviation sector. Diesel and aviation fuels will probably remain important for the foreseeable future. Transport and heavy industry are responsible for more than 45% of global CO_2 emissions (fig. 18 [18]).

How difficulty would it be to remove hydrocarbons from the energy mix by 2050? Answer: Very! (fig. 19).

Segment progress report

Sub-sector	% 2018 share emissions	What's being done	Challenge	Progress
Power	34%	Cost tailwinds allow RE to increase at pace	Long-duration storage a barrier	
Road Transport	18%	EV penetration growing in light-duty vehicles, policy support in cities	Cost, charging, heavy-duty transport and raw materials	
RCA (low temp heat)	10%	Fuel substitution (coal-to gas, or to electricity)	Economics and technology	
Aviation, Shipping, Other Ind.	20%	Pilot testing using electricity, Synfuels, Hydrogen, Ammonia	Economics and technology	
Steel	9%	Pilot testing using Hydrogen, HI-Smelt, biofuels and CCS	Economics and technology	
Cement	9%	Pilot testing using Hydrogen, biofuels and CCS	Economics and technology	
Agriculture and Farming	Included in RCA	Intelligent use of fertilizers, soil productivity, vertical farming	Economics and technology	

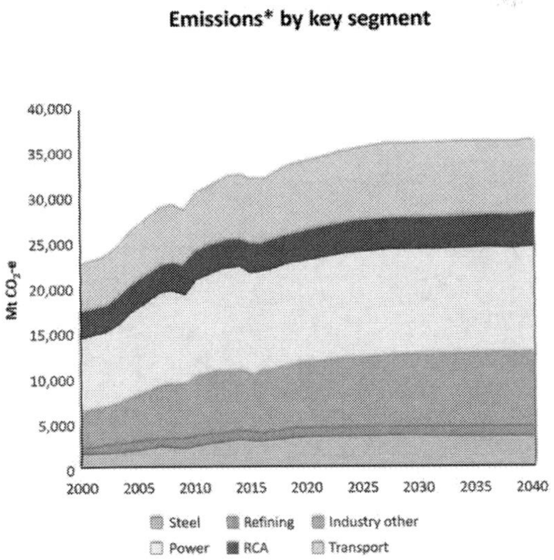

Fig. 18. – Why is it so challenging to decarbonise? Some segments of the global economy are hard to decarbonise (Source: Wood Mackenzie, 2019).

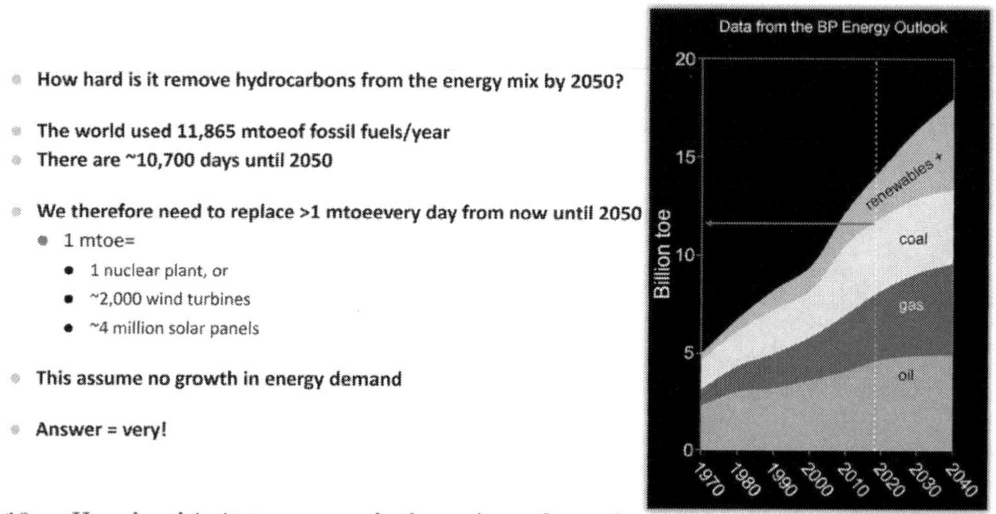

Fig. 19. – How hard is it to remove hydrocarbons from the energy mix by 2050? (Courtesy of Mike Simmons, Halliburton Landmark).

Fig. 20. – The end of hydrocarbons? Wall mural in Stavanger, Norway, May 2022.

Despite what some people would like to believe —even in Norway, it seems (fig. 20)— the hydrocarbon era will not be over soon!

10. – What are the big energy companies doing about their carbon emissions? What is their role in the "Energy Transition"

Collectively, the "Majors" (the world's largest independent integrated energy companies) account for only about 10% of global oil production. The largest, Exxonmobil, accounts for about 2%. The Majors cannot be expected to "lead the world into a low-carbon future", but they can, and must, "play their part". Like it or not, to billions of people around the world, they are the face of the global oil industry, they are instantly, emotive brands. They are integrated along the value chain in every part of the energy sector —including in low carbon energy and technologies, among the biggest companies on their respective equity markets and are woven into the fabric of everyday life.

Listed Energy companies are facing shareholder pressure and actively setting carbon reduction goals, but it would be quite naïve to think that large oil and gas companies can divest from billions of dollars of investment just because we need to decarbonise. Besides the fact that shareholders expect to have their investment renumerated, millions of people are employed in the oil and gas industry and all our economies still depend heavily on fossil fuels for materials, industry and transportation [18].

The Oil and Gas Climate Initiative (OGCI) —a group of 13 companies, including all the largest western International Oil Companies (IOCs) and some of the leading National Oil Companies (NOCs)— is "*dedicated to the ambition of the Paris Agreement to progress to net-zero emissions in the second half of this century*". At a meeting in October 2020, OGCI reported progress on curbing methane leakage, promising further support for investment in carbon capture, storage and use projects and pledging to set a common emissions intensity target in the year ahead. The CEOs of ExxonMobil, Chevron, Shell, Total, BP, Equinor, Eni, Repsol and Occidental were all present, together with high-level representatives from Saudi Aramco, CNCP and Petrobras. Methane leakage is one of the key areas of focus for OGCI —the group's members have set a collective target to cut methane leakage to 0.25% of their production by 2025.

There is an internationally accepted classification of Greenhouse Gas Emission into Scope 1, 2, and 3. In very simplified terms, Scope 1 emissions are those created by the company itself —the direct emissions from activities, Scope 2 are primarily indirect emissions produced by their suppliers and Scope 3 —the most difficult to address— are the indirect emissions produced by their customers (fig. 21).

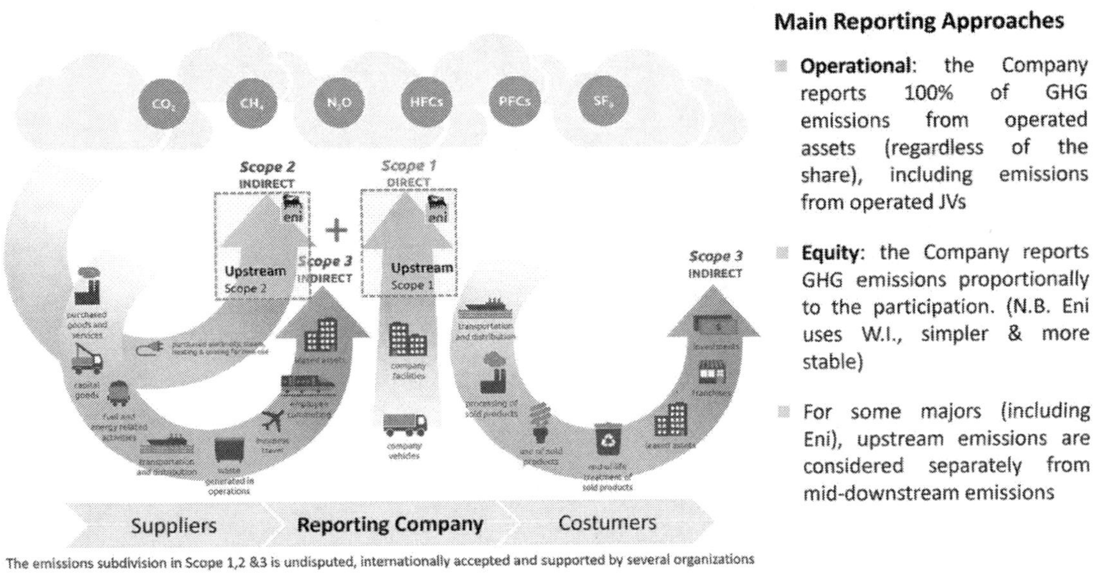

Fig. 21. – Greenhouse Gas (GHG) emissions classification, Scope 1, 2 and 3 (Source: Eni, 2020).

EPJ Web of Conferences **268**, 00004 (2022) https://doi.org/10.1051/epjconf/202226800004
LNES 2021

	ExxonMobil	Chevron	bp	eni	Shell	TotalEnergies	equinor
GHG Emissions Scope 1 – Scope 2	**reduction in Upstream carbon intensity** (vs 2016)* -15 to -20% @2025 -40 to -50% @2030 **Corporate emission** (vs 2016)* -20 to -30% @2030	**Net zero ambition** **reduction in Upstream carbon intensity** (vs 2016)* -35% @2028	**reduction in upstream GHG emissions*** (vs 2019) -20% @2025 -50% @2030	**Zero net carbon footprint** (equity) **Upstream @2030* Eni @ 2035**** **Upstream intensity -43% @2025 vs 2014** *(operated)*	**Net Carbon Footprint intensity** vs. 2016 (Scope 1+2+3) -6 to -8% @2025 -20% @2030 -100% @2050	**reduction in upstream GHG emissions*** (vs 2015) -15% @2025 -40% @2030	**Upstream CO2 intensity** <8 kgCO2/boe @2025 (100% op., Scope 1) **Carbon Neutral in Global Operations** @ 2030 (operated, Scope 1+2)
Scope 3	-	**reduction in carbon emissions intensity** (vs 2016) >5 % @2028	**Carbon intensity of sell products (vs 2019)** -5%@ 2025 -15 to -20%@ 2030 -100%@ 2050	**net carbon intensity** (equity, Scope 1+2+3) **-15% @2030 vs 2018**		**Carbon Intensity of energy products** (Scope 1+2+3) -20% @2030 vs 2015	**reduction of carbon intensity*** -20% @2030 vs 2019
Methane Emissions	**methane intensity** (vs 2016) -40 to -50%@2025 -70 to -80%@2030	**methane reduction** (vs 2016) >50% @2028 **2 kgCO2eq/boe**	**methane intensity*** 0.2% @2025	**Upstream fugitives** (vs 2014) **-80% @2025**	**methane intensity** (operated Upstream & Integrated Gas) <0.2% @2025	**methane reduction** (vs 2020) 50%@2028 80%@2030	**Keep methane emission intensity** near zero by 2030 (100% operated)
Gas Flaring	**flaring intensity** (vs 2016) -35 to -45% @2025 -60 to -70%@2030 **Zero routine flaring by 2030**	**flaring reduction** (vs 2016) >60% @2028 **3 kgCO2eq/boe** **Zero routine flaring by 2030**	**Zero routine gas flaring @2030**	**Zero routine gas flaring @2025**	**Zero routine gas flaring @2030**	**Zero routine gas flaring @2030**	0,2% @2020 **Zero routine gas flaring @2030**

*Operated assets — *Equity emissions — *Operational — *Scope 1, ** Scope 1+2 — *Operated — *Scope 1+2, 100% op. Scope 1+2+3, equity

Upstream carbon intensity: $kgCO_2eq/boe$ or $tCO_2eq/Kboe$

Scope 3 Carbon intensity: gCO_2eq/MJ

Fig. 22. – The approach of Eni and its peers to setting Greenhouse Gas (GHG) emissions targets (Source: Eni, 2022).

Not surprisingly, the different "Majors" have set a variety of different targets —covering not only GHG emissions, but also for methane emissions and gas flaring. It is quite difficult to compare between different companies because they use different metrics (fig. 22). There is, at least, a common ambition to have "Zero Routine Flaring" by 2030 —or in Eni's case by 2025.

In 2021, Eni launched a new strategy that will enable it to provide a variety of fully decarbonized products, through the development of bio-refineries and biomethane production, exploiting circular processes, the sale of low-carbon energy carriers, and the use of carbon sequestration and blue and green hydrogen technologies (fig. 23).

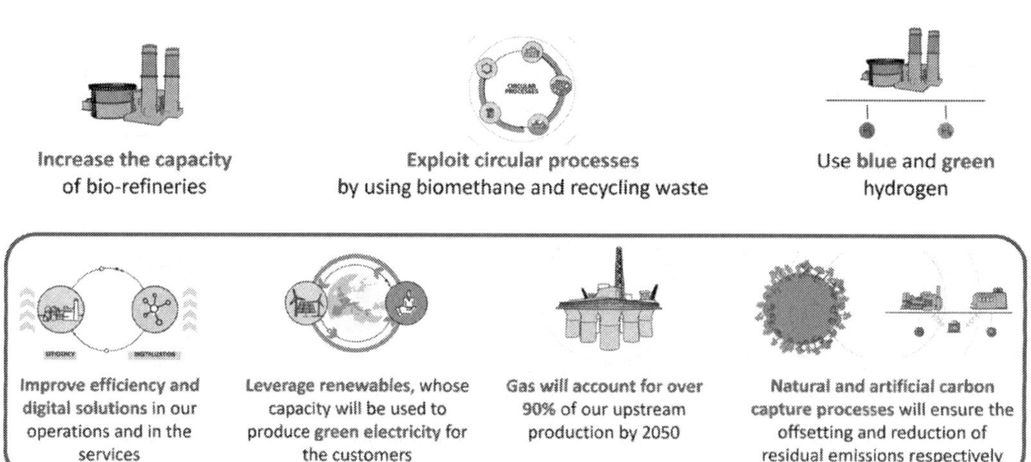

Fig. 23. – Eni's strategy for decarbonisation (Source: Eni, 2022).

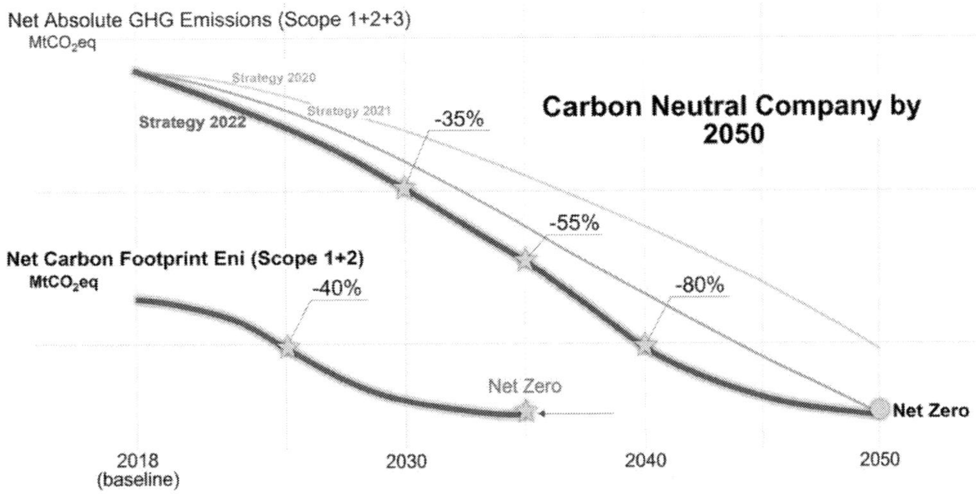

Fig. 24. – Accelerating Eni's Greenhouse Gas emission targets (Source: Eni, 2022).

One key target for Eni is that by 2050 gas will account for over 90% of our upstream production (fig. 23).

Eni aspires to contribute to the achievement of the UN's Sustainable Development Goals, supporting a just energy transition that meets the challenge of climate change with concrete and economically sustainable solutions while promoting efficient and sustainable access to energy resources, for all.

Today Eni has the most aggressive timeline of all the Majors for "net zero" Scope 1 and 2 emissions and, even more challenging has committed to reach "net zero" from Scope 3 on all products sold and reducing absolute emissions to net zero by 2050 (fig. 24) —one of the most ambitious emissions-reduction goals of any energy company in the world.

Eni has set an objective to achieve net-zero emissions in its upstream business by 2030, through increased efficiency to minimise direct upstream CO_2 emissions and offsetting residual upstream emissions through planting 20 million acres of forest in Africa. The forests planted in South Africa, Zimbabwe, Mozambique and Ghana should be removing 20 million tons a year of CO_2 from the atmosphere by 2030. Eni has also invested in converting two of its crude oil refineries in Italy —Gela and Venice— to use vegetable oils. The Venice refinery is the first in the world to be converted from processing fossil fuels to biofuels. In September 2021 the International Renewable Energy Agency (IRENA) finalised a 3-year partnership agreement with Eni to collaborate in the promotion of renewable energy and accelerate the energy transition, it is the first such agreement with an oil and gas company.

Carbon Capture and Storage (CCS) will undoubtedly play a critical role in decarbonising of society in the future, particularly in those areas of industry and energy supply where renewable energy solutions are more difficult to apply, or where CO_2 is a by-product of industrial processes such as the manufacture of cement or fertilizer [19, 20].

Today, there are two key issues —price and scale. The carbon price needs to be in the range of \$60–\$100 per ton of CO_2 for direct CCS to be a truly economic proposition and if we want to sequester enough CO_2 to follow the "2-degree pathway" we need a CCS industry that is about twice the size of today's oil and gas industry. Most forecasts of what it would take to get the world to "net-zero" by 2050 require CCS to increase essentially 100 times from today's levels, as well as eliminating global coal consumption and reducing global oil production by 50% or more from current levels.

To get to net zero, it is likely that between 1 billion and 2 billion tons of CO2 will need to be sequestered each year. In 2022, the world's 27 operating CCS plants scrubbed and stored just a few thousand metric tons of CO_2 [21]. Some authorities recommend that the ultimate target should be to return one tonne of CO_2 to geological storage for every tonne generated by the continuing burning of fossil fuels by 2050 —so called "geological net zero"— including using technologies such as direct air capture [22].

Eni has already made a major commitment to CCS with projects such as the "Hynet" project in Liverpool Bay, which will produce hydrogen from natural gas and store the resultant CO_2 in Eni's depleted fields in Liverpool Bay in the UK, and the "Adriatic Blue" project in Italy which will be a major European hub for CCS using depleted gas fields in the Adriatic Sea (see Della Moretta and Craig, in this volume).

Hydrogen can be used as a fuel and, if stored in the subsurface, can be an energy storage medium feeding from off-peak electricity from renewables. Hydrogen will be one of the main drivers of the energy transition and many countries have announced plans to increase the role of hydrogen in their future green economies, among them the United Kingdom and India. To achieve net zero by 2050, production of low-carbon hydrogen probably needs to reach 650 Mtpa, generating a US\$475 billion per annum international traded market [8]. The UK government has launched a plan for a world-leading hydrogen economy. The "Hynet" project uses "Grey hydrogen" from natural gas, but the ultimate goal is zero-emission "Green Hydrogen" created through electrolysis of water —a zero-emissions process. Given India's major dependence on energy imports (85% of its oil and 53% of its gas), green hydrogen is expected to be a "game changer" for the county, leveraging its large land mass and low solar and wind tariffs to produce low-cost green hydrogen and ammonia for export. In many countries the future is increasingly likely to take the form of a significantly scaled-up electricity network underpinned by a hydrogen gas-based distribution and energy storage network [23]. Hydrogen is expected to provide about 10% of global energy demand by 2040.

The world's major energy companies are a key driver for the "Energy Transition", but reputation management is a priority as the low-carbon world still needs some fossil fuels. There is a need for State and National oil and gas companies to adopt the same disclosures as the IOCs —lack of transparency will affect the whole industry, both operationally and reputationally. About 65% of the world's proven oil and gas reserves are under the control of NOCs. Many of these companies hold huge portfolios of assets, manage complex projects, employ many thousands of people and are often major cash generators for their countries [2]. Almost all these national oil companies will still be expected to provide revenues for their governments. Many are also investing in alternative energy resources to help them, and their countries, move away from unsustainable dependence on oil, but some are also seeking to increase their production of both oil and gas while they still can. Of the top 20 oil & gas companies with the highest emissions, 12 are NOCs [4].

11. – What role will geoscientists play in the "Energy Transition"?

Geoscientists bring unique skills to the "Energy Transition", given their deep knowledge of climate change through geological time and their deep experience of solving highly complex problems, dealing with complex Earth Systems and managing "big data". Geoscience skills will still be needed to provide the increasing amount of energy the world needs in the decades to come. This energy will come from a more diverse set of sources and many require geoscience skills (*e.g.*, geothermal, minerals for batteries, wind turbines, grid infrastructure and power equipment etc.). Even if the move to renewable energy resources occurs more rapidly than anticipated, subsurface skills will still be required for environmental solutions, such as carbon capture and sequestration.

Of course, renewable energy and batteries at scale come with environmental impacts of their own, including mining, manufacturing, deployment and landfill disposal. The low carbon "2-degree economy" will create huge demand for new metals and materials for batteries, wind turbines, solar panels etc. with a 200% increase in demand for aluminium, iron, lead and nickel and a staggering 1000% increase for cobalt, lithium and manganese. If it has not been grown, then it has got to have been mined. At present we have little idea where these new supplies will come from and very little geological data. The mining and minerals sector is, and will be, central to the "Energy Transition" story. Geoscience skills will be critical to providing them.

Geosciences are data-rich, so digitalization, Artificial Intelligence (AI) and Machine Learning (ML) will materially change the way we work but will not materially reduce the need for geoscience skills. There will, however, be even greater need for people who are multi-skilled and can integrate geological knowledge with computer science. This will require universities to change the curriculum and the focus of the geoscience training they provide.

Geoscientists can be proud, indeed must be proud, of the huge contribution they have made to the progress of society over the past decades. The challenge now is for geoscientists to step forward and engage with the wider society on solving the dilemmas associated with the Energy Transition. Transforming from a traditional hydrocarbon business model to a decarbonised net-zero culture with a flexible working environment, while still providing opportunities for growth and career development, will be critical if we are to attract new young talent into the energy industry. Geoscientists —and the Energy Companies they work for— must become "part of the solution" to the huge challenge of simultaneously meeting the world's growing energy demand, eliminating energy poverty and reducing global greenhouse gas emissions to slow global climate change.

12. – Conclusions and key messages

How are our energy systems changing today, and how should we respond?

The figure below brings together all the various issues that are causing major disruption in the Energy System today: from supply/demand imbalances, the rapid growth in renewables to the Covid pandemic (fig. 25). Steering a course through these factors is exceptionally difficult. Uncertainty is crippling!

Given all these issues, increasing government regulation, competition for capital, shareholder and public pressure to decarbonise, many traditional oil & gas companies are asking:

"Is it time to redefine the oil & gas industry and to find new directions for this new world?"

One of the most critical issues is, undoubtedly, Global Climate Change. The predictions made by a wide variety of organisations still suggest an increase in oil demand from ca. 100 million b/d today to perhaps 105 to 110 million b/d in 2040 —projections that would lead to a global temperature rise of 3.5 to 5.5 °C by 2100, if not countered by very significant reductions in greenhouse gas emissions. To keep the rise in global temperature below 2.0, and preferably below 1.5 °C, by 2100 we would need to cut global oil demand from 100 million b/d today to about 65–70 million b/d by 2040.

In recent years a startling difference in approach has emerged between the European

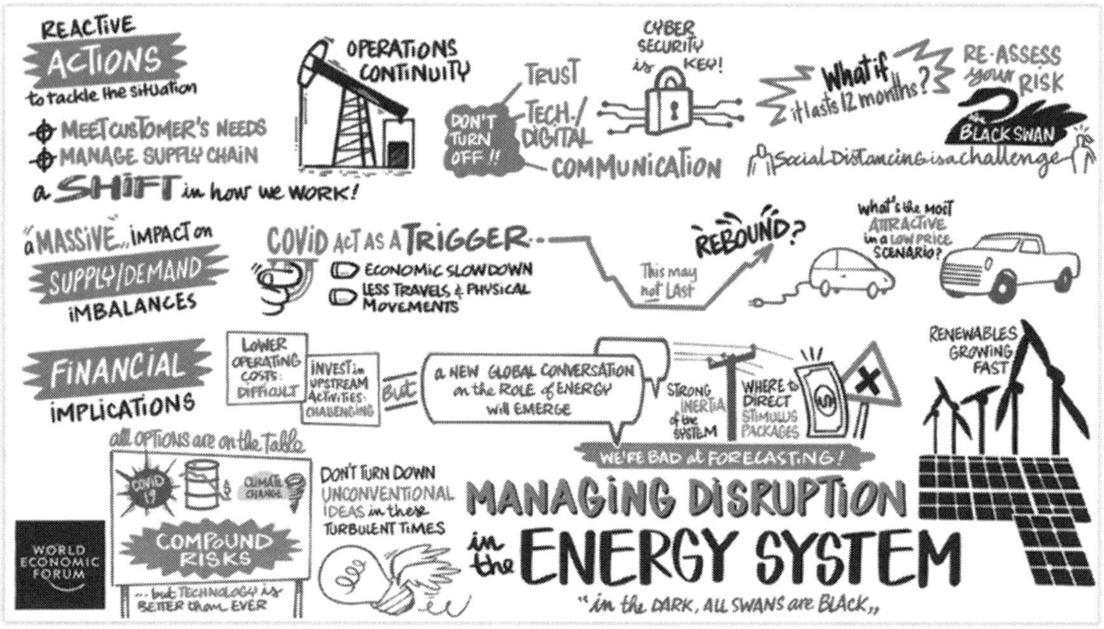

Source: World Economic Forum, 2020

Fig. 25. – How are our energy systems changing today? How should we respond? (Source: World Economic Forum, 2020).

"Majors" —Eni, bp, Repsol, Equinor, Shell, Total, who are all aiming for high levels of new energy diversification and a strong focus on carbon neutrality— and the U.S. "Majors", primarily ExxonMobil and Chevron, who, for now at least, are largely sticking to the old model of "hydrocarbons win". The European "Majors" are building up their renewable energy portfolios and looking for new income streams using their in-house knowledge and expertise, particularly in the broad area of "geoenergy".

But what is the right mix of traditional oil & gas, renewables —such as wind and solar— and new energy solutions, like hydrogen, for a company transitioning from an International Oil Company to an International Energy Company? What, indeed, is the recipe for success? (fig. 26).

Renewable energy has the power to change global politics! Wind, Solar and low-temperature geothermal are all available everywhere to some degree, so in the future many countries could be much less reliant for their energy supplies on, for example, the Middle East and Russia. Unfortunately, the materials needed to use renewable energy —the metals for wind turbines, solar panels, and rechargeable batteries, for example— are not so universally distributed. China controls almost the entire global market in the rare-earth elements used to make wind turbines. Cobalt and lithium are vital for making rechargeable batteries. A low-carbon "2-degree economy" will create huge demand for metals and materials. A World Bank study in 2017 estimated that the "Energy Transition" would result in a 200% increase in demand for aluminium, iron, lead and nickel, a 400% increase in graphite, a 600% increase in cobalt and a staggering 1000% increase in

Recipe for transition success? BP and chief executive Bernard Looney are cooking up something different for the supermajor's future

Fig. 26. – Transitioning from an International Oil Company (IOC) to an International Energy Company (Source: Rytis Daukantas/Upstream).

demand for lithium by 2050. Most of the known deposits of cobalt in the world are in the Democratic Republic of Congo where its production is linked to child labour. Chile is the world's largest producer of lithium, but indigenous communities are resisting continued lithium production there because of the environmental problems associated with consumption of scarce groundwater for solar evaporation. Lithium demand is expected to more than triple between 2020 and 2025, rising to a million tons by 2025 and outpacing supply by 200000 tonnes [24]. Demand for battery-grade nickel, another important component in electric vehicle batteries is expected to outstrip supply by 2024 [24]. There are substantial undeveloped nickel resources in Indonesia, but they are almost all in areas of extremely rich biodiversity. There were 10 586 active mines in the world in 2022 and a further 4 790 mining projects under construction. The largest mine in the world is La Escondida, a copper mine in the Atacama Desert in Northern Chile, but current global copper demand could sustain 20 additional mines the size of La Escondida [4]. It is estimated that 3 trillion new heavy-duty rechargeable batteries will be needed for electric vehicles over the next 20 years. Each electrical vehicle requires the mining of 250 tons of ore to produce the metals required for its batteries [16]. Imagine the possible environmental impact of the mining required if rigorous standards are not applied? There is a huge challenge facing the mining industry to provide the critical minerals required for the Energy Transition and a huge responsibility on mining companies to ensure that their operations have minimal environmental impact. Irresponsible, unsustainable and non-ESG compliant mining is a major issue and needs to be eliminated, but it will take time.

It is important to remember in any discussion about the "Energy Transition" that it is not the energy source that is the problem, it is the environmental impact of using it. This applies just as much to renewable energy resources as to fossil fuels.

Do not be fooled —there is no such thing as "clean energy", just more or less dirty forms of energy!

Most published predictions of World Energy Demand provide scenarios for the period through to 2040 or 2050. But what if we run a scenario that tries to meet the three great challenges of reducing greenhouse gas emissions, reducing energy poverty and satisfying growing world energy demand through to 2100? There are, of course, huge uncertainties and huge dangers in trying to predict this far into the future —we are notoriously bad at predicting the future— but one not completely unreasonable scenario looks like this (fig. 27) ...

There are some important assumptions involved in constructing this scenario —2 of the most critical are that:

- Global population will increase from 7.7 billion today to around 13 billion by 2100 with growth mainly in Asia-Pacific to 2050 and then in Africa from 2050 to 2100: this is the main driver of world energy demand.
- There will be social and political pressure in countries like India, China, Indonesia and across Africa to lift a substantial proportion of their populations out of energy poverty.

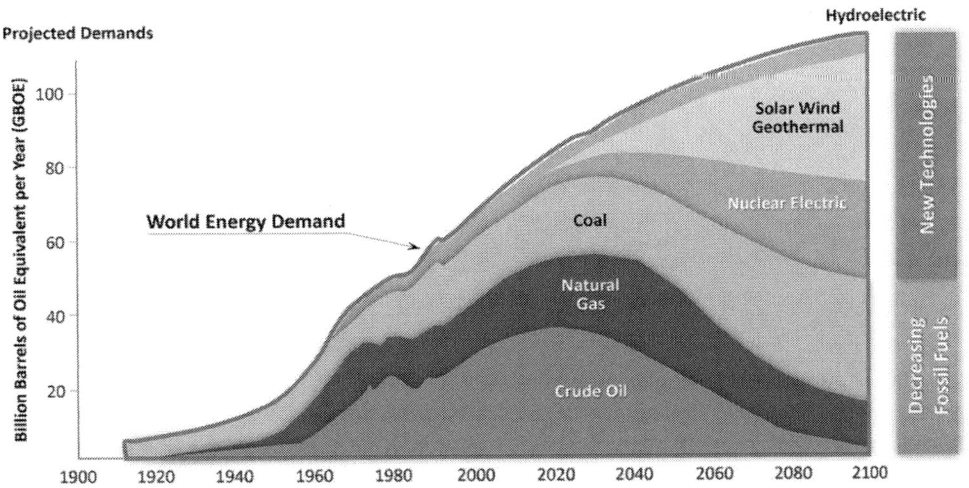

Fig. 27. – One possible forecast of sources of energy required to reduce greenhouse gas emissions, reduce energy poverty and satisfy growing world energy demand to 2100.

Based on these (and many other) assumptions, this model predicts peak oil demand will occur around 2025, with peak gas demand in about 2040. Coal will continue to be important for power generation until the end of the century because it is cheap and abundant —but we'll need to sequester the carbon dioxide produced— but by 2100, oil & natural gas will still be used to meet about 15% of global energy demand —and, on top of this, hydrocarbons will probably still be used as feedstock for at least some chemical processes. If you are wondering, this scenario gives about 2.5 to 3.0 °C temperature rise by 2100.

Ultimately, whether you consider the current "Energy Transition" to be a "myth" or a "reality" probably depends on your view about the timeframe over which the transition can and should occur and where you live in the world. If you live in Europe, Australia or North America and you believe the transition will take decades, then the "Energy Transition" is already very much a reality. If, however, you believe it can and should be completed in years rather than decades, or you live in large parts of Sub-Saharan Africa or parts of South America or the India Subcontinent where many people still have little or no access to affordable energy, it is certainly a myth and, worse, in the latter case, largely irrelevant... except.... except, that it is exactly these people, the poorest in the world, who will suffer the most from the effects of global climate change.

The worst possible scenario would be if the world cannot continue to meet its total energy demands, does not extend the availability of power and fuel to energy-starved communities, and also fails to avoid the most serious effects of climate change. Creating a new and sustainable energy system is certainly humanity's greatest challenge. The enormous and unprecedented challenge we face is to power the planet, reduced energy poverty and slash greenhouse gas emissions, simultaneously.

REFERENCES

[1] BIRDWELL J., "Complexities of the Energy Transition", *AAPG Explorer* (2022).
[2] WHALEY J., "A Difficult Energy Transition", *GEOExPro*, **18** (2021) 30.
[3] BURTON K. and MAITRE P., "Light and Power", *Geographical* (2021), 30.
[4] AMERICAN ENERGY SOCIETY, *Energy Matters*, Ser. 8, Vol. **17**, August 22, 2022.
[5] AMERICAN ENERGY SOCIETY, *Energy Matters*, Ser. 8, Vol. **2**, January 10, 2022.
[6] SPENCER B., "The futuristic fix for hot homes: shutters", *The Sunday Times*, 24th July 2022.
[7] BP, *Statistical Review of World Energy 2022*, 71st edition, 2022.
[8] WOOD MACKENZIE, *Accelerated Energy Transition 1.5° C scenario: Highlights. Global net zero 2050*, February 2022, `www.woodmac.com`.
[9] DAVIES A. and SIMMONS M. D., "Demand for 'advantaged' hydrocarbons during the 21st century energy transition", *Energy Rep.*, **7** (2021) 4483.
[10] AMERICAN ENERGY SOCIETY, *Energy Matters*, Ser. 8, Vol. **9**, April 18, 2022.
[11] AMERICAN ENERGY SOCIETY, *Energy Matters*, Ser. 8, Vol. **14**, June 27, 2022.
[12] AMERICAN ENERGY SOCIETY, *Energy Matters*, Ser. 8, Vol. **19**, September 26, 2022.
[13] AMERICAN ENERGY SOCIETY, *Energy Matters*, Ser. 8, Vol. **14**, July 11, 2022.
[14] INTERNATIONAL ENERGY AGENCY (IEA), World Energy Investment 2022.
[15] MCCARTY M., "OTC 2022: Reaching net zero will require more investment in oil, gas and renewables", *World Oil Online*, 3rd May 2022.
[16] CASIANO J., STERNBACH C.A., LUCHTEL-FERGUSON K. and DEMIS W. D., "It's a Good Time to Be in the Oil and Gas Industry", *AAPG Explorer online*, June 2022.
[17] HICKIE J. and TOVEY F., "Plastics in the hydrosphere", *Geoscientist*, **29** (2019) 22.
[18] ANDERSON H. and CASTIGLIONI R., "There is no energy transition without creation of value", *EnilitNews*, 29th June 2021.
[19] BAINES S. J. and WORDEN R. H., "Geological storage of carbon dioxide", *Geol. Soc., London, Special Publ.*, **233** (2004) 1.
[20] RINGROSE P. S. and YIELDING G., "Geoscience for CO_2 storage: an introduction to the thematic collection", *Pet. Geosci.*, **28** (2020) petgeo2022-003.
[21] AMERICAN ENERGY SOCIETY, *Energy Matters*, Ser. 8, Vol. **16**, August 8, 2022.
[22] ALLEN M., "Could Liz Truss allow new drilling for oil and gas and still strengthen the UK's next zero target?", *The Conversation*, 6th September 2022.
[23] SMALLBONE A., *Making net-zero easier by embracing solutions for heat, transport and electricity*, `https://openaccessgovernment.org/making-net-zero-easier-by-embracing-solutions-for-heat-transport-and-electricity/95591/`, 6th October 2020.
[24] BROWN I., "Critical minerals for electric vehicles", *GEO ExPro* **19** (2022) 32, issue No. 1.

EPJ Web of Conferences **268**, 00005 (2022) https://doi.org/10.1051/epjconf/202226800005
LNES 2021

Carbon capture and storage (CCS)

DAVIDE DELLA MORETTA and JONATHAN CRAIG
Eni Natural Resources - Milan, Italy

Summary. — Carbon Capture and Storage (CCS) is an important tool for the decarbonization of the energy system to achieve the mid-century global climate change targets. CO_2 is captured using different industrial processes that involve membrane filtering or enhanced combustion. The CO_2 is then transported, preferably by pipeline, to a storage site where it is injected into a permeable reservoir. Sealing capacity of the storage site is of paramount importance for safe CO_2 sequestration, to avoid any geological leakage. Each CCS project must have a dedicated MMV (Measurement, Monitoring and Verification) programme to ensure conformance with the expected evolution of the CO_2 plume and its containment within the storage site. Eni is committed to the implementation of CCS, with several ongoing projects.

1. – Introduction (CCS defined)

Carbon capture and storage (CCS) or carbon capture and sequestration is the process of capturing carbon dioxide (CO_2) before it enters the atmosphere, transporting it, and storing it (carbon sequestration) for centuries or millennia(1), normally in an underground geological formation (fig. 1). Alternatively, after capturing, the CO_2 can be used(2) to produce other substances through chemical transformation. In both cases its dispersion in the atmosphere is avoided.

(1) Excerpt from Wikipedia `https://en.wikipedia.org/wiki/Carbon_capture_and_storage`
(2) If CO_2 is utilized in the sequestration process, the term CCUS is used instead of CCS.

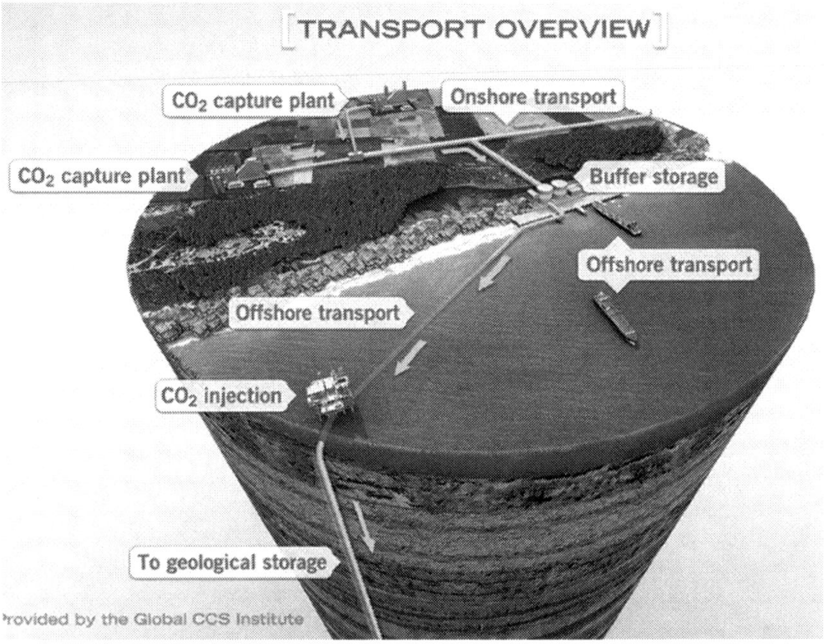

Fig. 1. – Schematic representation of a Carbon Capture and Storage project.

2. – Why CC(U)S is important

Internationally, CCUS is considered as an important tool for the decarbonization of the energy system. The International Energy Agency (IEA) stated in a recent report (CCUS in Clean Energy Transitions) that the capture, utilization, and storage of carbon dioxide must be a fundamental pillar of the efforts required to eliminate net greenhouse gas emissions within this century (fig. 2). CCUS is critical to address process emissions from cement, natural gas-based hydrogen, and biofuel production, to produce synthetic fuels, and to reach negative emissions from bioenergy with carbon capture and storage and direct air capture with storage.

CC(U)S is vital to reduce emissions to net-zero by mid-century and achieve global climate change targets.

3. – CCS in the world

CCS technologies are tried and tested as they have been in operation since the '70s. More than 260 Mt of anthropogenic CO_2 have been captured and stored to date, but there remains a massive gap between current CCS capacity and what is required to reduce global emissions to net zero. The global CCS industry must grow by more than a factor of 100 by the year 2050, to achieve the Paris Agreement climate targets. This means building 70 to 100 facilities a year. Most of the operational and planned facilities are located in USA, the rest is fragmented between 21 different countries (fig. 3).

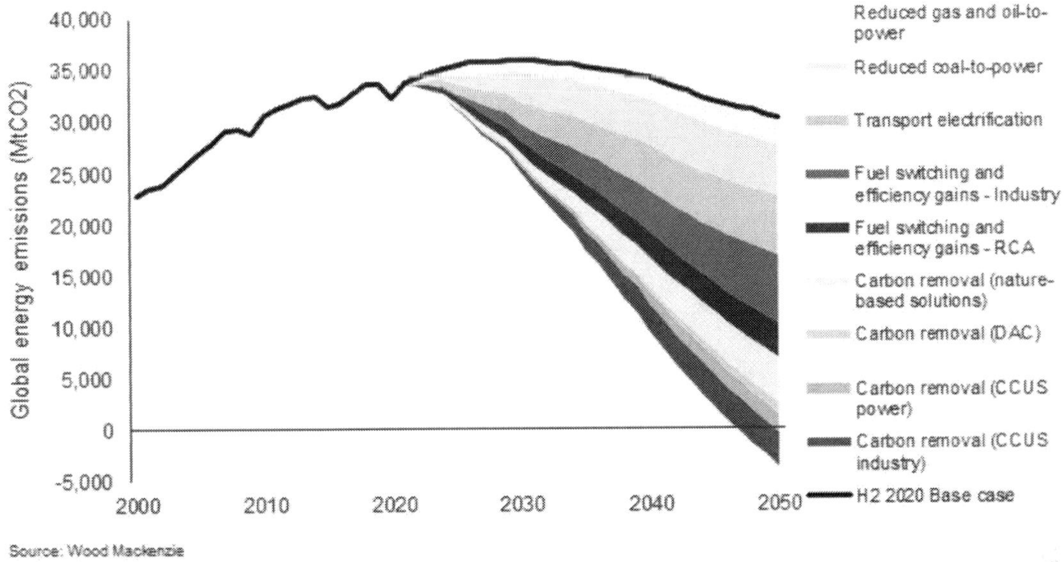

Fig. 2. – Change in global CO_2 emissions needed to reach net zero by 2050 [1].

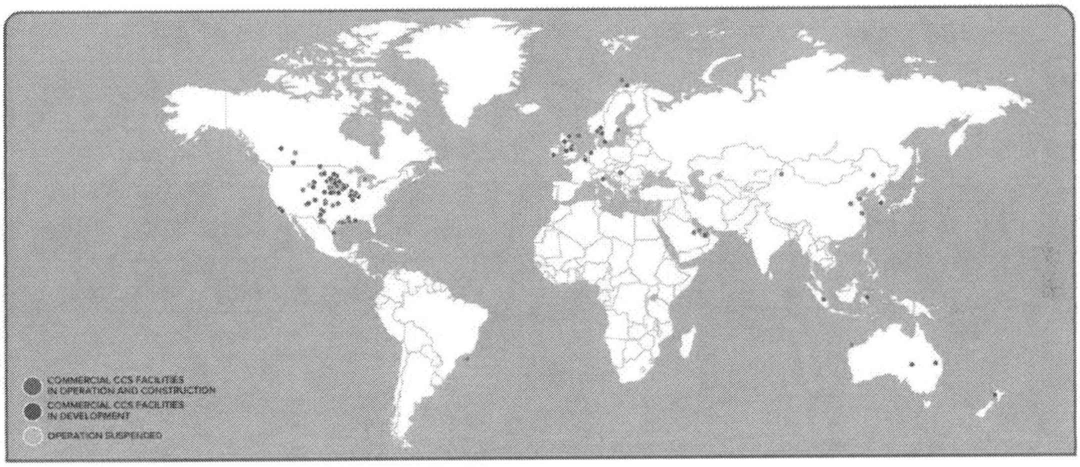

Fig. 3. – World map of CCS facilities at various stages of development [2].

The current operational capacity today is around 55 Mtpa([3]) (@July 2021), but the number of CCS projects is growing on a weekly basis (fig. 4).

The planned capacity has reached almost 300 Mtpa (@June 2021), up from 100 Mtpa at the turn of 2021 (fig. 5). Over 50 projects have been added in the first half of 2021, with a strong input from multi-user, multi-industry hubs, which makes up over half of the capacity of the planned projects.

([3]) Mtpa = Million Tons per Annum = Mt/y.

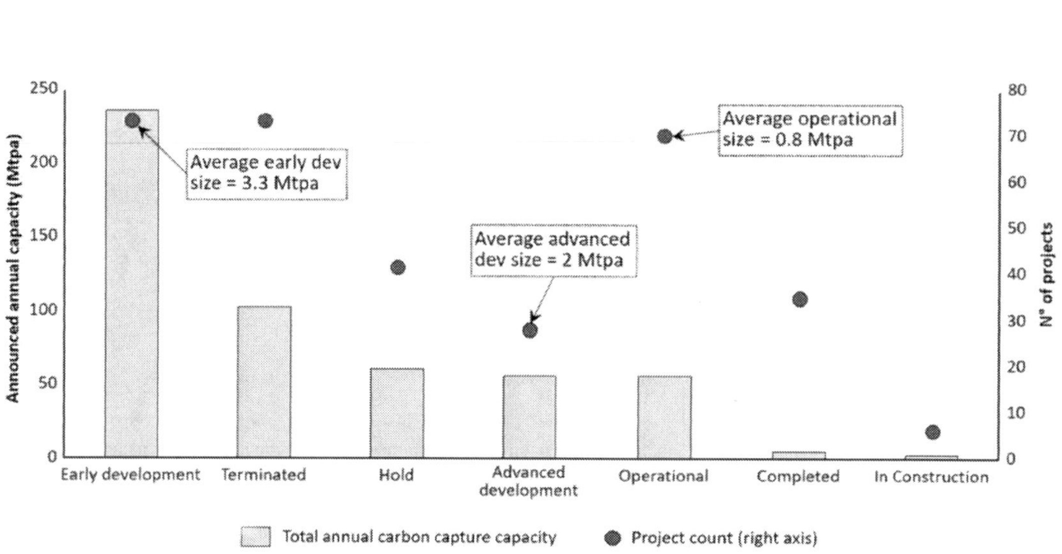

Fig. 4. – CCS projects development status in terms of capacity (left axis) and number (right axis). Please note the increase of average size per project moving from "Operational" to "Early development" [1].

The source of CO_2 to be reinjected comes for different industrial sectors, and it is evolving over the years (fig. 6). In the early stages of CCS, natural gas processing was one of the most important CO_2 sources, while for the future the "hard-to-abate" emissions will replace it.

The worldwide theoretical CO_2 storage capacity is huge and quite well distributed in most of the sedimentary basins (fig. 7). However, the screening for proper locations should take into consideration many local factors, both geological and logistical. In

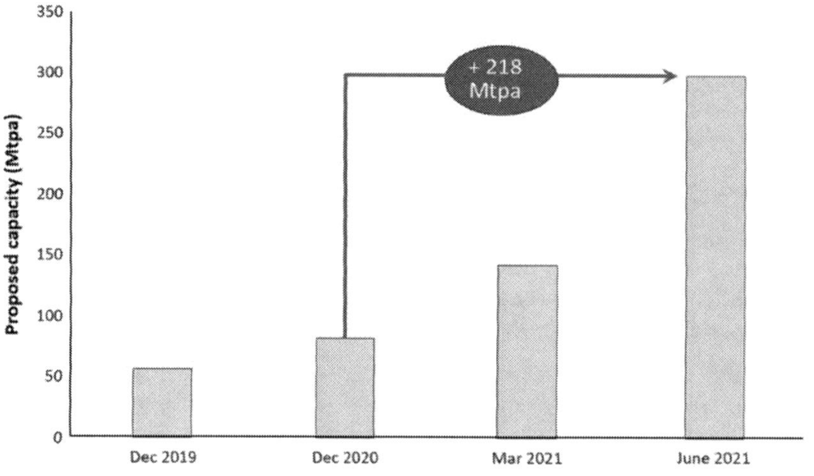

Fig. 5. – Total proposed CO_2 storage capacity in CCS projects [1].

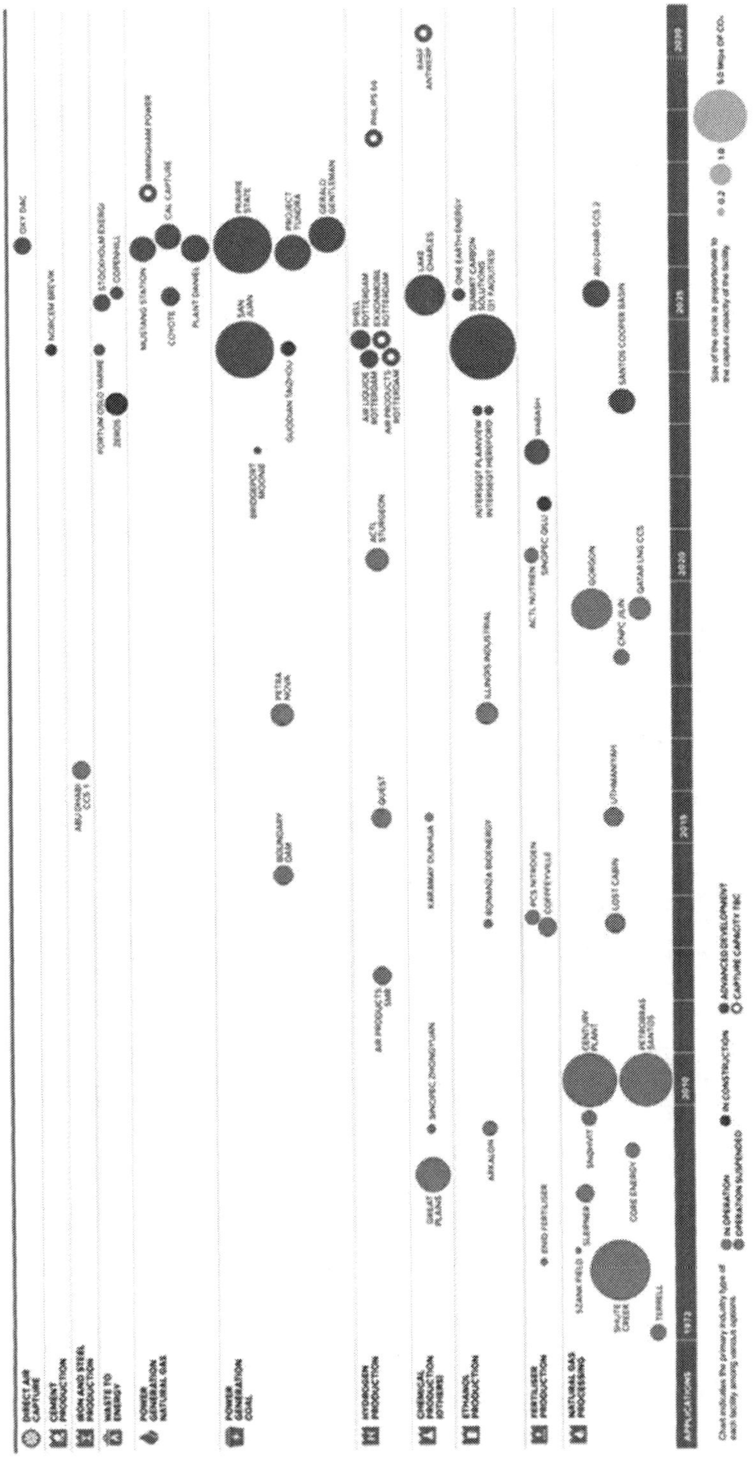

Fig. 6. – CCS projects by sector and scale over time [2].

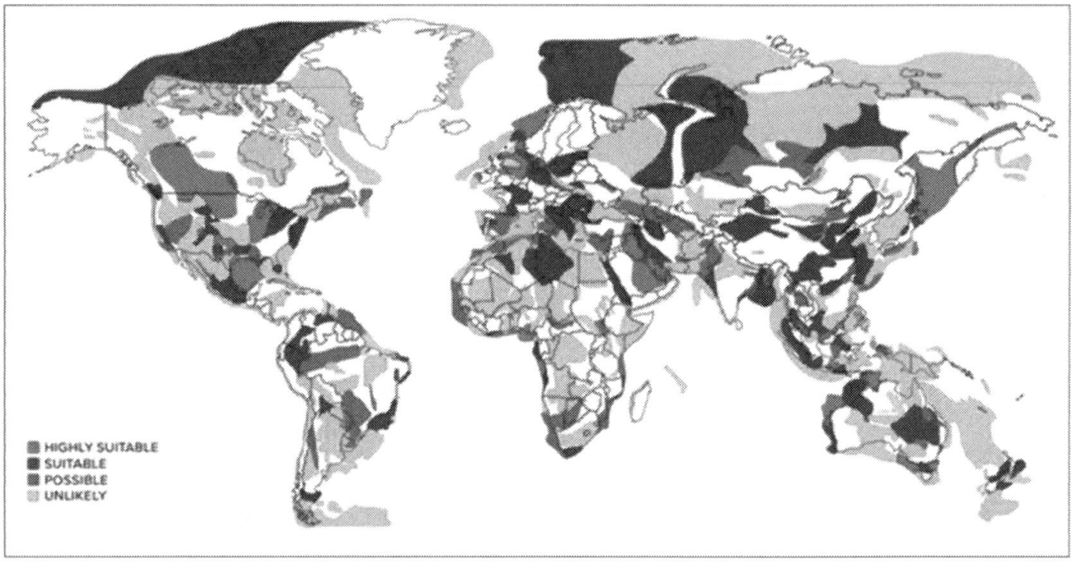

Fig. 7. – Suitable storage regions of the world based on the Global CCS Institute's storage basin assessment database [2].

particular, CO_2 sources should be close to the injection sites in order for the project to be economic/feasible. The carbon removal capacity should increase far beyond the current scenarios to meet climate targets (figs. 8, 9 and 10).

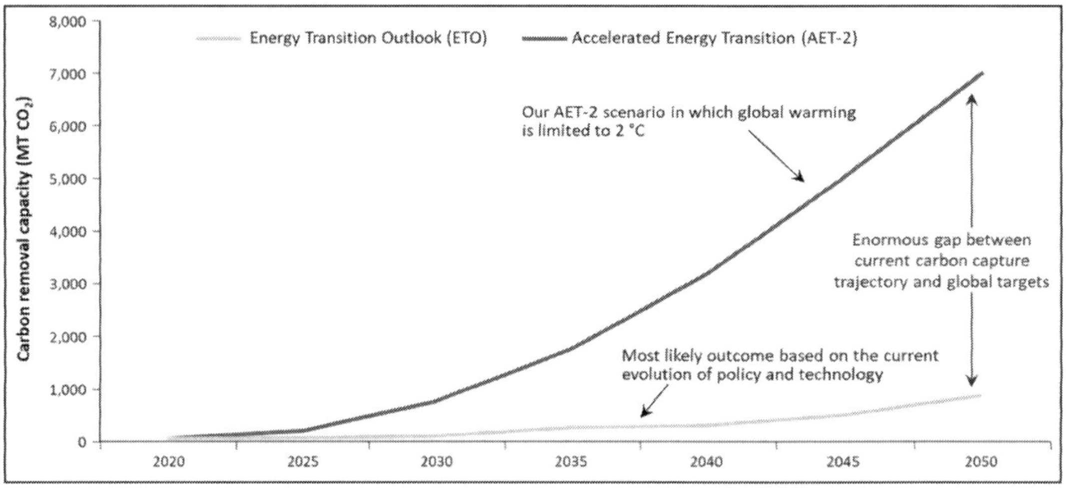

Fig. 8. – Comparison between carbon removal capacity needed for global targets and the current most likely scenario [1].

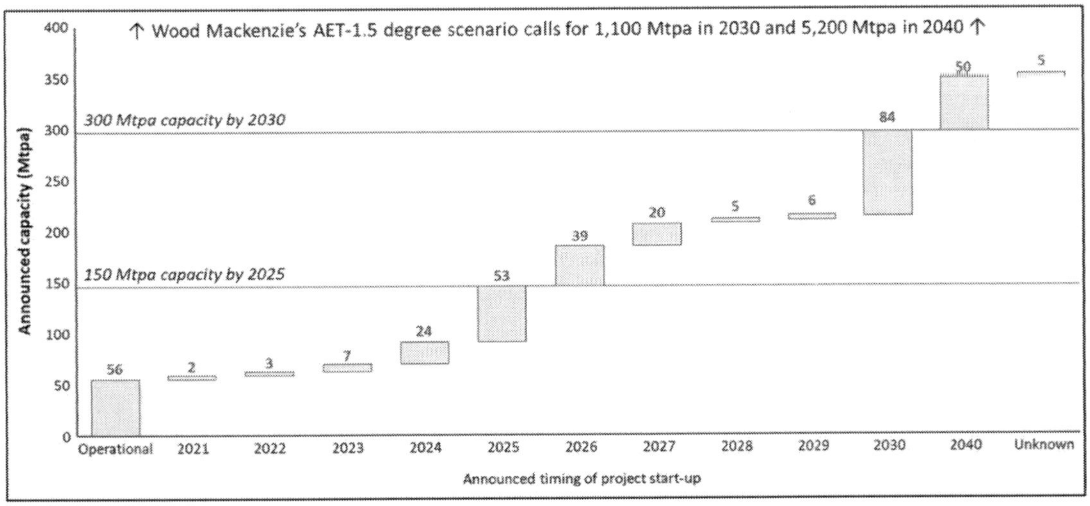

Fig. 9. – Planned carbon capture capacity additions by year of announced start-up [1].

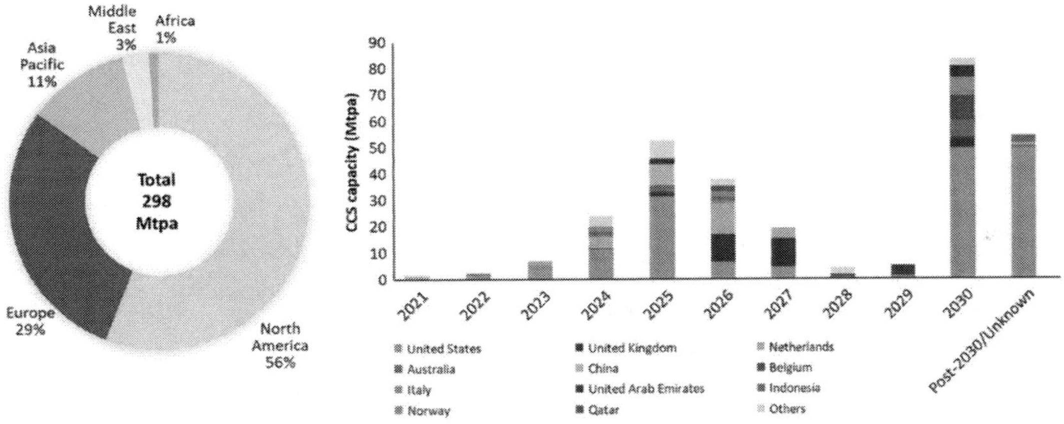

Fig. 10. – Geographical and time distribution of CCS capacity [1].

4. – CO_2 properties

CO_2 is a gas at standard temperature and pressure, and a solid if cooled and pressured sufficiently (fig. 11). If the temperature and pressure are both increased above the critical point for carbon dioxide (31 °C and 1070 psi, 73.8 bar), it can adopt properties midway between a gas and a liquid. More specifically, it behaves as a supercritical fluid above its critical temperature and pressure expanding in a container like a gas but with a density like that of a liquid.

CO_2 at deep reservoir depth is in a supercritical state, and to avoid phase changes the reservoir should be deeper than ≈ 800 m. Phase changes generate abrupt temperature

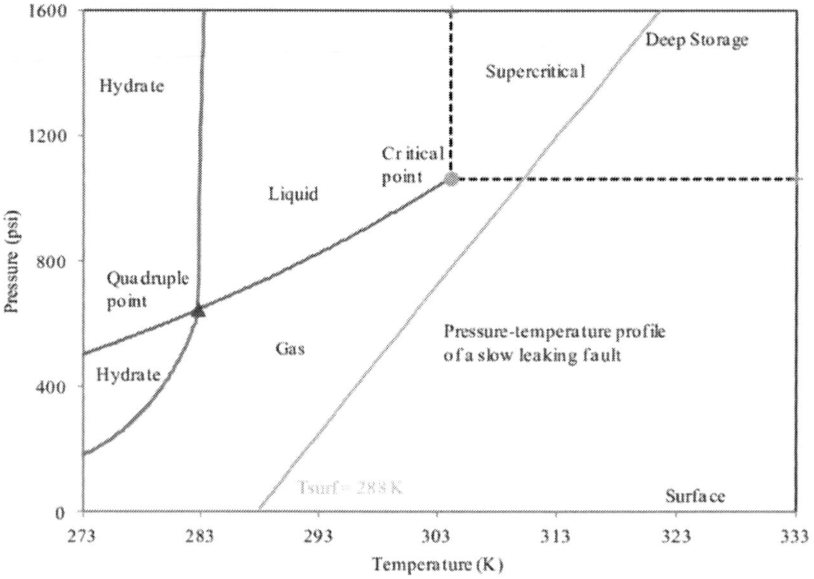

Fig. 11. – CO_2 phase diagram. The green line is a possible path of a CO_2 molecule moving from the reservoir up to the surface along a leaking fault. Depending on the temperature and thermal gradient, phase changes may occur [3].

variations, hence a proper planning of the processing design (*i.e.*, compressors, heaters) is needed. At supercritical state CO_2 is lighter than water, around 700 kg/m^3 at 300 bar and 90 °C.

CO_2 is not inert; it reacts with water and rocks leading to the acidification of the brines. A common reaction between CO_2 and water leads to carbonic acid formation.

5. – Carbon capture

Carbon capture is the process of extracting CO_2 from an industrial process. The following main types of capture processes are distinguished (fig. 12):

- Post-combustion: CO_2 captured from the exhaust gases is absorbed in a suitable chemical solvent. The CO_2 is then separated from the solvent and compressed in order to be transported and stored. Other post-combustion separation methods are by high-pressure membrane filtration or cryogenic separation.

- Pre-combustion: the fuel is converted into a mixture of H_2 and CO_2 before combustion using a process called gasification. CO_2 can then be transported and stored, while hydrogen can be used as a clean fuel.

- Oxy-combustion: this process involves the use of pure oxygen, or highly enriched air, in the combustion chamber. This type of combustion mainly produces steam and concentrated carbon dioxide, which is easier to process, to transport and store.

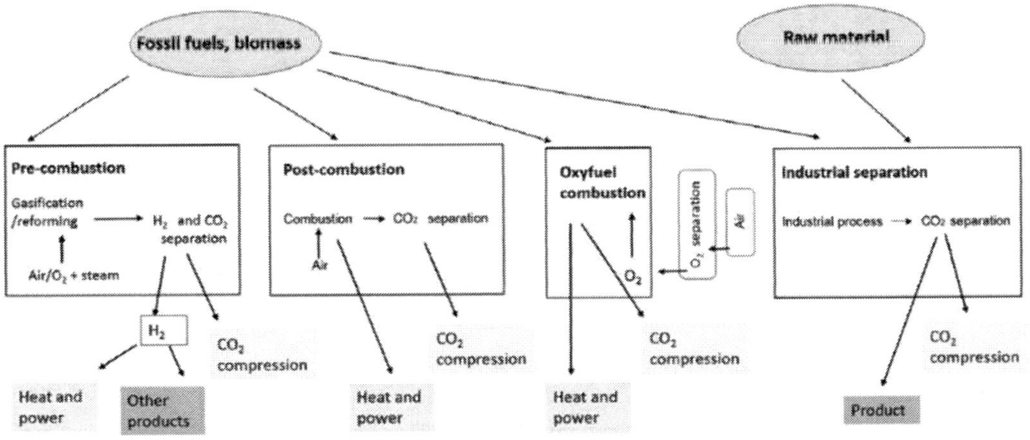

Fig. 12. – Schematic representation of carbon capture systems [4].

– Industrial separation: unwanted CO_2 from raw material is separated in different industry processes, such as natural gas sweetening, production of hydrogen and ammonia, etc.

6. – Carbon transport

Pipelines are by far the most used transport medium for CO_2. Special design considerations need to be implemented when constructing facilities for pipelining CO_2. The phase behaviour must be controlled during transportation, even in the case of multiphase flows. The best way of transporting CO_2 in the pipes is at ambient temperature and with a pressure above the supercritical one. Corrosion and fracture control should also be implemented, especially if CO_2 is transported together with water, which may lead to acidification. The pipes should be also kept at a safe distance from other infrastructure.

CO_2 is not flammable and inert at atmospheric conditions, but a sudden drop in pressure due to a pipe break could generate a similarly sudden drop in temperature around the broken area that could be dangerous and possibly mortal: a CO_2 pipe failure may lead to asphyxiation due to the very high CO_2 concentration around the break point. Since CO_2 is heavier than air, it stays on the ground for a longer time. CO_2 can also be transported using ships, in the solid phase at low temperature ($-35/-55\,°C$) and pressure (7/20 bar). Ships need to have heaters and compressors, which are heavy and add to the transportation cost, and CO_2 is not a valuable cargo at the moment. Therefore, there is no easy way of increasing the number of CO_2 cargo ships if the CO_2 pricing or governments incentives do not change.

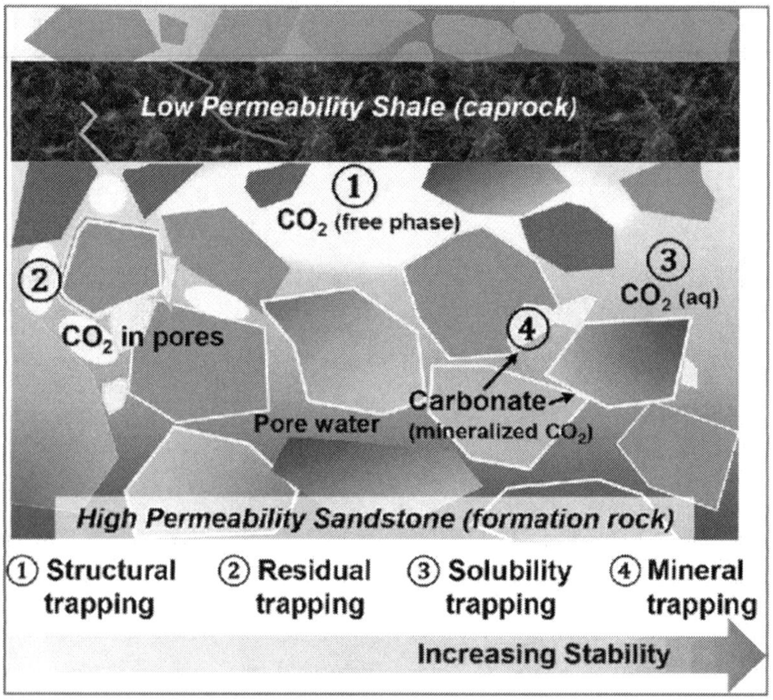

Fig. 13. – CO_2 trapping mechanisms [5].

7. – Carbon storage

The permanent storage of CO_2 can be done by injection in geological reservoirs. Deep saline aquifers and depleted oil/gas fields are most common, but other geological containments have been studied, such as coal beds and fractured basalts. CO_2 migrates inside the pore spaces driven by the balance between buoyancy and capillary threshold pressure. Different CO_2 trapping mechanisms contribute to the permanent sequestration of the CO_2 (fig. 13).

In depleted reservoirs the main trapping mechanism is structural trapping, where CO_2 replaces existing pore fluids (brine, residual hydrocarbons) filling the trap that previously contained hydrocarbons. Each trapping mechanism is associated with a time-dependent stability, with the mineral trapping (*in situ* carbonatation) as best option. Next to geological requirements, a CO_2 reservoir has to fulfil additional requirements, to be suitable for CO_2 storage:

Depth

The depth of the reservoir is usually larger than ≈ 800 m. In such conditions, CO_2 always remains in the supercritical phase.

Capacity
The reservoir must be able to accommodate the total planned CO_2 injected volumes. The capacity can be computed with simple formulas; one of the most common is the one introduced by [6] and it refers to depleted hydrocarbons (HC) reservoirs:

$$m_{CO_2} = \text{HOIP} \; * \; \text{URF} \; * \; \text{FVF}(T_{\text{res}}; pi) \; * \; \rho_{CO_2}(T_{\text{res}}; pi) \; * \; E,$$

where HOIP (Sm^3) is the hydrocarbon originally in place at standard conditions, URF is the ultimate recovery factor, FVF (rm^3/Sm^3) is the HC formation volume factor at initial reservoir conditions, ρ_{CO_2} (kg/m^3) is the CO_2 density at initial reservoir conditions, E is the efficiency factor accounting for the non-ideality of the process (aquifer strength, compressibility during re-pressurization, compartmentalization, mixing with the residual gas).

A robust estimation of the efficiency factor is very difficult to obtain, due to the geological uncertainties that characterize natural reservoirs, especially deep saline aquifers. Moreover, this is a "static" approach, since all the dynamics of the process (pressurization, CO_2 migration, dissolution, etc.) are not considered.

Seal
The reservoir must be isolated from the surface, without leakage of CO_2. Shales and evaporites are well-known to be seal rocks. A good seal is usually thick, not fractured and extremely impervious.

Transport & injectivity
A CCS project can only work if CO_2 is available and can be transported at economically sustainable costs. Therefore, the best storage locations are close to CO_2 emitters. In addition, the reservoir porosity/permeability should be able to accept the planned CO_2 flows to avoid project failure but, also and most importantly, it should withstand the required pressure ramp-ups that could lead to unwanted fracturing.

Monitoring & remediation
CO_2 movements in the subsurface must be monitored and remedial actions should be possible: not all storage sites can be properly monitored due to logistical/regulatory/environmental constraints. If a proper monitoring cannot be implemented, the storage site should be given up.

Stakeholders
Before starting any operations on an identified storage site, agreement among all involved parties (authorities, local communities, CO_2 providers, scientific advisors, etc.) should be reached.

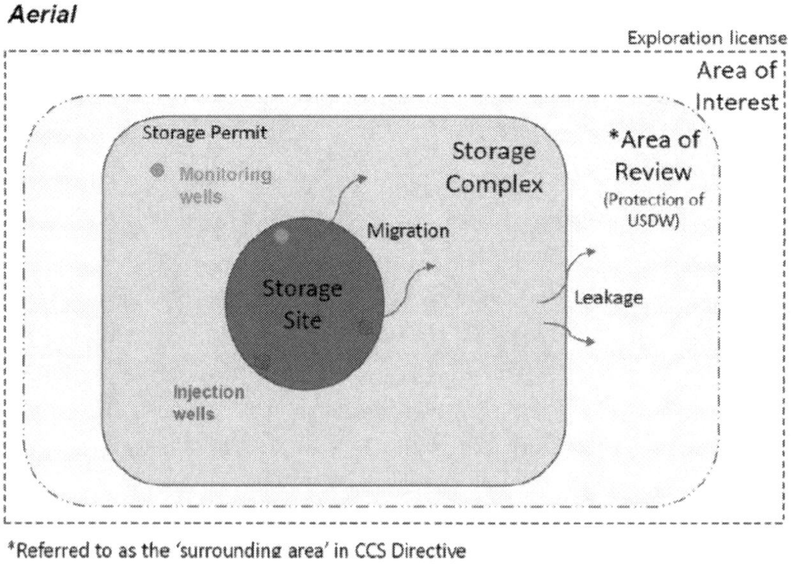

Fig. 14. – Conceptual map view of the CO_2 storage site/complex definitions.

7·1. *The "storage site" and "storage complex" concepts.* – International laws that rule CCS projects attach major importance to permanent CO_2 storage in the subsurface. In particular, the CCS European directive stresses the importance of analysing larger areas than the doublet reservoir-caprock (fig. 14), and introduce the following definitions:

- *storage site*: the defined volume within a geological formation used for the geological storage of CO_2 and associate surface and injection facilities;
- *storage complex*: the storage site and surrounding geological domain which can influence overall storage integrity and security, that is, secondary containment formations.

Therefore, a more comprehensive approach to subsurface characterization must be applied compared to the reservoir-oriented analysis that are usually carried out in oil & gas exploration/exploitation (fig. 15).

7·2. *Leakage prevention.* – In real storage systems many possible leakage pathways may jeopardize a CCS project (fig. 16). Leakage may occur due to the inefficiency of the geological containment (*i.e.*, top-, fault-seal failure) or it could be related to existing infrastructure, mainly legacy wells.

A risk analysis should be performed, in order to set up a risk register with all possible sources of leakage. Dedicated studies must be performed to find the best actions to mitigate the identified risks. In depleted reservoirs, the presence of many wells without CO_2-proof cement poses considerable risks during the project execution: a closed system is required in order to proceed with CO_2 injection. In deep aquifers, most leakage risks are associated with geologic uncertainties because of inadequate documentation, in contrast

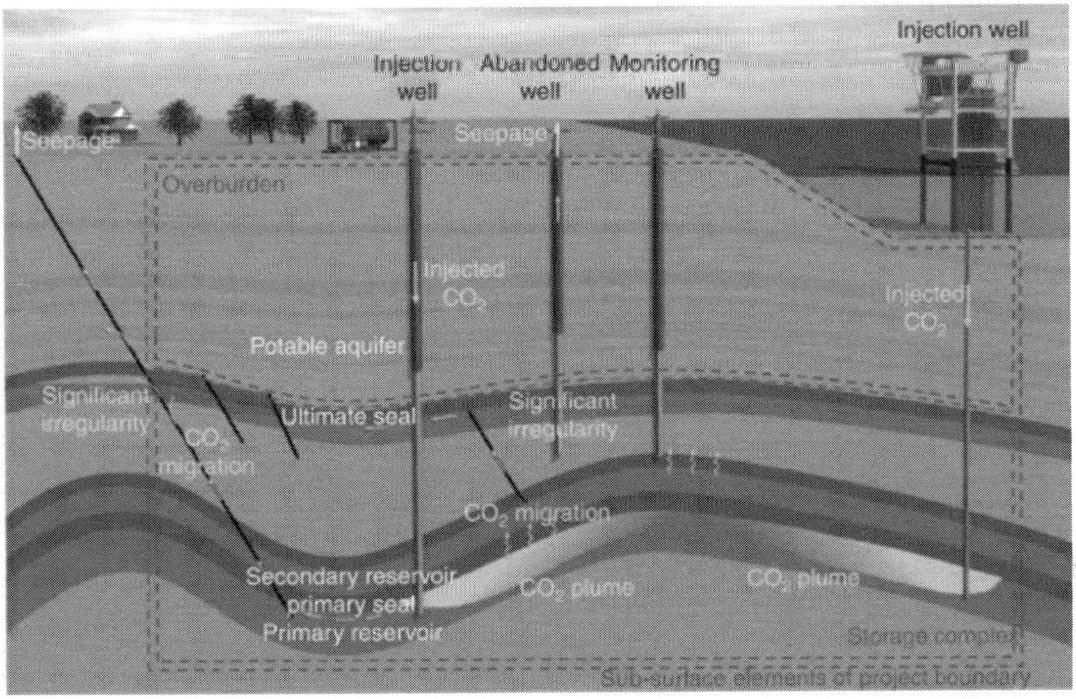

Fig. 15. – Conceptual cross-section of the CO_2 storage site/complex definitions. Highlighted in green, the outline of the "overburden" is also shown [7].

to oil wells. The geometry of the reservoir, its internal permeability distribution and the effectiveness of the seal are the major risks.

7·3. *Comparison between saline aquifers and depleted reservoirs.* – Saline aquifers have the largest capacity compared to the other reservoir types because they are available worldwide. Nevertheless, estimates of their CO_2 capacity are subject to a high degree of uncertainty, mainly because the knowledge of saline formations is quite limited.

Depleted reservoirs are well-known, structural trapping is proven (although related to different fluids) and volumetrics can be accurately computed with more confidence. In addition, depleted reservoirs are located in areas where existing infrastructure could be repurposed for CO_2 injection rather than hydrocarbon production. The pore pressure regime of the reservoir plays a crucial role for the injection strategy into depleted reservoirs or aquifers. When a rock reaches the fracture pressure, cracks are generated and permeability is enhanced, allowing the fluids to escape. The fracture pressure represents the highest limit that can be reached in injection operations. In depleted reservoirs, hydrocarbon production is likely to reduce the virgin reservoir pressure, allowing a larger "safe window" for injection operations compared to aquifers, that did not experience previous depletion. On the other hand, an excessive pressure reduction could move the CO_2 PVT([4]) conditions to the hydrate phase, which is not compatible with fluid injection.

([4]) PVT = Pressure, Volume, Temperature.

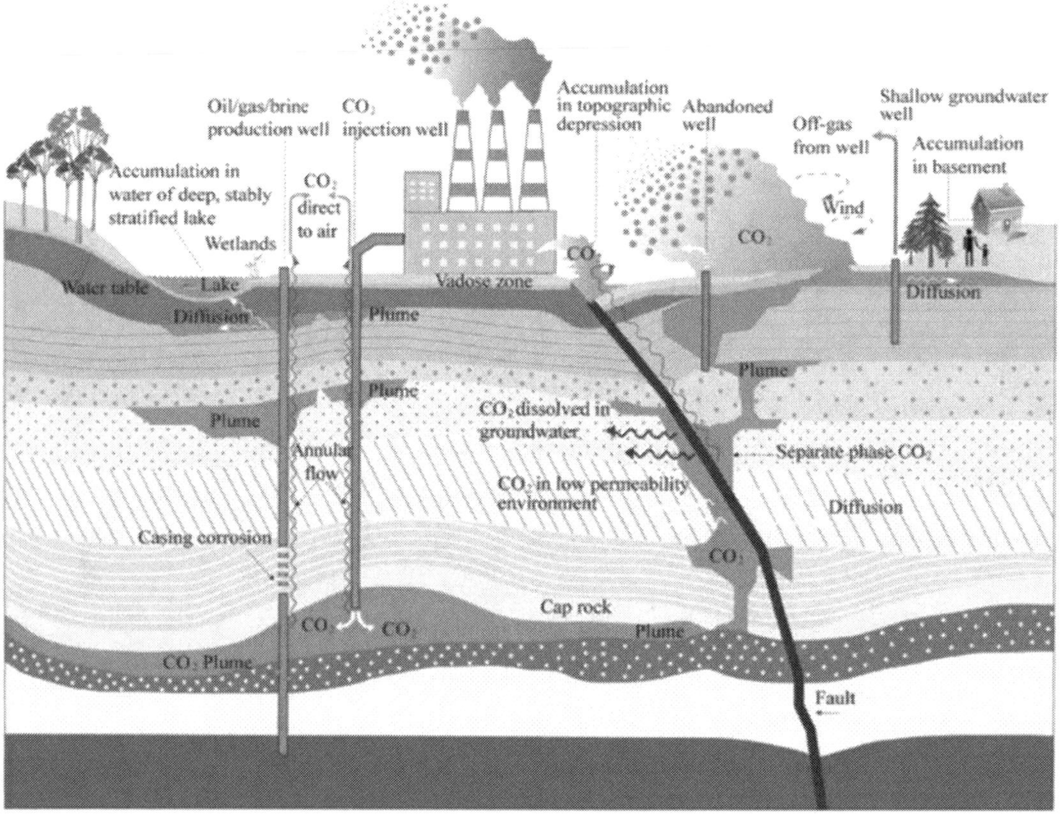

Fig. 16. – Possible leakage pathways of sequestered CO_2 [8].

8. – Monitoring

Each CCS project must have a dedicated MMV (Measurement Monitoring and Verification) program, which consists of a set of direct and indirect (mainly geophysical) measures to ensure conformance and containment [9]:

- *Containment*: the current security of CO_2 storage by verifying geological containment, well integrity, and the absence of any environmental effects outside the storage complex. Also, by detection of early warning signs of any unexpected loss of containment trigger, if necessary, additional safeguards to prevent or remediate any significant environmental impacts.
- *Conformance*: pressure and CO_2 accumulation inside the storage site are consistent with model-based forecasts and, if necessary, calibrate and update these models. Evaluate and, if necessary, adapt injection and monitoring to optimize storage performance. Provide the monitoring data necessary to support CO_2 inventory reporting.

Monitoring can be broadly divided in:

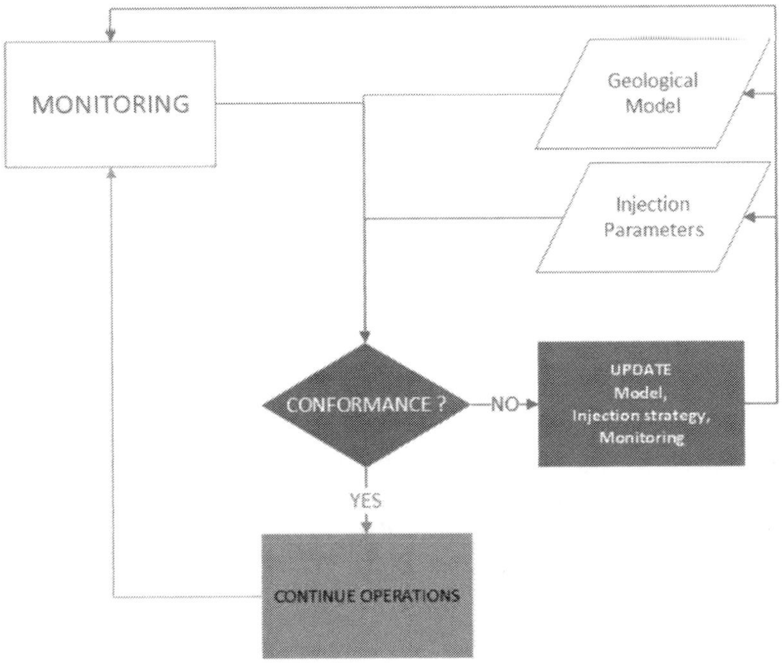

Fig. 17. – Block diagram of the MMV workflow. The geological model and the selected injection parameters are compared to the monitoring data. If there is conformance, operations can continue as planned. If not, an update of the geological model, injection strategy and monitoring should be implemented.

- Site storage monitoring—related to the injection management.
- Environmental monitoring—related to data acquisition for leakage detection.

Storage monitoring is aimed at the detection of the evolution of the CO_2 plume inside and, possibly, *outside* the reservoir. This is usually done by repeating geophysical surveys, such as 4D seismic, which records changes in seismic response due to the CO_2 that is replacing brine or hydrocarbons in the reservoir. Environmental monitoring is aimed at ensuring the environmental sustainability of the operations. It includes leakage detection, micro seismic monitoring and ground deformation monitoring.

The MMV workflow (fig. 17) should have strong feedback on the geological model and the injection strategy. The verification of the conformance between the expected system response and the actual data acquired during monitoring should guide the continuous update of the model, during and after operations.

8'1. *Geophysical 4D monitoring.* – Repeating the geophysical measurements over the years allows monitoring the evolution of the CO_2 plume. The periodicity of the data collection depends on the threshold of detectability of the changes in the properties, of course, on the costs.

https://doi.org/10.1051/epjconf/202226800005

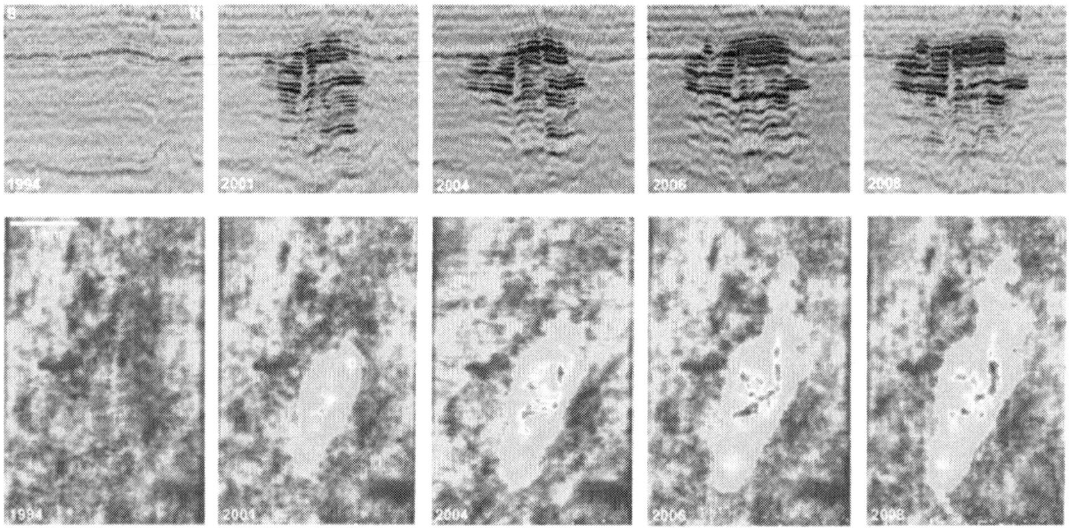

Fig. 18. – A spectacular example of 4D seismic from the Sleipner CO_2 storage field (Norway). From left to right the baseline survey and periodic monitoring survey are shown, documenting the evolution of the CO_2 plume over the years in cross-section (top) and plane view (bottom) [10].

8·1.1. 4D seismic. The reflection seismic method is recognized as the most effective monitoring technology because it can recognize subtle variations in CO_2 saturation with a 3D coverage. The physical principle of reflection seismics is that any acoustic impedance changes in the subsurface can produce a reflected sound wave when struck by an artificially generated seismic pulse. Acoustic impedance is related to the density and the velocity of the rocks according to the following simple formula:

$$I = \rho V_p,$$

where ρ is density and V_p acoustic velocity. The injection of CO_2 displaces the fluids in the pore spaces changing both velocity and density of the rock. Such changes can be detected by carefully analysing the seismic data acquired at different times. A baseline survey is required just before injection begins, and if the subsurface has a good 4D response, time-lapse seismic can easily be used to track the evolution of the CO_2 plume over the years, including potential leakage paths (fig. 18).

8·1.2. 4D gravity/electromagnetic/borehole seismic. Since CO_2 is lighter than brine (and heavier than natural gas), density variations can be detected by repeated gravimetric measurements. The (small) difference in CO_2 density compared to water can be used to track the plume during injection through 4D gravimetric surveys. The feasibility of such monitoring depends on the depth of the reservoir, the saturation and the amount of CO_2 injected, so it is not always applicable. In particular, the spatial resolution of the method is not comparable with 4D seismic.

Another non-seismic method that can be used for monitoring is electromagnetics (EM). When CO_2 replaces some of the saline pore fluid in an aquifer, the bulk electrical resistivity increases significantly, and this change can be monitored with electromagnetic measurements in the wells and at the surface. EM methods are very effective when surface and borehole data are combined, but they require a high density of boreholes with short distances between them. The same is true for borehole seismic techniques, which can provide very high-resolution 4D images but are geometrically limited due to the distribution and number of boreholes.

8·2. *Well monitoring.* – Geophysical monitoring is "indirect", since CO_2 saturation is derived from proxy measurements (*i.e.*, density, velocity, resistivity). Wells, instead, allow direct CO_2 measurements by sampling the formation fluid. They represent an important factor of any MMV program, especially in depleted reservoirs. One or more dedicated measuring holes are used for monitoring: they provide information about the CO_2 plume with unparalleled resolution.

Fluid sampling is not the only information that can be provided by newly drilled or repurposed wells: petrophysical parameters, caprock mechanical properties, cores, cement status are examples of data that can be collected in wells.

8·3. *Microseismic monitoring.* – The injection of CO_2 increases reservoir pore pressure, and thus it changes the stress distribution within the reservoir. Such changes may generate unwanted fractures in the weakest parts of the reservoir that should be detected and analysed to avoid induced seismicity. A microseismic network is a group of seismic sensors that are evenly spaced around the storage site, aimed at detecting any microearthquake that could be related to injection operations. Moreover, in seismic active areas, it can be used to identify naturally occurring earthquakes with better accuracy.

Building a microseismic network (fig. 19) is easier for onshore reservoirs, while offshore stations are costly and difficult to manage (fig. 20). If the storage project makes use of depleted fields, the existing platforms can be used to deploy microseismic stations. The platform provides power and communication facilities, while the sensors are placed on the sea bottom.

8·4. *Ground deformation monitoring.* – The pressure increase due to CO_2 injection into a reservoir usually has only insignificant effects on the surface. However, especially for shallow reservoirs, a measurable bulge right above the storage site can occur. In order to detect possible ground deformations, a monitoring strategy must be implemented. Many technologies are available for such monitoring:

- High precision leveling
- Settlement gauge/piezometer
- LiDAR/Photogrammetry
- InSAR
- Hi-res bathymetry
- Continuous-GPS.

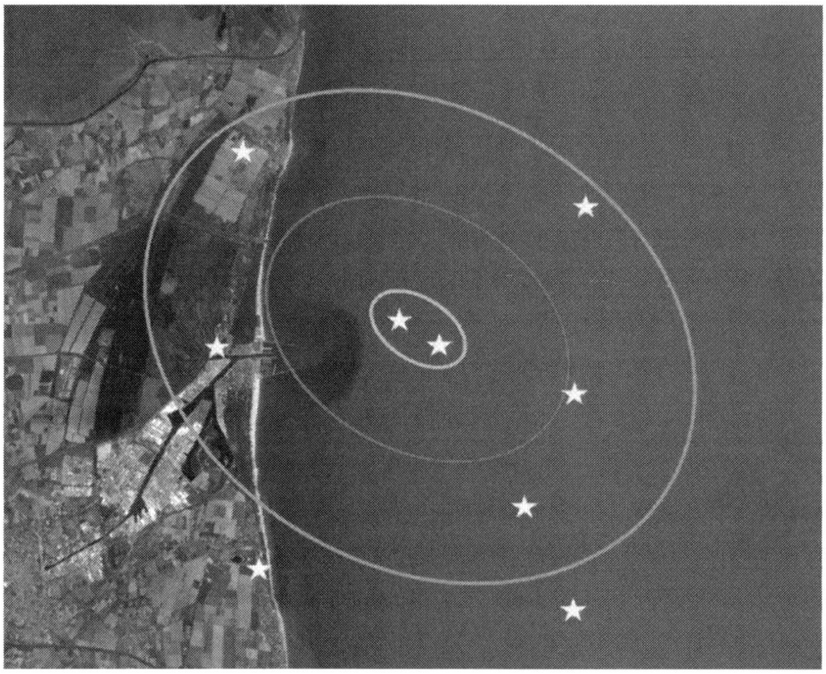

Fig. 19. – Example of a hybrid (onshore/offshore) microseismic network. Stars indicate the locations of seismic stations. Their number, azimuthal distribution and characteristics must comply with local guidelines, and are the result of a dedicated feasibility study. The green contour indicates the area of the depleted field, while the red and blue contours are at a distance of 5 and 10 km from the field.

Each of them has its own accuracy and applicability, so they are usually applied in combination. The site location (onshore or offshore) plays a fundamental role: leveling, piezometers and InSAR cannot be used in offshore environments. In addition, measurements should be combined according to the characteristics of each technology. For example, GPS is extremely precise, but provides only point measurements (fig. 21), while InSAR is less accurate, but can cover a large area.

8·5. *Surface leakage monitoring.* – Measuring CO_2 leakage is very difficult because of the high natural CO_2 "noise" in the biosphere. A baseline survey aimed at detecting annual, seasonal and daily variations of CO_2 natural emissions must be included in the monitoring plan. CO_2 can be detected using *p*H measurements, variations of CO_2 *vs.* O_2 concentrations, vegetation changes, tracers and direct CO_2 detectors.

8·6. *Geophysical and geological studies for CCS projects.* – Despite some differences, the regulatory framework for CCS is quite homogeneous regarding containment. The importance of preventing any leakage at the surface and ensuring permanent containment of CO_2 in the reservoir is recognized worldwide.

Fluid-dynamic, geomechanical and geochemical (G&G) modeling are indispensable to assess the containment of a reservoir for CO_2 storage. However, prior to modeling,

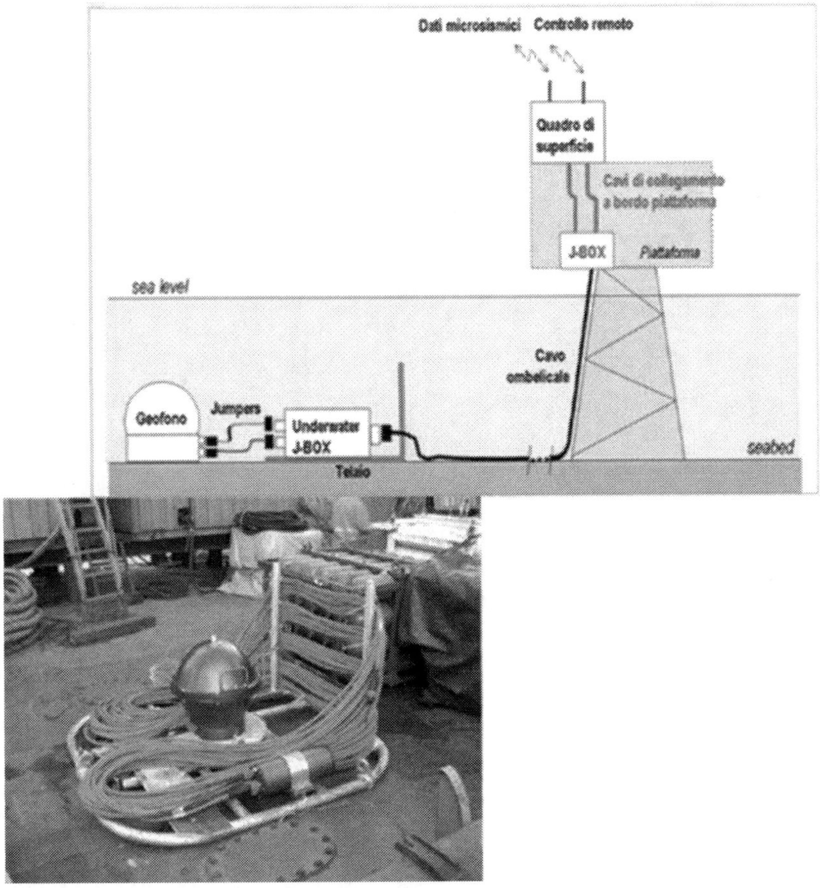

Fig. 20. – Schematic view of an offshore micro seismic station (top) and photograph of the underwater sensor (bottom).

a comprehensive characterization of the subsurface should be carried out. Integrated G&G studies provide geological/geophysical analysis of both the storage complex and overburden successions, in order to better assess their properties at various scales, and mitigate problems associated with injection and storage. Such studies can be summarized as follows:

- *Geophysical characterization*
  - seismic inversion (acoustic, elastic, Extended Elastic Impedance)
  - rock physics modeling
  - calibration to rock properties
  - classification/segmentation
- *Structural characterization*
  - structural interpretation

Fig. 21. – Continuous GPS onshore station.

- structural modelling
- fracture studies for caprock and reservoir
- fault seal analysis

– *Stratigraphic/sedimentological characterization*

- biostratigraphic analyses on core/cuttings (with different tools based on lithology)
- stratigraphic definition/revision of well-tops
- well top homogenization on neighbouring fields
- Paleoenvironmental reconstructions
- characterization of the depositional model
- seismo-stratigraphic/sedimentological analysis (on both seismic and attributes)
- integration with petrographic/mineralogical data
- analysis of the carbonate pore network

– *Basin-scale fluid migration analysis*

- qualitative analysis for monitoring purposes

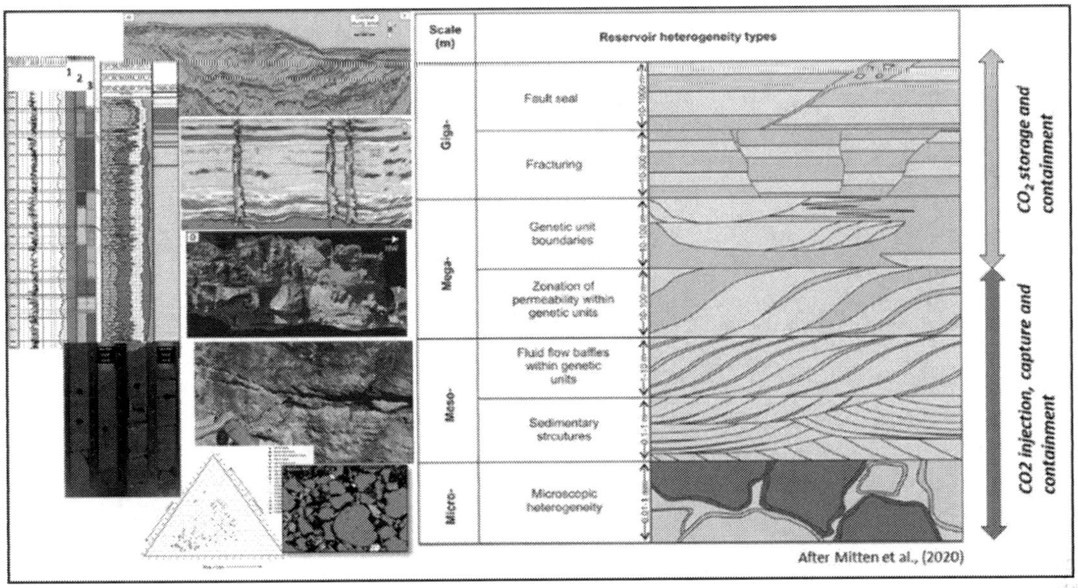

Fig. 22. – "Scale factor" in subsurface characterization. Different data are used at different scales.

The study area must be larger than the extent of the sites used for injection purposes to allow an adequate geologic evaluation. The geological characteristics of the basin should drive the selection of the area to be investigated for both the storage complex and the overburden, and a comprehensive G&G dataset is required (fig. 22).

The result of the integrated studies is the correct definition of the storage complex and the overburden, in terms of geometries and petrophysical parameters within a 3D "geological shared model" (fig. 23). This model is the basis for any subsequent dynamic modelling. Depleted reservoirs are less prone to geological uncertainties than saline aquifers, but the fact that the reservoir is "known" should not preclude the importance of reviewing data and updating models, within and outside the reservoir.

8·7. *Modelling.* – Similar to oil & gas exploitation, fluid-dynamics modelling is the reference study for the evolution of the CO_2 plume. Efficiently and accurately solving the equations governing fluid flow in subsurface reservoirs is very challenging because of the complex geological environment and the properties of fluids (especially CO_2) at high pressure. Complex multiphase and multicomponent systems should be modelled to describe miscible gas injection processes during CO_2 sequestration.

With respect to confinement, geomechanical modelling is of paramount importance. It determines if and how the pressure changes in the reservoir during injection may affect the integrity of the caprock or ground level movements (*i.e.* subsidence, swelling). The results of fluid-dynamic modeling in terms of pressure variations are the inputs for the geomechanical modelling, although sometimes a coupled approach is used to understand the changes in effective stress dynamics with the decrease in reservoir pressure due to

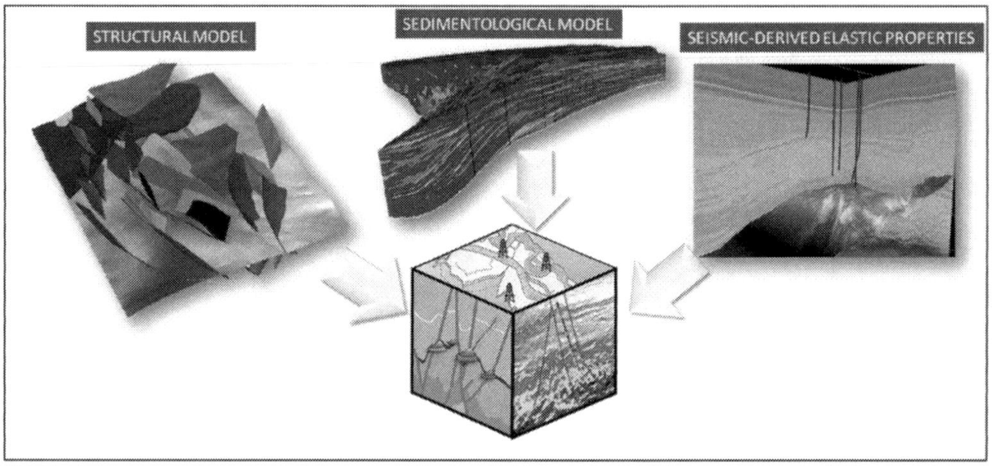

Fig. 23. – A "shared model" approach to integrated G&G studies. All results from each discipline are combined into a single 3D geological model, that can be used as input for all further modelling.

injection. Geomechanical modelling also provides an indication about the stability of faults around the reservoir, which is an important information when to evaluate induced seismicity.

As CO_2 is not inert at reservoir pressure and reacts with brines and rocks, reactive transport modelling (RTM) is very important in CCS projects. RTM in porous media refers to the creation of models that link chemical reactions with transport of fluids in the reservoir. Such models predict the of the spatial and temporal distribution of chemical reactions that occur along a flow path.

9. – CCUS in Eni

Eni is committed to the implementation of CCUS because:

- It is a mature and scalable technology that is considered the only viable option for energy-intensive industries.

- When combined with other measures, CCUS is a cost-effective way to rapidly reduce greenhouse gas emissions.

- It is an enabling technology for the future elimination of carbon dioxide.

- It promotes the development of a blue economy and, through utilization, a circular economy of CO_2.

- It enables synergies with existing facilities with minimal footprint, boosting local economies, employment levels and creating long-term value.

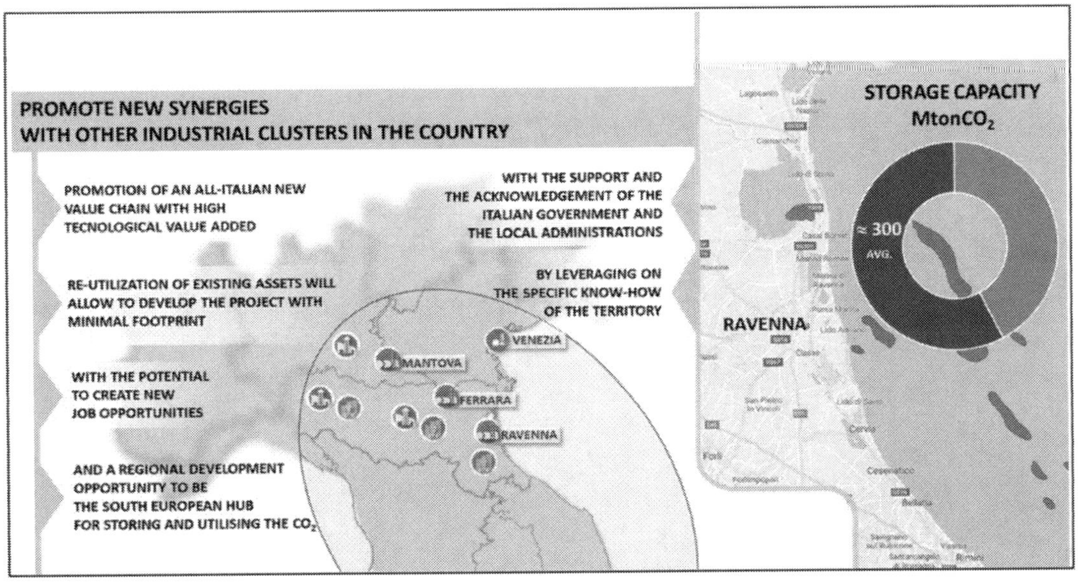

Fig. 24. – Adriatic Blue CCS project. The aim is to de-carbonize Italian industrial districts by means of CO_2 injection in offshore depleted gas reservoirs. Existing facilities (pipelines, platforms) will be re-purposed.

"Without CCUS as part of the solution, reaching our climate goals is practically impossible. We simply cannot afford another decade of muted progress". IEA, Fatih Birol - November 2018.

Eni has several ongoing projects in the world of CCS (Carbon Capture and Storage) and CCU (Carbon Capture and Utilisation) technology.

Eni is optimizing every stage of the process, from transport to fluid-rock interaction and field monitoring systems, to make the technology more efficient and facilitate its large-scale application. On the operational side, Eni plans to build one of the world's largest CO_2 storage hubs —the first in the Mediterranean— off the coast of Ravenna. The conversion of depleted gas fields in the Adriatic Sea into exclusive permanent CO_2 storage sites and the reuse of a small part of the existing infrastructure will provide a quick and concrete solution for reducing emissions in the Italian industrial sector at very competitive prices. CCS is in particular the only readily available option for so-called hard-to-abate sectors such as cement, steel, chemical plants, etc., where a significant share of carbon dioxide emissions is linked to the industrial process and therefore cannot be avoided by *e.g.*, electrification or renewables. Internationally, Eni is also a partner in HyNet North West in the United Kingdom, in the Liverpool Bay area on the north-west coast of the country. Outside Europe, we are also assessing the feasibility of a CO_2 capture project in the United Arab Emirates at Ghasha and investigating a CCS application in Libya, for the Bahr Essalam project. in addition, Eni is exploring opportunities to develop CCS projects in Australia and East Timor.

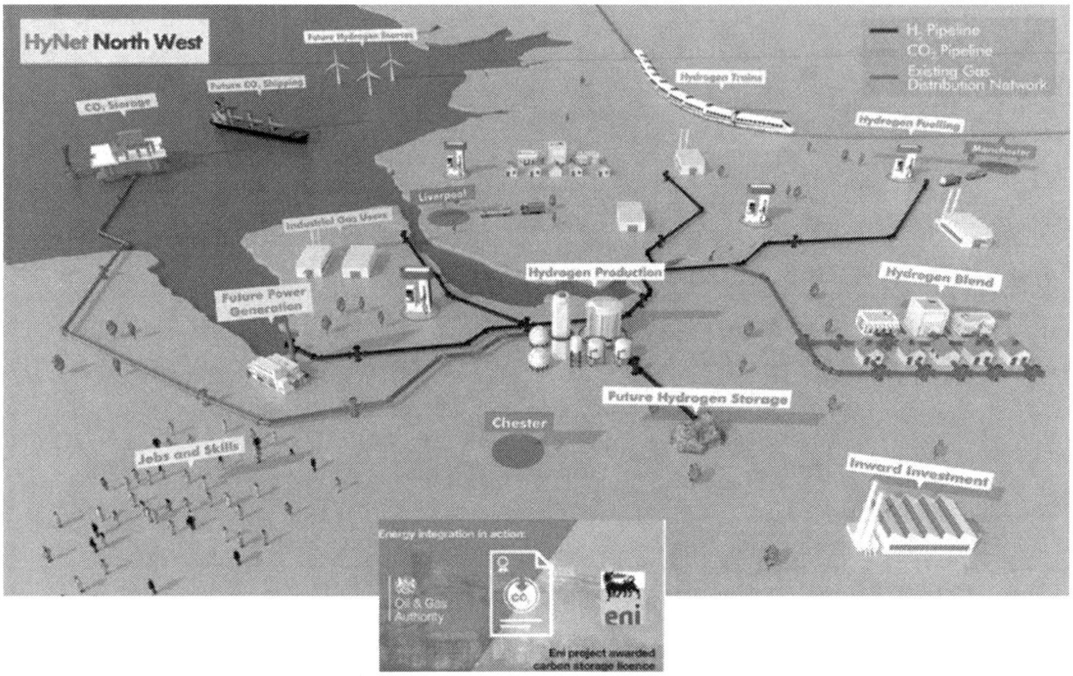

Fig. 25. – Overview of the HyNet project (Eni operator): decarbonisation of refineries, fertiliser and cement plants, H_2 production with Autothermal Reforming (ATR) in Northern Wales. Storage licence awarded in October 2020. Planned start-up: 2025.

The presence of depleted gas fields and decommissioned plants at the Ravenna offshore site provides Eni with a unique opportunity to create a large hub for the storage of CO_2 generated by onshore production (fig. 24). The Ravenna project is not only aimed at reducing Eni's emissions. By combining our extensive knowledge of reservoir dynamics with new technologies, we aim to create the leading centre for carbon dioxide capture and storage for Italy and the Mediterranean in the Middle Adriatic. Above all, we aim to provide a concrete solution for all of Italy's strategic industrial sectors, such as cement, steel, paper and chemicals, which account for about 20% of Italy's emissions and for which there are currently no viable technological alternatives that could be rapidly implemented. Part of the expected initial 4 Mtpa of the project will be reserved for these industrial sectors, which, may then be rapidly increased, to over 10 Mtpa of CO_2 in the following years, thanks to the availability of large storage capacities (500 Mt).

Regarding activities in the UK, in October 2020 Eni received a licence from the UK Oil and Gas Authority for the HyNet North West project, which aims to build a storage hub in Liverpool Bay. This project also involves the reuse of Eni's depleted offshore fields, with Eni taking operational responsibility for CO_2 storage and transport activities, to reduce emissions generated by the region's major industrial centre, where a large 4 Mtpa hydrogen production plant will also be built. The project is scheduled to start in 2025, with the possibility of expanding the capture capacity to

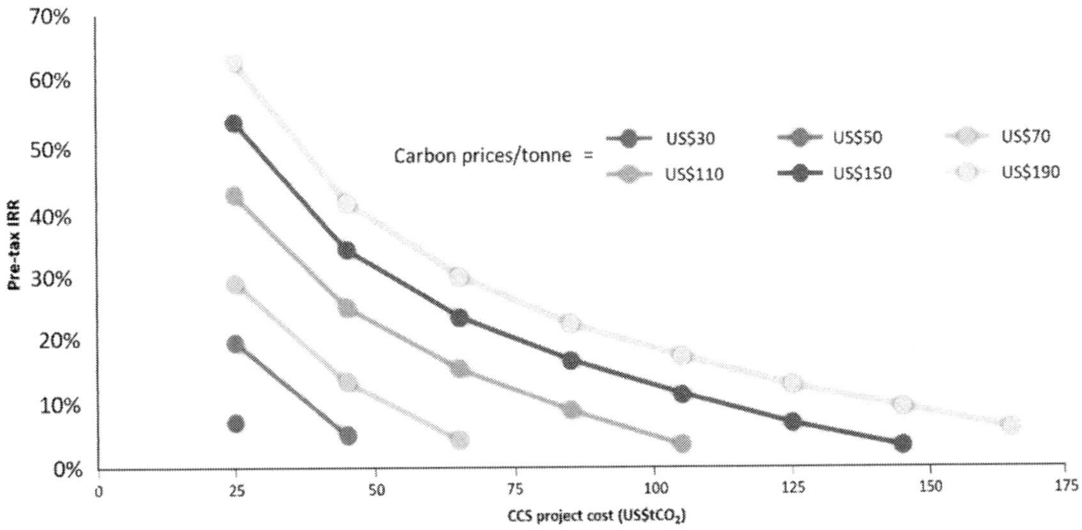

Fig. 26. – CCS project costs *vs.* pre-tax-Internal Rate of Return. The current carbon price (< 30 \$/ton) hinders the economic viability of any CCS project [1].

10 Mtpa by 2030. The business plan was submitted to the UK authorities on July 9, 2021 and, in October 2021 Eni UK, on behalf of the HyNet Consortium Cluster, was able to announce that its application to the Cluster Sequencing process had been accepted as a Track 1 project. This will enable Eni UK and its supporting Hynet companies to become one of the first UK industry clusters to apply carbon capture and storage (CCS), to significantly reduce carbon emissions in the UK. Meanwhile, several Memoranda of Understanding have already been signed with companies interested in capturing, transporting and storing their emissions in Eni UK's depleted fields. The agreements signed so far cover the hard-to-abate sectors and will play a critical role in decarbonising the industrial area in Northwest England and North Wales (fig. 25).

10. – Remarks on the economic viability of CCS

No CO_2 storage project is currently economically viable. Governmental incentives are essential to drive existing and planned CCS project. The CO_2 price is far below the break-even point, regardless of the project type or the amount of CO_2 injected. The carbon price needs to be at least \$ 50/ton for most stand-alone CCS projects to be economically feasible. The average global carbon price is currently around 23 \$/ton. At a carbon price of \$ 190/ton, even a CCS project costing \$ 150/ton can generate a 10% return on investment (fig. 26).

REFERENCES

[1] Wood McKenzie Report, 2021.
[2] Global Status of CCS 2021 - CCS Institute, `https://www.globalccsinstitute.com/resources/global-status-report`.
[3] Ramachandran Hariharan, Pope Gary and Srinivasan Sanjay, "Effect of Thermodynamic Phase Changes on CO_2 leakage", *Energy Proc.*, **63** (2014) 3735.
[4] Međimurec N. G., Novak-Mavar K. and Majic M., *Carbon Capture and Storage (CCS): Technology, Projects and Monitoring Review* (The Mining-Geology-Petroleum Engineering Bulletin) 2018.
[5] Jun Young-Shin, Zhang Lijie, Min Yujia and Li Qingyun, "Nanoscale Chemical Processes Affecting Storage Capacities and Seals during Geologic CO_2 Sequestration", *Acc. Chem. Res.*, **50** (2017) 1521.
[6] Bachu S. and Shaw J. C., "CO_2 storage in oil and gas reservoirs in western Canada: effect of aquifers, potential for CO_2-flood enhanced oil recovery and practical capacity", in *Proceedings of the 7th International Conference on Greenhouse Gas Control Technologies* (Elsevier) 2005, pp. 361–369.
[7] Jagger E. and Drosin E., "8-Risk assessment of CO_2 storage complexes and public engagement in projects", in *Geological Storage of Carbon Dioxide* (CO_2), edited by Jon Gluyas and Simon Mathias (Woodhead Publishing) 2013.
[8] Vishal V. and Singh T. N. (Editors), *Geologic Carbon Sequestration – Understanding Reservoir Behavior* (Springer) 2016.
[9] Dean M. and Tucker O., "A risk-based framework for Measurement, Monitoring and Verification (MMV) of the Goldeneye storage complex for the Peterhead CCS project", *Int. J. Greenhouse Gas Control*, **61** (2017) 1–15.
[10] Chadwick A., Williams G., Delepine N., Clochard V., Labat K., Sturton S., Buddensiek M. L., Dillen M., Nickel M., Lima A. L., Arts R., Neele F. and Rossi G., "Quantitative analysis of time-lapse seismic monitoring data at the Sleipner CO_2 storage operation", *The Leading Edge*, **29** (2010) 170.

Sb_2Se_3: a possible future for thin-film photovoltaics?

G. Spaggiari(*)

Dipartimento di Scienze Matematiche, Fisiche e Informatiche, Università di Parma - Parma, Italy

Istituto dei Materiali per l'Elettronica ed il Magnetismo, CNR - Parma, Italy

S. Rampino

Istituto dei Materiali per l'Elettronica ed il Magnetismo, CNR - Parma, Italy

D. Bersani

Dipartimento di Scienze Matematiche, Fisiche e Informatiche, Università di Parma - Parma, Italy

Summary. — Antimony selenide (Sb_2Se_3) is today one of the most promising alternative materials for p-type absorbers in thin-film photovoltaics, with an optimal band-gap and a very high absorption coefficient. However, its crystal structure is extremely anisotropic and its natural carrier density is generally very low. Sb_2Se_3 thin films have been deposited by two different high-energy techniques: magnetron RF-sputtering (MS) and low-temperature pulsed electron deposition (LT-PED). Their dominant crystallographic orientations have been studied as a function of deposition parameters and of the different used substrates, while complete solar cells have been subsequently made with the obtained samples to confirm the dependence of conversion efficiencies on the observed $(Sb_4Se_6)_n$ ribbon orientation. Cu-doped Sb_2Se_3 thin-films have been also preliminary prepared in order to evaluate a possible route to further improve the free charge-carrier density and the cell performance.

(*) Corresponding author. E-mail: giulia.spaggiari1@unipr.it

1. – Introduction

Today the world is in strong need of economic and efficient renewable energy sources. While commercial silicon-based solar panels are already available for large-scale photovoltaic energy production, at lower and lower cost and with efficiencies closer and closer to the maximum theoretical value, thin-film solar cells are instead making their way in the parallel growing markets of building-integrated photovoltaics, semitransparent and bi-facial solar panels, flexible and/or light-weight applications. The leading materials in thin-film photovoltaics are today amorphous-silicon (a-Si), cadmium telluride (CdTe) and copper-indium-gallium selenide (CIGS), but each of them has a weakness: efficiency of a-Si modules is languishing since several years between 11 and 12% [1]; CdTe modules sales are threatened by the carcinogenicity of cadmium-related compounds after decomposition [2], despite their good efficiency and solid production process; CIGS modules are more expensive than the other two because of the high prices of indium and gallium [3]. For all these reasons, in the last years researcher are intensively seeking for possible alternative absorber materials to be exploited in thin-film solar cells, possibly featuring high efficiency, low toxicity, and low competitive cost, all at the same time.

Antimony selenide (Sb_2Se_3) is currently one of the most studied materials for this purpose. It is indeed characterized by very promising physical properties: a 1.22 eV direct band gap (1.03 eV indirect) [4], a very high density of states in the proximity of the valence band maximum, a very high absorption coefficient (more than $10^5\,cm^{-1}$) [5], and a 32.2% theoretical power conversion efficiency (PCE) limit (calculated by the Shockley-Queisser model) [6]. At the same time, other properties are strongly related to its highly anisotropic crystal structure, dominated by 1D ribbons made of (Sb_4Se_6) stacked units along the c-axis of the orthorombic cell ($Pbnm$ space group symmetry, $a = 11.62$ Å, $b = 11.77$ Å, $c = 3.962$ Å), as clearly visible in fig. 1. Within each ribbon covalent bonds are present, while van der Waals forces are predominant between different ribbons. So, for example, hole mobility drops from $42\,cm^2V^{-1}s^{-1}$ to $0.7\,cm^2V^{-1}s^{-1}$ when the charge motion direction is changed from along the c-axis to along a perpendicular one [7]. This means that arranging these ribbons perpendicular to the substrate, in order to maximize the carrier transport properties, is a key challenge in the deposition of Sb_2Se_3 thin films for photovoltaic applications. Once researchers focused on this goal, PCE values reported in the literature rapidly grew from less than 1% to 10% [5, 8].

In this work we report the comparison between different Sb_2Se_3 thin films, deposited by two different high-energy techniques: low-temperature pulsed electron deposition (LT-PED) and Magnetron RF-Sputtering (MS). The process parameters and the substrates that produced the films with the best average crystallographic orientation are here highlighted. Complete cell structures were also realized with some of the deposited thin films and the cells performance was then correlated to the previously observed crystallographic properties. Finally, preliminary results on the production of Cu-doped Sb_2Se_3 thin films are also reported and discussed, in order to assess the effects of a possible increase in the charge carrier density, which is intrinsically low in this material.

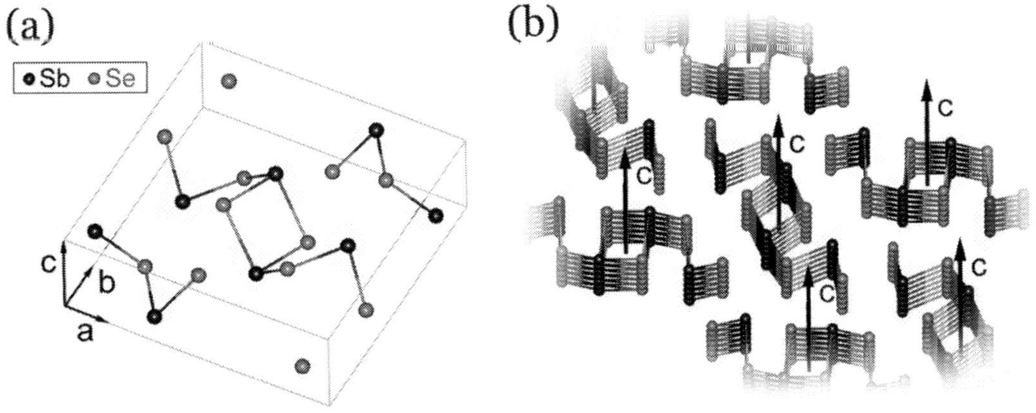

Fig. 1. – Crystal structure of Sb_2Se_3 is characterized by an orthorombic cell in the *Pbnm* space group: (a) unit cell; (b) (Sb_4Se_6) ribbons visualization, here obtained by the stacking of $n = 6$ planes along the *c*-axis.

2. – Experimental

Sb_2Se_3 thin films were deposited using both LT-PED and MS techniques. At first, bare soda-lime glass was used to determine the best growth conditions for Sb_2Se_3 crystallization, using different substrate temperatures: between 200 °C and 400 °C for LT-PED and between 50 °C and 330 °C for MS. Other parameters were previously optimized and reported (see ref. [9] for LT-PED and ref. [10] for MS). Then, depositions were also performed with both techniques on different substrates, *i.e.* molybdenum (Mo), fluorine tin oxide (FTO), cadmium sulfide on FTO (CdS/FTO), and undoped zinc oxide (ZnO), at 300 °C to verify how crystallization was influenced by them.

The structural properties of the obtained films, with a special focus on preferential orientations, were characterized by a X-ray diffraction system (XRD, Siemens D500), in Bragg-Brentano geometry and with CuKα radiation. Morphology and composition were instead controlled by a field-emission gun scanning electron microscope (FESEM-FIB, Zeiss Auriga Compact) equipped with an energy dispersive X-ray spectrometer (EDX, Oxford) optimized with a Co standard. The current density *vs.* voltage (J-V) characteristics of the solar cells were instead measured under standard test conditions (solar simulator, ABET SUN 2000) by means of a multimeter (Keithley 2614B).

3. – Results and discussion

Both LT-PED and MS are high-energy techniques that can be successfully employed in the deposition of Sb_2Se_3 thin films. We previously demonstrated that LT-PED can be successfully used for the deposition of different layers for photovoltaic applications [9, 11-14], while MS is widely recognized as one of the leading techniques in industrial applications. The use of high-energy techniques generally enables the use of low temperatures for the substrates, because the generated ions have enough energy to move and/or rearrange on the substrate surface, hence resulting in a good crystallization without the need of high-temperature kynetic energy. This is a very important request when industrial application is considered and/or temperature sensitive substrates are used (glass, polymers, etc.).

In the following analysis we are going to address the effect of deposition conditions on thin-films quality in terms of "good" or "bad" preferential crystalline orientations of their grains. Since charge transfer is much better along the $(Sb_4Se_6)_n$ ribbons, the best possible preferential crystalline orientation for thin-film solar cells is the one providing vertically aligned ribbons on the substrate (*i.e.*, when $00l$ peaks are dominant in XRD spectra), while the worst preferential crystalline orientation is the one with ribbons lying on the substrate (*i.e.*, when $hk0$ peaks are dominant in XRD spectra). Generally also ribbons oriented along an intermediate (hkl) crystalline direction with $l \neq 0$ and a high inclination angle to the substrate plane are considered "good", because also in that case charge carriers can run through a rather short oblique path towards the cell contacts.

Then the obtained structural results are compared with the performance of photovoltaic cells made with some of the deposited Sb_2Se_3 thin films.

3·1. *Crystallization vs. temperature.* – The effect of substrate temperature on Sb_2Se_3 films crystallization was controlled at first on bare glass substrates, in order to exclude any effect due to the presence of a matching/unmatching templating lattice. The XRD spectra of selected films are reported in fig. 2 to highlight the main results.

Thin films deposited by LT-PED have a poor crystallization below 200 °C, which then improves with increasing temperature. However, while at 300 °C films with a partially "good" orientation are obtained, at higher temperature "bad" orientations become dominant (fig. 2(a)). On the contrary, thin-films deposited by MS crystallize above 80 °C but start to have grains with "good" orientation only above 260 °C (fig. 2(b)). At higher temperature (> 330 °C), and for long deposition times, SEM imaging revealed the formation of elongated nano/micro-wires over the surface [10], which may be interesting for the realization of 3D cell structures, but not desired for this study.

Hence, a common substrate temperature of 300 °C was chosen as the best one to make comparisons between the two techniques and the films obtained on a different type of substrates.

3·2. *Crystallization vs. substrate.* – Since the substrate texture and lattice can act as a template, that favours different crystallizations, the average preferential crystalline orientation was compared for Sb_2Se_3 films that where deposited at 300 °C on different substrates. In order to simplify the comparison, the following values can be defined [9]:

$$TC(hkl) = n \times \frac{\frac{I(hkl)}{I_0(hkl)}}{\sum_n \frac{I(hkl)_n}{I_0(hkl)_n}}, \qquad (1a)$$

$$RCT = \sum_n \left(TC(hkl)_n \times \cos\alpha(hkl)_n\right), \qquad (1b)$$

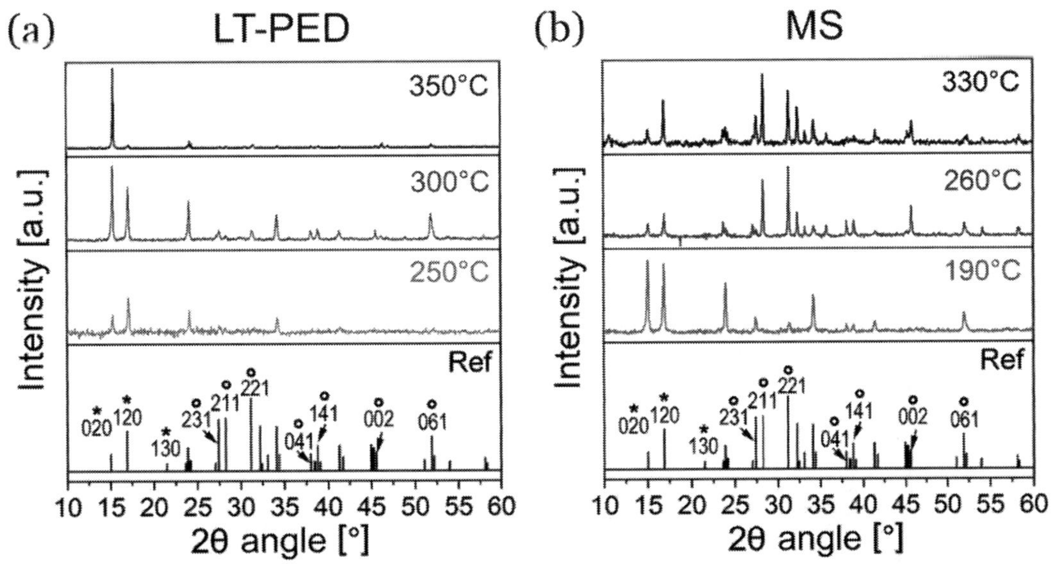

Fig. 2. – X-ray diffraction patterns of Sb_2Se_3 films deposited at different temperatures on bare glass substrates: (a) by LT-PED; (b) by MS. JPCDS 15–0681 reference is reported at the bottom, where a few selected reflections are highlighted with * symbol, if related to ribbons laying parallel to the substrates, or with ° symbol, if related to ribbons with an orientation having a non-zero perpendicular component (hkl with $l \neq 0$).

where $I(hkl)$ is the relative intensity of the (hkl) peak in the collected XRD pattern and $I_0(hkl)$ is the relative intensity for the same peak in the JCPDS 15–0681 reference; summation are calculated for $n = 10$ reflections, chosen among the more intense ones that have been experimentally observed in the collected spectra and labeled in the reference pattern in fig. 2: (020), (120), (130), (231), (211), (221), (041), (141), (002) and (061); $\alpha(hkl)$ is the angle between the ribbon axis and the normal to the surface for hkl orientation.

$TC(hkl)$ is the normalized "texture coefficient", while RCT is the "ribbon carrier transport" coefficient. According to the definition, a preferential orientation of grains in ASe films along one of these directions is present when the corresponding $RTC > 1$; so the larger is the RCT value, the better is the average preferential crystalline orientation of the Sb_2Se_3 film.

The comparison of the obtained RCT values obtained for the different substrates, using both techniques at 300 °C, is reported in fig. 3(a). While the best preferential orientation is achieved on FTO for LT-PED deposition, CdS provides the best result in the case of MS deposition. Good results are also obtained with ZnO, while Mo, which is often used in other cell structures reported in the literature, is the worst substrate for both techniques. As a general consideration, the plot in fig. 3(a) shows that MS deposition generally performs better than LT-PED. The details of XRD spectra of Sb_2Se_3 films on CdS and Mo are reported in fig. 3(b) to highlight the differences.

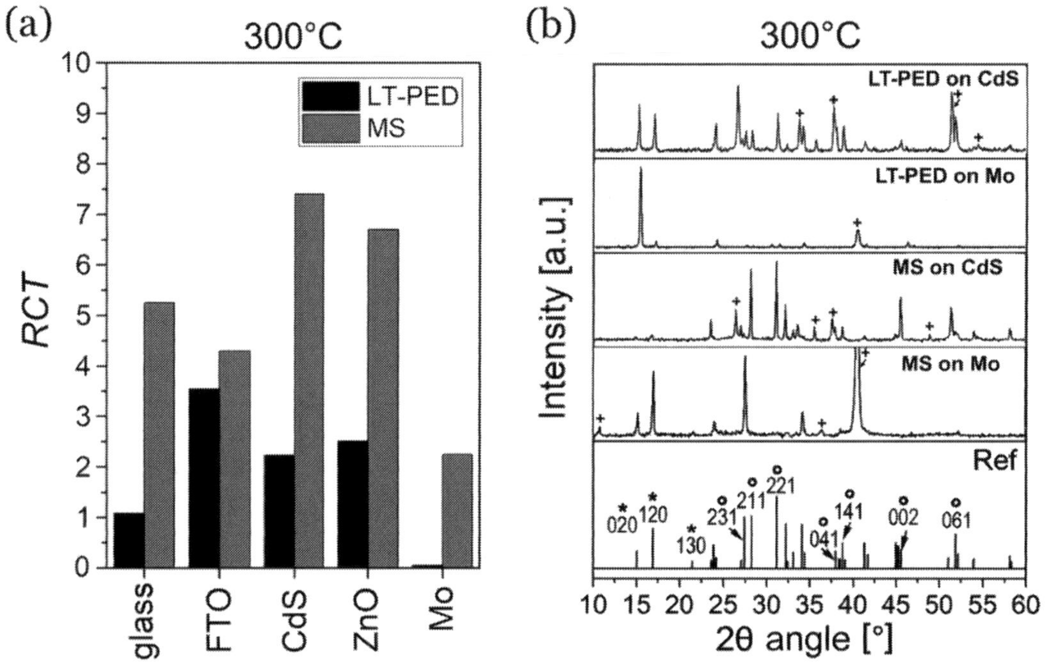

Fig. 3. – (a) *RCT* values for Sb_2Se_3 films deposited on different substrates by LT-PED and MS at 300 °C. (b) X-ray diffraction patterns of Sb_2Se_3 films deposited on Mo and CdS at 300 °C using LT-PED and MS; peaks from the used substrates are marked with + symbol; JPCDS 15–0681 reference is reported at the bottom, where a few selected reflections are highlighted with * symbol, if related to ribbons laying parallel to the substrates, or with ° symbol, if related to ribbons with an orientation having a non-zero perpendicular component (*hkl* with $l \neq 0$).

3·3. *Cell performance and Cu-doping.* – In order to confirm the correlation between the preferential crystalline orientation of grains in Sb_2Se_3 films and the photovoltaic performance, complete cells were made with some of the samples that showed the best and the worst orientations: Mo and FTO by LT-PED deposition, as well as Mo, FTO and CdS by MS deposition (see refs. [9, 10] for details on cells production). The *J-V* characteristics of these cell are reported in figs. 4(a) and (b) respectively.

From these plots it is easily visible that the cell performance is somehow proportional to the correct or uncorrect alignment of ribbons in the film (*i.e.*, proportional to the *RCT* value). Both cells with a Sb_2Se_3 film deposited on Mo show very low short-circuit current density (J_{SC}) values. The cell with a Sb_2Se_3 film deposited by LT-PED on the FTO substrate has $J_{SC} = 20.28\,\mathrm{mA\,cm^{-2}}$, open-circuit voltage ($V_{OC}$) = 256 mV, fill factor (FF) = 39.2%, PCE = 2.1%. The cell with a Sb_2Se_3 film deposited by MS on FTO substrate does not perform as well, having a poor FF, but the one on CdS substrate, that is related to the highest *RCT* value, has the best result: $J_{SC} = 27.06\,\mathrm{mA\,cm^{-2}}$, $V_{OC} = 266\,\mathrm{mV}$, FF = 32.7%, PCE = 2.36%.

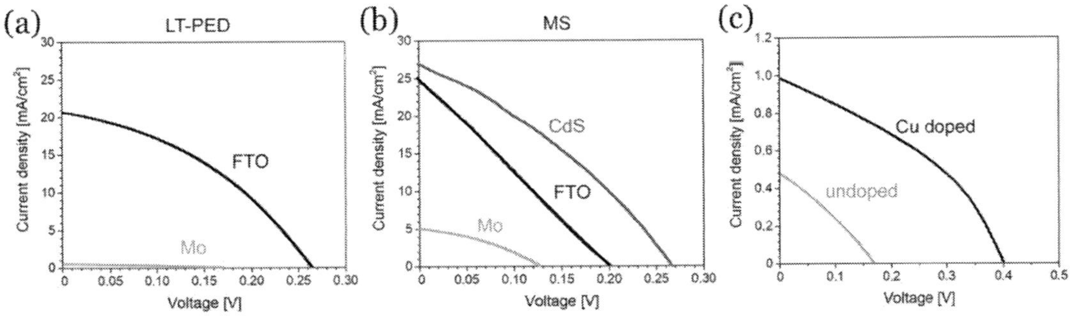

Fig. 4. – (a) *J-V* plot for the two cells made with Sb_2Se_3 films deposited by LT-PED: ZnO:Al/ZnO/CdS/(Sb_2Se_3/Mo)/glass and ZnO:Al/ZnO/CdS/(Sb_2Se_3/FTO)/glass. (b) *J-V* plot for the three cells made with Sb_2Se_3 films deposited by MS: ZnO:Al/ZnO/CdS/(Sb_2Se_3/Mo)/glass, ZnO:Al/ZnO/CdS/(Sb_2Se_3/FTO)/glass, and superstrate Au/(Sb_2Se_3/CdS)/FTO/glass. (c) *J-V* plot for the undoped and the Cu-doped cells made with Sb_2Se_3 films deposited by LT-PED.

All these results, however, are clearly affected by a very low V_{OC}. One of the possible explanation for such a low value may be found in the very low free charge-carrier density (namely as low as $10^{13}\,cm^{-3}$) that generally characterize Sb_2Se_3 [5]. In order to proof this point, a Cu-doped Sb_2Se_3 film was deposited on the Mo substrate by LT-PED, using a Cu-doped Sb_2Se_3 target (custom synthesis). The preliminary result is shown in fig. 4(c) for the samples on the Mo substrate: thanks to the doping V_{OC} increases up to 0.4 V, nearly doubling the value of the undoped films.

4. – Conclusions

Two industrial-production–oriented techniques, LT-PED and MS, have been tested for the deposition of Sb_2Se_3 thin films for photovoltaic application. This study clearly demonstrates the importance of a "good" crystallografic orientation of Sb_2Se_3 grains in the deposited films, avoiding $(Sb_4Se_6)_n$ ribbons to lay on the same plane of the substrate, to build a working photovoltaic cell based on a Sb_2Se_3 absorber. The best orientation in LT-PED and MS depositions, evaluated by means of their *RCT* values, have been respectively obtained on FTO and CdS substrates, heated at 300 °C. Despite the measured PCE values, that are the results of the contributions from all the cell layers and interfaces (to be optimized in the next future), the characterization of the cells, made with the deposited Sb_2Se_3 films, clearly confirms the correlation between the cell performance and its average crystallographic orientation. Finally, the preliminary test of a cell, made with a Cu-doped Sb_2Se_3 film, showed a possible way to improve the V_{OC} value in this kind of devices.

REFERENCES

[1] Green M. A., Dunlop E., Hohl-Ebinger J., Yoshita M., Kopidakis N. and Hao X., *Prog. Photovolt.*, **29** (2021) 3.
[2] Nain P. and Kumar A., *Renew. Sustain. Energy Rev.*, **119** (2020) 109592.
[3] Anctil A. and Fthenakis V., *Prog. Photovolt.*, **21** (2013) 1253.
[4] Vadapoo R., Krishnan S., Yilmaz H. and Marin C., *Phys. Status Solidi B*, **248** (2011) 700.
[5] Mavlonov A., Razykov T., Raziq F., Gan J., Chantana J., Kawana Y., Nishimura T., Wei H., Zakutayev A. Minemoto T., Zu X., Li S. and Qiao L., *Sol. Energy*, **201** (2020) 227.
[6] Shockley W. and Queisser H. J., *J. Appl. Phys.*, **32** (1961) 510.
[7] Cao Y., Zhu X., Jiang J., Liu C., Zhou J., Ni J., Zhang J. and Pang J., *Sol. Energy Mater. Sol. Cell.*, **26** (2020) 110279.
[8] Tang R., *Nat. Energy*, **5** (2020) 587.
[9] Pattini F., Rampino S., Mezzadri F., Calestani D., Spaggiari G., Sidoli M., Delmonte D., Sala A., Gilioli E. and Mazzer M., *Sol. Energy Mater. Sol. Cell*, **218** (2020) 110724.
[10] Spaggiari G., Pattini F., Bersani D., Calestani D., De Iacovo A., Gilioli E., Mezzadri F., Sala A., Trevisi G. and Rampino S., *J. Phys. D: Appl. Phys.*, **54** (2021) 385502.
[11] Rampino S., Pattini F., Bronzoni M., Mazzer M., Sidoli M., Spaggiari S. and Gilioli E., *Sol. Energy Mater. Sol. Cell.*, **185** (2018) 86.
[12] Cavallari N., Pattini F., Rampino S., Annoni F., Barozzi M., Bronzoni M., Gilioli E, Gombia E., Maragliano C., Mazzer M., Pepponi G., Spaggiari G. and Fornari R., *Appl. Surf. Sci.*, **412** (2017) 52.
[13] Mazzer M., Rampino, S., Spaggiari G., Annoni F., Bersani D., Bissoli F., Bronzoni M., Calicchio M., Gombia E., Kingma A., Pattini F. and Gilioli E., *Sol. Energy Mater. Sol. Cell.*, **166** (2017) 247.
[14] Bronzoni M., Stefancich M. and Rampino S., *Thin Solid Films*, **520** (2012) 7054.

Electrocatalysis with quantum chemistry

KALISHANKAR BHATTACHARYYA(*)

Department of Molecular Theory and Spectroscopy, Max-Planck-Institut für Kohlenforschung, Kaiser-Wilhelm-Platz 1, 45470 Mülheim an der Ruhr, Germany

Summary. — The following article presents a brief introduction to modeling an electrochemical reaction. Two crucial concepts, oxidation-reduction and acid-base reactions, are briefly illustrated to understand the structural changes of the electrocatalyst. These two concepts are applied to compute the stability of catalysts for electrochemical reactions from the density functional theory calculations.

1. – Introduction

Electrocatalysis is one of the forefront leading technologies to store intermittent energy and its transformation to valuable chemicals [1]. Significant interest from academic and industrial research efforts has been increased to utilize the full potential of electrocatalysts for maximum productivity, sustainability, and stability of the catalyst materials [2]. Electrochemical reactions, including water splitting reactions (oxygen and hydrogen evolution reaction), carbon dioxide reduction, and nitrogen reduction, are among the most important reactions for developing sustainable energy storage solutions [3-7]. Proton

(*) E-mail: ksb@kofo.mpg.de

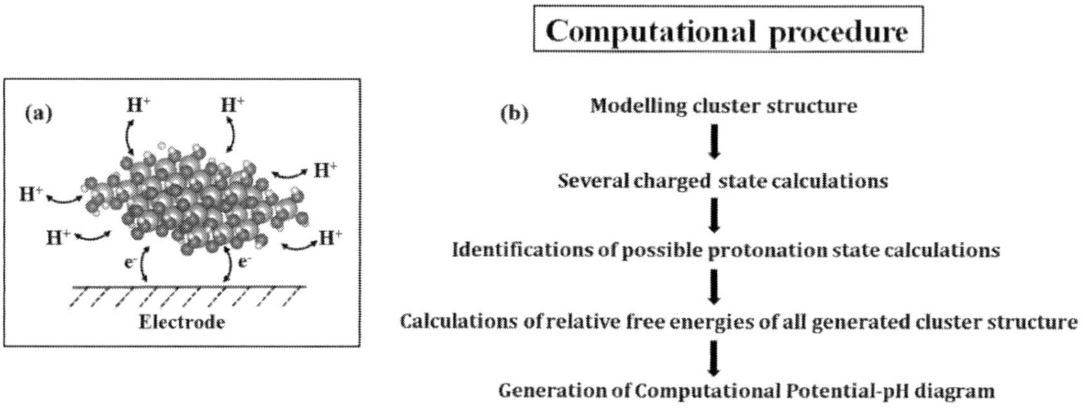

Fig. 1. – (a) Modeling of electrocatalytic materials, including the effect of potential and pH. An electrode is held for the electron transfer, and proton transfer is studied using an implicit solvation model. (b) Computational protocol for the calculations of the potential and the effect of pH of an electrocatalytic reaction.

transfer (PT), electron transfer (ET), and successive proton-coupled electron transfer (PCET) are the fundamental basis of these electrochemical reactions [8, 9]. Therefore, unique aspects of the electrocatalysts are based on controlling these proton/electron transfer kinetics and thermodynamics, which are regulated by the reaction conditions, including electrode potential, temperature, concentrations, and electrolyte [10, 11]. In this regard, computational modeling based on a predictive-quality framework in addition to experimental results has represented a pivotal role for catalysts design and optimization with atomic-scale mechanistic insights [12, 13].

2. – Computational modeling of electrocatalytic reactions for finite-size systems

As shown in fig. 1, the effect of potential and pH on the amorphous structures considering the exchange of electrons with an electrode and protonation/deprotonation with a solvent are considered simultaneously. Upon changing the potential and pH, charged states (electron transfer with the electrode) and protonation states (proton transfer with the water around the cluster) of the catalysts also change. We provide a detailed computational modeling to incorporate the effect of potential and pH in the next section.

3. – Influence of potential: determination of E^0

This section outlines our methodology to compute the effect of potential that is conceptually different from the widely applied constant charge method [14]. In the standard DFT calculations, the number of electrons is kept constant during the simulations, and the electrochemical potential can be tuned by varying the proton-coupled electron transfer (constant charge method). On the other hand the electrode potential in our approach is a pre-set variable and can be kept constant through changing the number of electrons independently of the proton transfer (constant potential method) [15, 16].

The main reason for applying the potential to the cluster structures in exchange with an external electron reservoir (electrode) is to determine the charges state of the system that has to be the lowest energy at a specific potential. Therefore, the electrode potential can be tuned by varying the number of electrons as shown in eq. (1). On the other hand, the applied potential can thus alter the direction of equilibrium either towards A^q or A^{q-1}. Therefore, other charge states also need to be considered to find the most stable charged structures at a given potential:

$$A^q + e^- \rightleftharpoons A^{q-1}. \tag{1}$$

Computationally, we optimize all possible charged structures with an implicit solvation model and compute their energy difference. Note that the energy difference between the two charged states, A^q and A^{q-1} is equivalent to the energy requirement of the electron transfer that corresponds to the potential of an electron, *i.e.*, to the electron in *a* vacuum. It is essential to mention that reference electrodes (RHE or SHE electrode) are considered during an experimental measurement. Henceforth, reference values should be regarded throughout the potential calculations to compare the computational data with experimental results. Trasatti proposed $-4.44\,\text{eV}$ as a reference value for SHE electrodes [17]. Recently, Isse and Gennro reported $-4.28\,\text{eV}$ by measuring the solvation free energy of a proton [18]. Although the reference value so far has been reported ranging from $-4.05\,\text{eV}$ to $-4.85\,\text{eV}$ in the literature, and it is still debated which value is most reliable. Taken $-4.28\,\text{eV}$ as a reference value, the equilibrium potential *vs.* SHE can be expressed as

$$E^0{}_{V\ vs.\ SHE}(A^q/A^{q-1}) = -(\mu(A^{q-1}) - \mu(A^q)) - 4.28\ [\text{eV}]. \tag{2}$$

Equation (2) is the general expression of the electrode potential on the scale of SHE, and these values will be defined as the stable boundary region of the respective charged states as shown in fig. 2(a).

4. – Influence of pH: determination of pK_a

Now, we discuss how the effect of pH is modeled in our computational methodology to understand the impact of electrolytes. Like the electrode potential computations, the effect of pH of a given electrolyte on the cluster structure is to identify the protonation sites with the lowest free energy at a specific pH. In this way, the thermodynamic equilibrium of the acid-base reactions will be established as follows:

$$AH_{n-1}^{q-1} + H^+ \rightleftharpoons AH_n^q, \tag{3}$$

where AH_n^q is acid and its conjugated base is AH_{n-1}^{q-1}. Therefore, several protonation sites will be computed to estimate the stability region by computing the pK_a values.

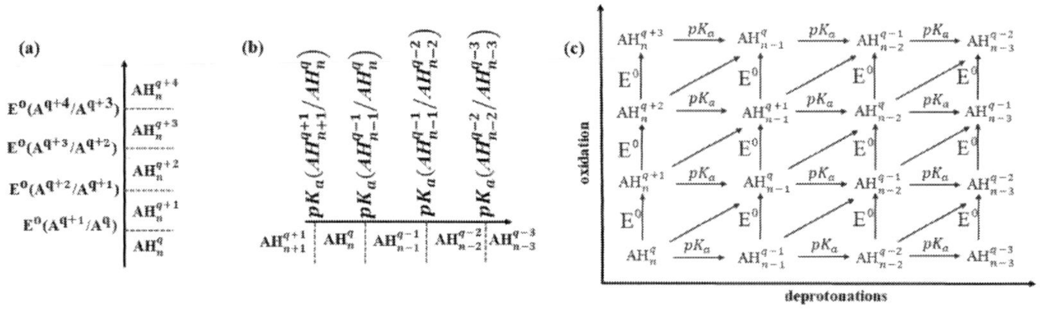

Fig. 2. – (a) Schematic description of each charge state for a potential domain with the corresponding reduction potentials. (b) Schematic depiction of each protonation state for the pH region of stability with the corresponding pK_a values as boundaries. (c) Schematic description of the two-dimensional phase diagram based on the oxidation-reduction and protonation-deprotonation reactions.

pK_a values of cluster models are calculated according to

$$pK_a = \frac{\mu(AH_{n-1}^{q-1}) + \mu_{\text{eff}}(H^+) - \mu(AH_n^q)}{RT\ln(10)} . \tag{4}$$

These values are applied to set the boundaries of the pH domain of each protonation site for the cluster, as shown in fig. 2(b).

5. – Computational potential-pH diagram: relative stability of the electrocatalyst

Based on the above methodology for the computations of potential and pK_a values for cluster structures, we now started adding and removing electrons and protons from the cluster independently and computing the potential for each electron transfer and pK_a for the proton transfer step. Hence, a two-dimensional phase diagram will be generated based on the boundaries of potential and pK_a values, as shown in fig. 2(c). This two-dimensional phase diagram will provide a stable regime based on potential and pK_a. However, to find the thermodynamic stabilities of the cluster at a particular potential and pH, we need to compute their relative stabilities with respect to the reference state: on eq. (5):

$$\begin{aligned}\Delta_r G_{AH_m^x}(AH_n^y) = {} & \mu(AH_n^y) - \mu(AH_m^x) + (m-n)\mu_{\text{eff}}(H^+) \\ & - (m-n)RT\ln(10)p\text{H} - (y-x+m-n)(U+4.28)\ [\text{eV}].\end{aligned} \tag{5}$$

Based on eq. (5), the lowest energy structures for a given potential and pH can be generated, and stable charged structures with available protonation sites of the cluster are easily identified. For a range of potential and pH, if these computations are carried out, the most stable cluster structure is specified in the three-dimensional computational

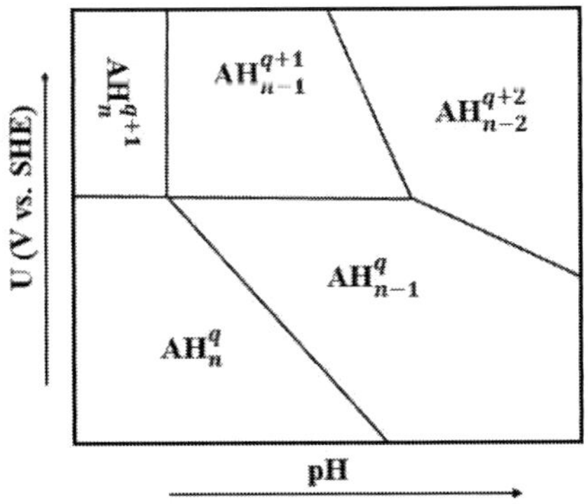

Fig. 3. – Computational potential-*p*H diagram describing the most stable structures based on the potential and *p*H scale.

potential-*p*H diagram as depicted in fig. 3. The interested reader can look at refs. [19] and [20] for detailed applications of the proposed methodology into the electrocatalytic reactions.

6. – Summary

This article presented a brief primer about incorporating the potential and the effect of *p*H on the electrocatalyst materials using electronic structure calculations. Based on the explicit consideration of the potential and the implicit consideration of the solvent effect for *p*H in our methodology, a computational framework is employed to investigate the structural changes of the model cluster structure. Consequently, the effect of the potential after the computations of the energetics of the neutral systems in *a posteriori* way is not required. Investigating the electron transfer through the explicit consideration of the potential and protonation-deprotonations process using the implicit solvation model will allow us to explore the electrocatalytic reactions.

* * *

The author thanks Professor Alexander A. Auer for his supervision to work on this problem and acknowledge the support from Alexander Von Humboldt Foundation for post-doctoral fellowship and Max Planck Society for computational facilities.

REFERENCES

[1] Bard A. J. and Faulkner L. R., *Electrochemical Methods: Fundamentals and Applications*, 2nd edition (John Wiley & Sons, Inc., New York) 2001.
[2] Hagen J., *Industrial Catalysis: A Practical Approach*, 2nd edition (Wiley-VCH, Weinheim, Germany) 2006.
[3] Trasatti R. S., *J. Electroanal. Chem.*, **39** (1972) 163.
[4] McCrory C. C., Jung S., Peters J. C. and Jaramillo T. F., *J. Am. Chem. Soc.*, **135** (2013) 16977.

[5] Rong X., Parolin J. and Kolpak A. M., *ACS Catal.*, **6** (2016) 1153.
[6] Montoya J. H., Tsai C., Vojvodic A. and Nørskov J. K., *ChemSusChem*, **8** (2015) 2180.
[7] Bhattacharyya K. and Datta A., *Phys. Chem. Chem. Phys.*, **21** (2019) 12346.
[8] Solis B. H. and Hammes-Schiffer S., *Inorg. Chem.*, **53** (2014) 6427.
[9] Koper M. T., *Chem. Sci.*, **4** (2013) 2710.
[10] Horvath S., Fernandez L. E., Soudackov A. V. and Hammes Schiffer S., *Proc. Natl. Acad. Sci. U. S. A.*, **109** (2012) 15663.
[11] Man I. C., Su H. Y., Calle-Vallejo F., Hansen H. A., Martinez J. I., Inoglu N. G., Kitchin J., Jaramillo T. F., Nørskov J. K. and Rossmeisl J., *ChemCatChem*, **3** (2011) 1159.
[12] Nørskov J. K., Bligaard T., Rossmeisl J. and Christensen C. H., *Nat. Chem.*, **1** (2009) 37.
[13] Seh Z. W., Kibsgaard J., Dickens C. F., Chorkendorff I., Nørskov J. K. and Jaramillo T. F., *Science*, **355** (2017) 146.
[14] Norskov J. K., Rossmeisl J., Logadottir A., Lindqvist L., Kitchin J. R., Bligaard T. and Jonsson H., *J. Phys. Chem. B*, **108** (2004) 17886.
[15] Lozovoi A., Alavi A., Kohanoff J. and Lynden-Bell R., *J. Chem. Phys.*, **115** (2001) 1661.
[16] Hörmann N. G., Andreussi O. and Marzari N., *J. Chem. Phys.*, **150** (2019) 041730.
[17] Trasatti S., *Pure Appl. Chem.*, **58** (1986) 955.
[18] Isse A. A. and Gennaro A., *J. Phys. Chem. B*, **114** (2010) 7894.
[19] Poidevin C. and Auer A. A., *Carbon*, **171** (2021) 618.
[20] Bhattacharyya K., Poidevin C. and Auer A. A., *J. Phys. Chem. C*, **125** (2021) 4379.

Nuclear energy basics

S. Leray

CEA Paris Saclay - Irfu, France

Summary. — In the first part of the paper, the basic features of nuclear fission and fusion energy are addressed. In the second part, an overview of energy consumption in the world is presented, which highlights the growing need for electricity from non-CO_2 emitting sources, among which nuclear power is expected to play a role. The perspectives for the future of nuclear energy are then discussed, showing that it develops mostly in emerging countries and the small modular reactors are likely to have an increasing role. The third part is devoted to the question of nuclear waste management and to the non-radioactive environmental impact of nuclear energy in comparison with other sources of energy.

1. – Basics on nuclear fission and fusion

Nuclear energy originates from the release of part of the energy that binds together the constituents of the atomic nucleus. Nuclei are composed of nucleons (neutrons and protons) held together by the strong nuclear force, which overcomes the Coulomb repulsion between the protons. The mass of the nucleus is less than the sum of the masses of its constituents. For a nucleus of mass A with Z protons, the mass is given by

$$M(A,Z) = Zm_p + (A-Z)m_n - B(A,Z), \tag{1}$$

where B is the binding energy.

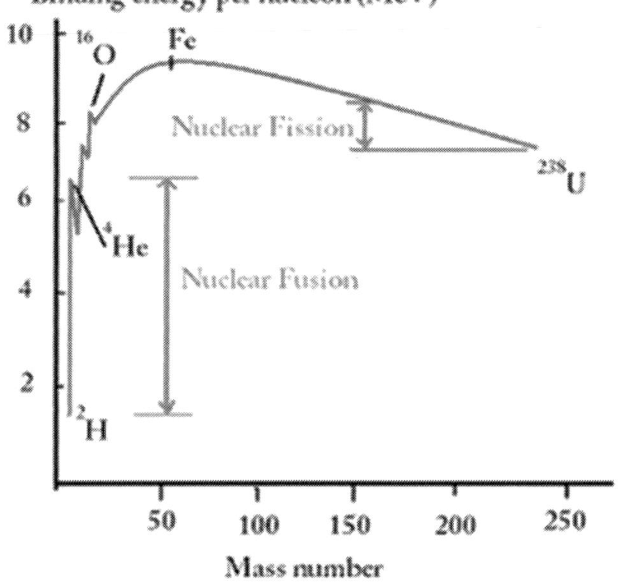

Fig. 1. – Binding energy as a function of the nucleus mass.

Actually, the binding energy per nucleon, B/A, is of the order of several MeV and is maximum around Fe, as can be seen in fig. 1 where it is plotted as a function of the mass of the nucleus. The consequence is that part of this binding energy can be released in nuclear reactions in which the constituents of the initial nuclei are redistributed into different final nuclei. In a reaction between two initial nuclei, A_1 and A_2, leading to A_3 and A_4 in the final state, the energy released, Q, is just the difference between the binding energies:

$$Q = B(A_1) + B(A_2) - B(A_3) - B(A_4). \tag{2}$$

A large amount of energy can be released, in particular in two cases corresponding to the two ends of the curve of fig. 1: either by inducing the splitting of heavy nuclei, such as uranium, into two more or less equal ones with neutrons —this is fission, in which around 1 MeV per nucleon is released, *i.e.* of the order of 200 MeV in total— or by merging two very light nuclei into a larger one — this is fusion, in which several MeV per nucleon are liberated.

1˙1. *Fission.* – Fission reactions can be induced by capture of very slow neutrons on some nuclei such as ^{235}U, which are said to be fissile, producing two fission fragments and 2.4 neutrons in average. The produced neutrons, after slowing down, can initiate further reactions leading to what is called a chain reaction. In a reactor, the chain reaction must be controlled so that it is self-sustained without the risk of diverging. This implies that for each fission reaction, one neutron is used to induce another fission and the remaining neutrons are absorbed or escape from the system. This is criticality.

Natural uranium is composed of 0.7% ^{235}U, 99.3% ^{238}U and traces of ^{234}U. ^{235}U is fissile by neutron capture regardless of the energy of the neutron, but the probability that a reaction can take place decreases with increasing neutron energy. Therefore, it is much easier to ensure the criticality with very slow neutrons, actually neutrons that are in thermal equilibrium with the medium they are interacting with, called thermal neutrons. This implies that the neutrons produced by fission are slowed down in a moderator material. ^{238}U also can fission but only with high-energy neutrons and with a small probability. In a thermal reactor, it has a high probability to absorb neutrons making it difficult to reach criticality with natural uranium. This is why it is necessary to enrich uranium by increasing the proportion of ^{235}U in the fuel. Typically, the enrichment is around 3% in light-water reactors.

^{235}U is the only naturally occurring fissile isotope but other isotopes are fissile, such as ^{239}Pu and ^{233}U that can be produced in reactors through the absorption of neutrons by ^{238}U and ^{232}Th, respectively. In both cases, a neutron is first captured, generating a nucleus which then undergoes two subsequent β-decay leading to the long-lived fissile isotope

$$(3) \qquad {}^{238}\text{U}\ (n,\gamma) \longrightarrow {}^{239}\text{U}\ (23\,\text{min}) \xrightarrow{\beta} {}^{239}\text{Np}\ (2.3\,\text{d}) \xrightarrow{\beta} {}^{239}\text{Pu}\ (2.4\ 10^4\,\text{y}),$$

$$(4) \qquad {}^{232}\text{Th}\ (n,\gamma) \longrightarrow {}^{233}\text{Th}\ (22\,\text{min}) \xrightarrow{\beta} {}^{233}\text{Pa}(23\,\text{d}) \xrightarrow{\beta} {}^{233}\text{U}\ (1.6\ 10^5\,\text{y}),$$

^{238}U and ^{232}Th are said to be fertile. In fact, in light-water reactors, up to one third of the fissions come from ^{239}Pu over the lifetime of the fuel. It can be seen from eq. (4) that natural thorium, which is composed only of the ^{232}Th isotope, can also be used as fuel in thermal reactors but it requires either that ^{232}Th is first irradiated in another reactor to provide ^{233}U or that plutonium is added to initiate the process.

It is also possible to burn ^{238}U in fast neutron reactors that use directly the neutrons produced by fission without moderation. This has the advantage that it multiplies the energy that can be extracted from a given mass of natural uranium by a factor of about 100. However, this requires a higher neutron flux in the reactor and the mobilization of a much larger inventory of fissile material, which can be made up of highly enriched uranium, or plutonium.

More details on how a nuclear fission reactor works are given in Ripani's contribution to these proceedings [1].

1·2. *Fusion.* – As said above and shown by fig. 1, the fusion of light isotopes can produce a huge amount of energy. These reactions are in fact what drives the nucleosynthesis of light elements and are the primary source of energy for stars. This is possible thanks to the gravitational attraction that leads to the formation of a hot and very dense plasma of light isotopes. To replicate this process on earth to produce energy in a controlled way is very difficult and requires the confinement of a very hot and dense plasma of hydrogen isotopes, with a temperature high enough to overcome the Coulomb barrier between the

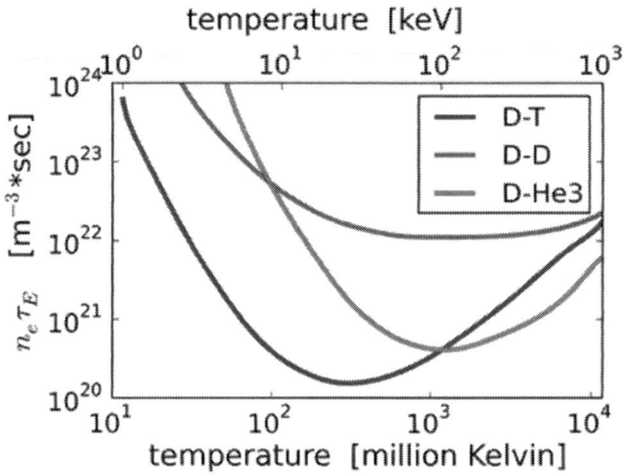

Fig. 2. – Lawson criterion for three possible fusion reactions: *d-d*, *d-t* and *d*-^{3}He. From [3].

ions, during a sufficient time, and being able to recover the generated energy.

The reactions that can be considered to produce energy in a fusion reactor are

- Deuterium-deuterium reactions

(5) $$^2\mathrm{H} + {}^2\mathrm{H} \to {}^3\mathrm{He} + n + 3.27\,\mathrm{MeV},$$

(6) $$^2\mathrm{H} + {}^2\mathrm{H} \to {}^3\mathrm{He} + {}^1\mathrm{H} + 4.03\,\mathrm{MeV}.$$

- Deuterium-tritium reaction

(7) $$^2\mathrm{H} + {}^3\mathrm{H} \to {}^4\mathrm{He} + n + 17.6\,\mathrm{MeV},$$

  of which 14 MeV are carried out by the neutron.

- Deuterium-^{3}He reaction

(8) $$^2\mathrm{H} + {}^3\mathrm{He} \to {}^4\mathrm{He} + {}^1\mathrm{H} + 18.3\,\mathrm{MeV},$$

  which is interesting as it does not produce neutrons.

The conditions for a sustained fusion plasma producing more energy than is required for its heating are generally expressed as the Lawson criterion [2] shown in fig. 2 for three different fusion reactions. In the figure T is the temperature, τ_E is the confinement time, n_e, is the ion density.

It can be seen that the most favorable reaction, *i.e.* the one requiring the lowest values of temperature and $n_e\tau_E$ product, is by far the deuterium-tritium reaction, on which current fusion energy projects are based. Deuterium is rather abundant in nature and easy to obtain by separation from ordinary water. On the contrary, tritium is a radioactive isotope, with a half-life of 12 years, which naturally exists only in trace amounts. Therefore, tritium breeding is a key issue for fusion energy economics. Actually, tritium can be produced from lithium which is largely abundant in nature. Natural lithium is composed of 7% of ^{6}Li and 93% of ^{7}Li and tritium can be produced using the reactions

$$^6\mathrm{Li} + n_{\mathrm{slow}} \rightarrow {}^4\mathrm{He} + T + 4.8\,\mathrm{MeV}, \tag{9}$$

$$^7\mathrm{Li} + n_{\mathrm{fast}} \rightarrow {}^4\mathrm{He} + T + n_{\mathrm{slow}} + 2.5\,\mathrm{MeV}. \tag{10}$$

Most fusion reactor designs involve surrounding the main vessel with a cover made of a lithium-based material. It remains to demonstrate that it is possible to attain tritium self-sufficiency.

Two main technologies are considered to realize the confinement of the hot plasma at a temperature of the order of 10^8 K:

- inertial confinement, using laser beams heating and compressing small fuel pellets, allowing to reach very high densities (10^6 times the air density) but during short characteristic time (10^{-11} s);
- or magnetic confinement, in which a low-density plasma (10^{-5} times the air density) is confined in a vessel, a tore in the case of tokamaks, by a high magnetic field, during a rather long characteristic time (10 s).

More information about fusion energy can be found in the contribution to these proceedings by Ongena [4] and Romanelli [5].

1˙3. *Nuclear energy content compared to fossil fuels.* – As seen above, the energy released in nuclear reactions is of the order of tens of MeV. In the combustion of fossil fuels, it is the chemical energy that is at stake, which is of the order of eVs. We compare below the amount of heat released during the combustion of 1 g of various fuels and the amoun¡t that would correspond to the consumption of a 1000 MWe power plant per day (which can feed around 1 million homes):

- Coal:

  $\mathrm{C} + \mathrm{O}_2 \rightarrow \mathrm{CO}_2 + 4\,\mathrm{eV}$;

  energy released by 1 kg of coal: 32 MJ;

  fuel comsumption of a 1000 MWe plant: 6750 ton/day (assuming a 40% efficiency).

- Natural gas (CH4):

  $\mathrm{CH}_4 + \mathrm{O}_2 \rightarrow \mathrm{CO}_2 + 2\mathrm{H}_2\mathrm{O} + 8\,\mathrm{eV}$;

  energy released by 1 kg of natural gas: 48 MJ;

  fuel comsumption of a 1000 MWe plant: 3600 ton/day or 64 m^3/s (assuming a 50% efficiency).

- Nuclear fission (U):

  $^{S23}5\mathrm{U} + n \to 2$ fission fragments $+ 2.7n + 200\,\mathrm{MeV}$

  energy released by 1 kg of ^{235}U: 8.2 10^4 GJ and by 1 kg of natural uranium in a thermal reactor: 570 GJ;

  fuel comsumption of a 1000 MWe plant: 3 kg/day of ^{235}U or 460 kg/day of natural uranium (assuming a 30% efficiency).

- Nuclear fusion:

  $d + t \to {}^4\mathrm{He} + n + 17.5\,\mathrm{MeV}$ (of which 80% are carried by neutrons),

  energy released by 1 g of D-T mixture = 3.4 10^5 GJ;

  fuel comsumption of an hypothetic 1000 MWe plant per day: 1.25 kg/day (assuming a 20% efficiency).

2. – Nuclear energy in the world: status and perspectives

2·1. *Total primary energy and electricity supply.* – At the beginning of the 21st century, the nuclear power industry was expecting a continued growth due to the increase in the world's population, the rapid industrialisation of emerging countries and the growing awareness of the effects of CO_2 emissions. This has not been the case. The Fukushima accident in 2011, which led to the shutdown of Japanese reactors and the decision by some countries to reduce or even abandon nuclear power, had of course an important impact but it was not the only cause. In fact, the decline had begun earlier, partly because of the shale oil "revolution" —the discovery of large oil and gas resources in the US, which provided cheap and easy to produce electricity— and partly because of the economic crisis of 2008. This is reflected in fig. 3, taken from the 2020 BP statistical review of world energy [6], which shows the nuclear energy consumption by region of the world over time. In recent years, however, nuclear power production had begun to grow again, at least before the COVID-19 crisis that led to a general decline in energy consumption.

The question of nuclear energy has to be considered in the context of the global increase in energy demand. This increase is driven by the growth of the world's population, the accelerated development of emerging countries aiming to reach the standard of living of developed countries, and the aspiration of developing countries to improve their living conditions. Figure 4 (left) from [7] presents the total primary energy supply by region over time. It clearly shows that while the energy supply in OECD countries has been stagnating for years, it increases rapidly in middle east, Asia and in particular in China. As regards electricity (see fig. 4 (right)) the global demand increases even faster, and again mainly in non-OECD countries, driven by the needs from industry and residential sector in these countries, and more recently also by the development of smart electronic devices and electric cars. It is quite unlikely that the trend will be reversed in the near future.

If one looks (fig. 5) at the supply of total primary energy (left) and electricity (right) per source, it can be noticed that: i) the vast majority of total primary energy is still

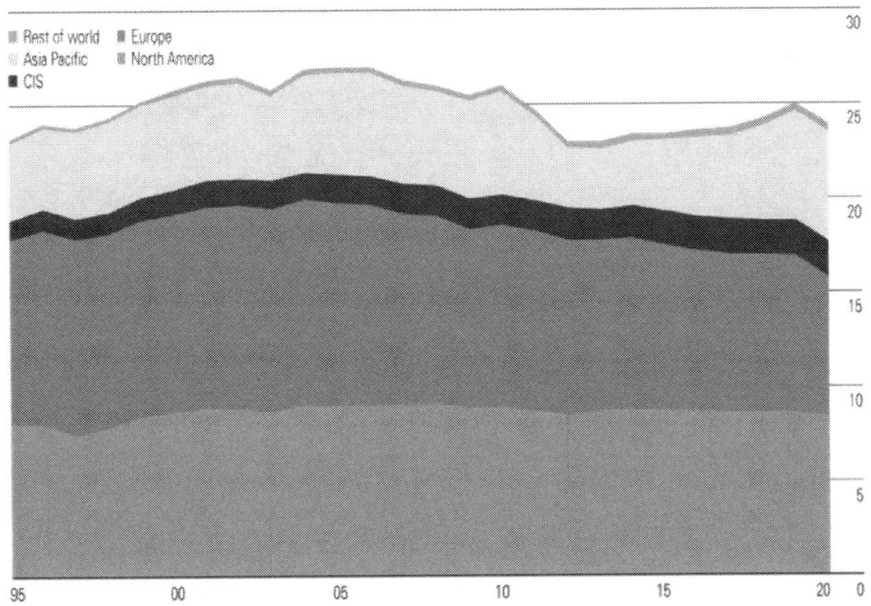

Fig. 3. – Nuclear energy consumption by region. From [6].

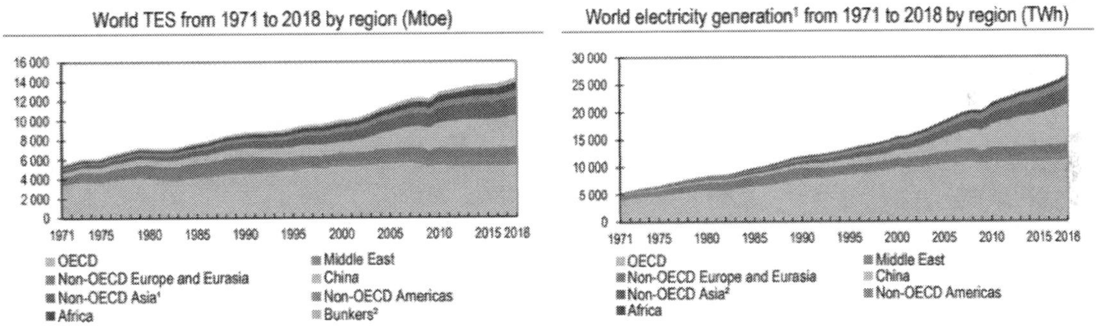

Fig. 4. – Total primary energy (left) and electricity (right) supply by region. From [7].

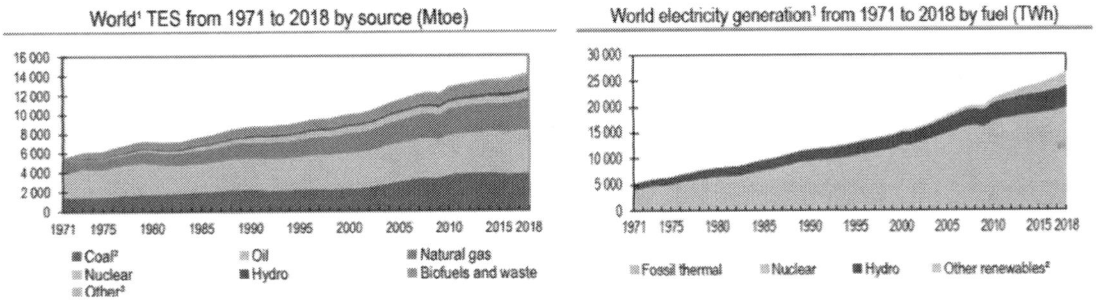

Fig. 5. – Total primary energy (left) and electricity (right) supply by source. From [7].

produced by fossil fuels; ii) the share of nuclear energy in electricity has not been increasing in recent years and remains around 10%; iii) the share of renewable energies, other than hydroelectricity, is rising significantly but is still quite low.

2·2. CO_2 *emissions.* – It is now well understood that CO_2 emissions contribute to global warming and that we must try to limit them as much as possible. Up to now, although some efforts have been done in Europe and in US, CO_2 emissions continues to grow, especially in emerging and developing countries, as can be seen in fig. 6 as estimated by the Global Carbon Project [8,9], which shows the evolution of the emissions of the 2020 four top emitters and of the rest of the world. This has led, after 2 years of stagnation, to a steady increase since 2017, with the exception of 2020 due to the COVID-19 pandemic.

In order to reduce CO_2 emissions, there are different possible options:

- Reducing the use of fossil fuels, in particular in electricity production and transportation, which is of course the first priority;
- Energy savings and increased energy efficiency need to be amplified, but they are limited and counterbalanced by increasing demand, especially in developing countries;
- Carbon capture and storage should be developed but it is expensive and profitable only if close to the emission site;
- Renewable energies, which are developing rapidly and getting cheaper, but solar and wind energies are intermittent and therefore need to be coupled to backup

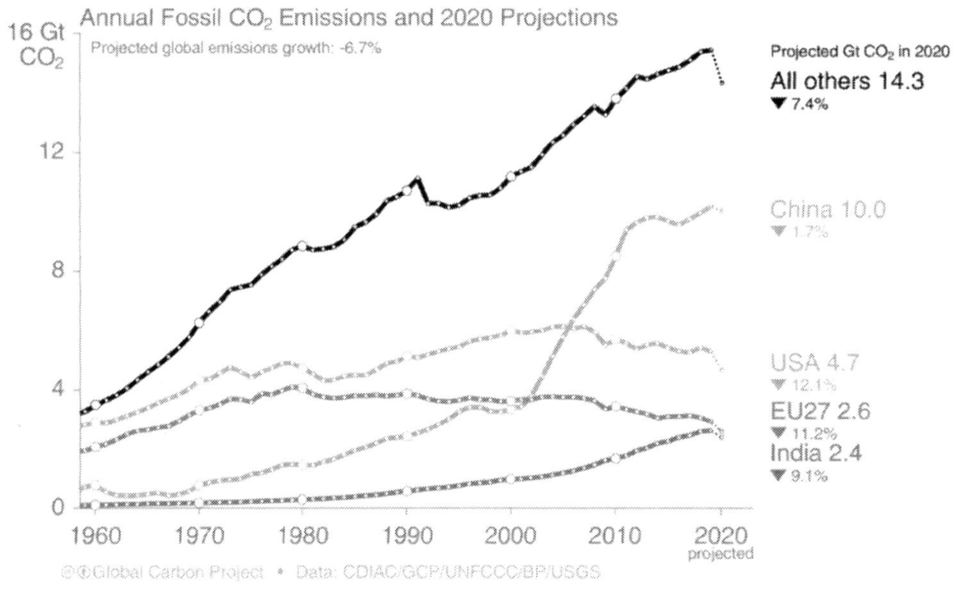

Fig. 6. – CO_2 emissions of the top 4 emitters. From [8].

resources, as far as possible CO_2 free, capable of responding to changes in their availability. Battery storage technology, up to now, has not reached a sufficient maturity and may face a problem in rare-earth element supply;

- Nuclear energy, which suffers from the rejection of a part of public opinion due to the fear of an accident and the question of waste.

It is clear that there is no single solution that can solve the problem. Instead, a combination of all these options will be needed to significantly reduce CO_2 emissions and thereby tackle global warming.

2.3. *The IEA net zero by 2050 scenario.* – This is precisely the conclusion of the International Energy Agency that has recently issued a report proposing a scenario for reducing the CO_2 emission down to zero by 2050 (NZE) [10]. It is stated that this would require a further rapid deployment of available technologies as well as a widespread use of technologies that are not on the market yet. In the NZE scenario, among the various hypotheses, the share of fossil fuel is reduced to about 20%, the electric car sales is assumed to be multiplied by 18, and an annual average reduction of 2.2% in energy intensity, *i.e.* energy use per unit of GDP is used.

One of the results of the NZE study is that the share of electricity in the total energy supply must increase drastically. The evolution of the distribution of total primary energy and electricity between the different sources of supply by 2050 is presented in fig. 7 and fig. 8, respectively.

The share of nuclear energy in the electricity supply will still represent only about 10%. However, because of the increase of the electricity demand, nuclear power capacity has to be at least doubled. It is advocated that this could be done differently in advanced economies and in emerging and developing countries. In the first case, the nuclear power capacity could be increased by extending the lifetime of existing reactors and building 4.5 GW per year of new reactors from 2021 to 2035, with an increasing emphasis on small modular reactors. In emerging and developing economies, the fleet of reactors would have to be multiplied by 4 by 2050, which would represents two thirds of the new nuclear power capacity, mainly in the form of large-scale reactors.

2.4. *Perspectives.* – The previous section shows that, after years of decline, the prospects for nuclear power are quite favorable. This is confirmed by the fact that the number of reactors under construction or planned is increasing again. New countries are joining the group of nuclear nations, such as Poland and the Czech Republic in Europe and Turkey, and other countries are planning to join, notably in Africa and the Middle East. To date, 56 reactors are under construction, including 16 in China, 6 in Russia and 7 in India, and 152 reactors are planned, including 43 in China, 25 in Russia and 14 in India, as can be seen in fig. 9 taken from [11], demonstrating the growing importance of Asia in the nuclear domain.

Most of the reactors currently in operation are based on light water technology and belong to the so-called Generation II, while the newly built reactors or those under con-

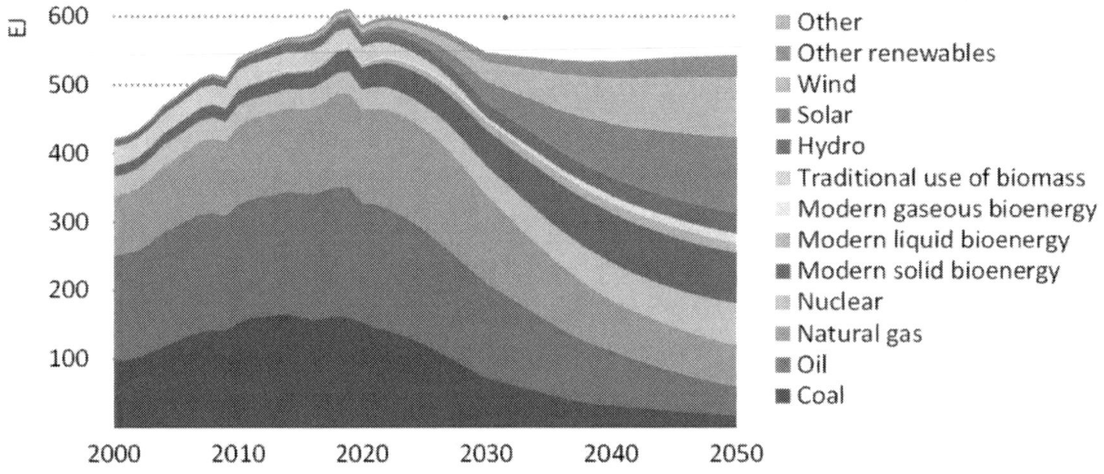

Fig. 7. – Evolution of the total energy supply by source in the IEA Net Zero Emission scenario up to 2050. From [10].

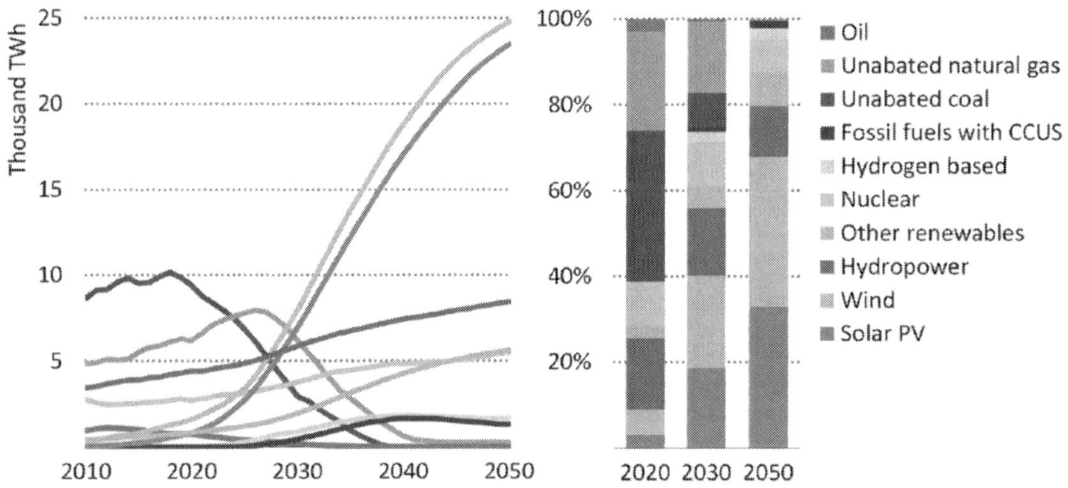

Fig. 8. – Evolution of the electricity supply by source in the IEA Net Zero Emission scenario up to 2050. From [10].

struction belong to Generation III, which is essentially an evolution of the Generation-II technology towards greater safety and competitiveness. There is however an important effort in the world to study more innovative designs, mainly driven by the Generation-IV International Forum [12], established in 2000, which brings together 13 countries, as well as Euratom —representing the 27 European Union members— to co-ordinate research and development on these systems. The goal is to design systems that would satisfy the following criteria: i) sustainable energy generation and effective fuel utilization; ii) long-term availability, iii) minimization and management of their nuclear waste; iv) economical competitiveness; v) high level of safety and reliability; vi) Proliferation-

COUNTRY	NUCLEAR ELECTRICITY GENERATION 2020		REACTORS OPERABLE June 2021		REACTORS UNDER CONSTRUCTION June 2021		REACTORS PLANNED June 2021		REACTORS PROPOSED June 2021	
	TWh	% e	No.	MWe net	No.	MWe gross	No.	MWe gross	No.	MWe gross
Argentina	10.0	7.5	3	1641	1	29	1	1150	2	1350
Armenia	2.6	34.5	1	415	0	0	0	0	1	1060
Bangladesh	0	0	0	0	2	2400	0	0	2	2400
Belarus	0.3	1.0	1	1110	1	1194	0	0	2	2400
Belgium	32.8	39.1	7	5942	0	0	0	0	0	0
Brazil †	13.2	2.1	2	1884	1	1405	0	0	4	4000
Bulgaria	15.9	40.8	2	2006	0	0	1	1000	2	2000
Canada	92.2	14.6	19	13,624	0	0	0	0	2	1500
China	344.7	4.9	51	49,569	17	18,616	38	41,785	168	196,86
Czech Republic	28.4	37.3	6	3934	0	0	1	1200	3	3600
Egypt	0	0	0	0	0	0	4	4800	0	0
Finland	22.4	33.9	4	2794	1	1720	1	1170	0	0
France	338.7	70.6	56	61,37	1	1650	0	0	0	0
Germany	60.9	11.3	6	8113	0	0	0	0	0	0
Hungary	15.2	48.0	4	1902	0	0	2	2400	0	0
India	40.4	3.3	23	6885	6	4600	14	10,5	28	32
Iran	5.8	1.7	1	915	1	1057	1	1057	5	2760
Japan †	43.0	5.1	33	31,679	2	2756	1	1385	8	11,562
Jordan	0	0	0	0	0	0	0	0	1	1000
Kazakhstan	0	0	0	0	0	0	0	0	2	600
Korea RO (South)	152.6	29.6	24	23,15	4	5600	0	0	2	2800
Lithuania	0	0	0	0	0	0	0	0	2	2700
Mexico	10.9	4.9	2	1552	0	0	0	0	3	3000
Netherlands	3.3	3.9	1	482	0	0	0	0	0	0
Pakistan	9.6	7.1	6	2332	1	1100	1	1170	0	0
Poland	0	0	0	0	0	0	0	0	6	6000
Romania	10.6	19.9	2	1300	0	0	2	1440	1	720
Russia ‡	201.8	20.6	38	28,578	2	2510	25	23,89	21	20,1
Saudi Arabia	0	0	0	0	0	0	0	0	16	17
Slovakia	14.4	53.1	4	1837	2	942	0	0	1	1200
Slovenia	6.0	37.8	1	688	0	0	0	0	1	1000
South Africa	11.6	5.9	2	1860	0	0	0	0	8	9600
Spain	55.8	22.2	7	7121	0	0	0	0	0	0
Sweden	47.4	29.8	6	6882	0	0	0	0	0	0
Switzerland	23.0	32.9	4	2960	0	0	0	0	0	0
Thailand	0	0	0	0	0	0	0	0	2	2000
Turkey	0	0	0	0	3	3600	1	1200	8	9500
Ukraine †	71.5	51.2	15	13,107	2	1900	0	0	2	2,4
UAE	0	0	1	1345	3	4200	0	0	0	0
United Kingdom	45.9	14.5	15	8923	2	3440	2	3340	2	2300
USA	789.9	19.7	93	95,523	2	2500	3	2550	18	8000
Uzbekistan	0	0	0	0	0	0	2	2400	2	2400
WORLD*	2553	c 10.1**	444	395,267	54	61,219	100	102,437	325	353,812
	TWh	% e	No.	MWe	No.	MWe	No.	MWe	No.	MWe
	NUCLEAR ELECTRICITY GENERATION		OPERABLE		UNDER CONSTRUCTION		PLANNED		PROPOSED	

Fig. 9. – Panorama of world nuclear power reactors: in operation, in construction and planned as of July 2021. From [11].

resistance. Six different concepts, at different levels of development, are studied or optimized by one or several countries: Gas-cooled Fast Reactor (GFR), Lead-cooled Fast Reactor (LFR), Molten Salt Reactor (MSR), Supercritical Water-cooled Reactor (SCWR),

Sodium-cooled Fast Reactor (SFR) and Very High Temperature Reactor (VHTR).

In recent years, there has been a growing interest in small modular reactors (SMRs), which offer advantages such as a lower initial investment and construction time, the possibility of powering smaller markets and power grids or remote areas, greater flexibility and the possibility of combining non-electricity applications, such as heat or hydrogen production, or desalination.

Actually, there are many different types of SMRs. Most of them are reactors between 50 and 200 MWe, generally based on GEN-III technology (PWR, BWR, sometimes HTR). Some of them, sometimes called Advanced Modular Reactors (AMR) are based on of GEN-IV type systems. There are also Micro Modular Reactors (MMR) or Very Small Modular Reactors (vSMR) of a range power from 1 to 20 MWe.

The dynamism of the SMR field is clearly evidenced in the recent report on the Advances in Small Modular Reactor Technology Developments published by IAEA [13], which lists more than 50 SMR designs under development for different application, many of them proposed by newly created companies. A few of them are in an advanced stage of construction in Argentina, China and Russia.

3. – Nuclear waste management and environmental impact

3·1. *Nuclear waste.* – After irradiation during 3 years, the fuel unloaded from pressurized water reactor, which originally was composed of about 3.3% of ^{235}U and 96.7% of ^{238}U, contains around 35 kg of fission products, 9 kg of plutonium and 1 kg of minor actinides (MA), *i.e.* isotopes of neptunium, americium, and curium, per ton of fuel. Many isotopes of fission products are highly radioactive but most of them are rather short lived or with a half life shorter than 30 years. On the contrary, minor actinides are mostly long-lived isotopes, some with half-lives of million years. To evaluate the potential danger of spent fuel for humans, rather than radioactivity, radiotoxicity is used, which takes into account the nature of the radioactivity and the biological effects of a possible ingestion. The evolution of spent fuel radiotoxicity over a period of 1 million years is presented in fig. 10 with the different contributions. It can be seen that in the short term, radiotoxicity is dominated by fission fragments, and then plutonium takes over. The time required to reach the level of the initial uranium ore is of the order of several hundred thousands of years.

There are actually two options for the management of spent fuel, which are illustrated in fig. 11, either what is called open or one-through cycle, which consists simply in a direct disposal in a deep underground repository —this is the choice for instance of US, Sweden, Finland— or a partially closed or twice-through cycle, in which the spent fuel is reprocessed to extract plutonium. Plutonium is then used to make a new type of fuel composed of a mixture of uranium and plutonium oxides, the MOX fuel, which can be easily burnt in current commercial reactors. This option has been adopted for instance by France, Japan, Russia, China.

As plutonium is the main contributor to the radiotoxicity after a few tens of years, as shown in in fig. 10 and also to the residual heat, reprocessing reduces the amount,

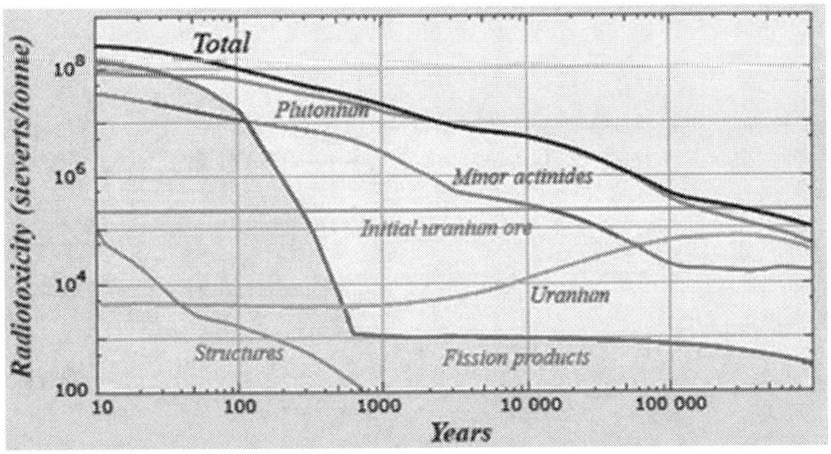

Fig. 10. – Evolution of spent fuel radiotoxicity over a period of 1 million years. © Source: CEA.

volume and radiotoxicity of the high-level waste to be stored, and of the size of the deep underground storage since the waste packages can be stored closer together. However, it leads to additional volumes of intermediate wastes during the reprocessing and fuel fabrication processes. In any case a final deep geologic disposal of the remaining long-lived high-level wastes will be necessary.

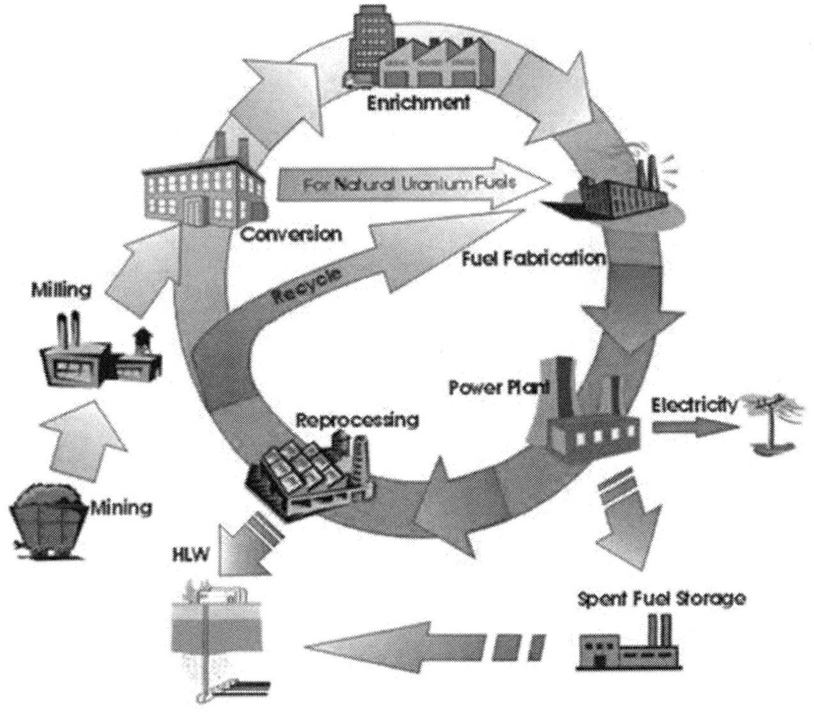

Fig. 11. – The open and closed spent fuel cycles. From [14].

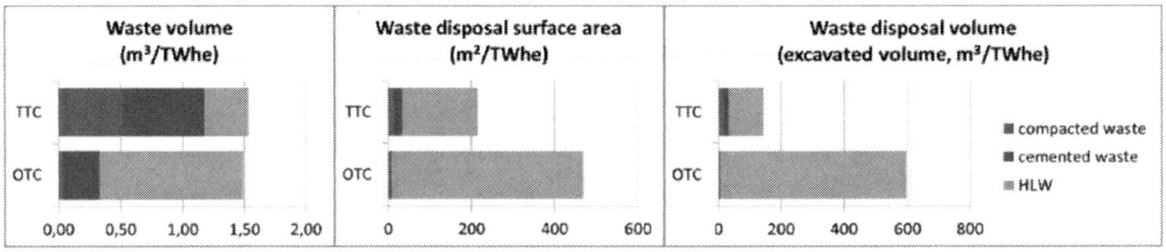

Fig. 12. – Comparison of the waste volumes, waste disposal surface areas and waste disposal excavated volumes for the Once-Through (OTC) and Twice-Through Cycles (TTC). From [16].

There are different types of nuclear waste, which are generally classified according to their activity [15]:

- High-level waste (HLW): used fuel or separated waste from reprocessing of used fuel, with a decay heat higher than 2 kW/m^3, which leads to significant temperature increase. They account for only 3% of the volume, but 95% of the total radioactivity of produced waste and contain both long-lived and short-lived isotopes.
- Intermediate-level waste (ILW): comprises resins, chemical sludges, and metal fuel cladding, as well as contaminated materials from reactor decommissioning. They are highly radioactive but have a low decay heat. They represent 7% of the volume and 4% of the total radioactivity.
- Low-level waste (LLW): waste with radioactive content not exceeding 4 GBq/t of alpha activity or 12 GBq/t beta-gamma activity such as tools, clothing, filters, etc., which contain small amounts of mostly short-lived isotopes. They constitute 90% of the volume, but 1% of total radioactivity.
- Very low-level waste (VLLW): in fact, radioactive materials at a level not considered harmful to people or the surrounding environment, essentially material produced during rehabilitation or dismantling operations.

3˙2. *Nuclear energy environmental impact.* – An interesting and in-depth comparison has been done, in the French case, by Poinssot *et al.* in [16] between the Once-Through (OTC) and Twice-Through Cycles (TTC), which correspond, respectively, to the open and closed cycles presented in fig. 11. It uses a simulation tool compiling and relating together all the energy and matter fluxes along the nuclear fuel cycle. In fact, the analysis compared the actual situation of the French TTC to a virtual scenario in which France would produce the same amount of electricity with the actual reactor park but without operating any recycling operations. Figure 12 shows the results concerning the total volume of waste, which is very similar in the two cases, and the repository (for long-lived waste) excavated volume and surface area, which are significantly lower for high-level waste (HLW) in the case of TTC.

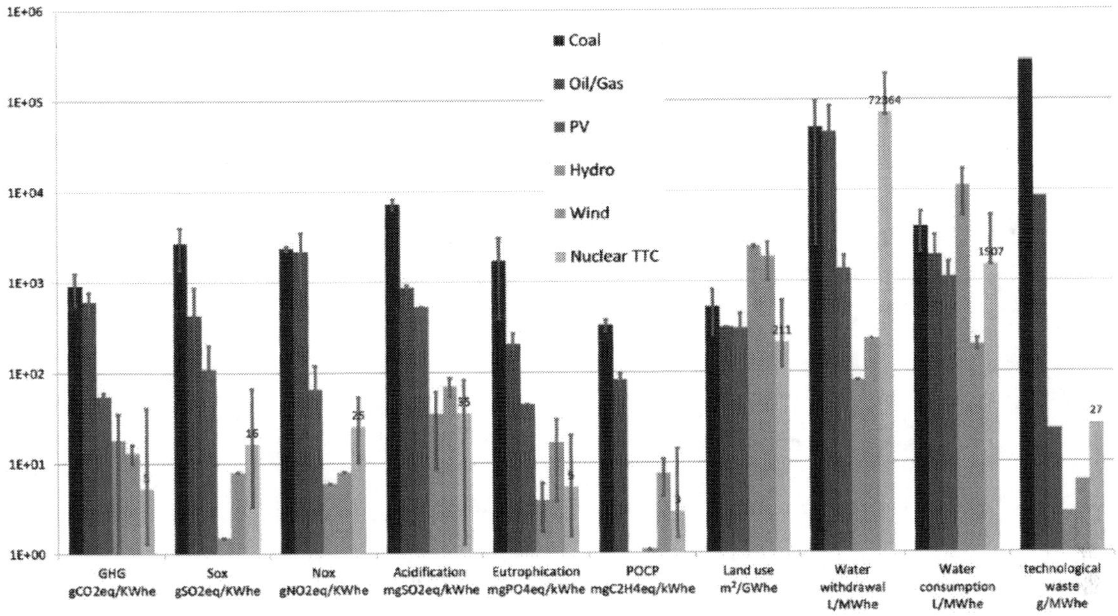

Fig. 13. – Comparison of the selected indicators between the French TTC and other energy sources. The error bars represent the gap between the minimum and maximum values found in the literature. From [16].

In the same paper, the non-radioactive impacts of nuclear energy, through the example of the French TTC, was also analysed and compared to other sources of energy. Some indicators have been selected to perform a comprehensive life cycle analysis of energy production from coal, oil or gas, photovoltaic, hydro, wind and nuclear power. These indicators are:

- green-house-gases emissions (GCG, gCO2eq/kWhe);
- atmospheric pollution (mg/kWhe), in particular SOx and NOx;
- water pollution (mg/kWhe), *i.e.* acidification, eutrophisation and POCP (photochemical ozone creation potential);
- land-use (m^2/GWhe);
- water consumption (l/MWhe);
- water withdrawal (l/MWhe);
- production of technological waste (g/MWhe).

Figure 13 shows the results, which clearly indicate that nuclear energy has the lowest impact regarding GHG emission and land use (which is in fact dominated by mining), has an impact close to renewable energies regarding atmospheric and water pollution and technological waste generation, obviously better than fossil energies. On the opposite, in terms of water consumption and withdrawal, nuclear energy is in the range of fossil fuels.

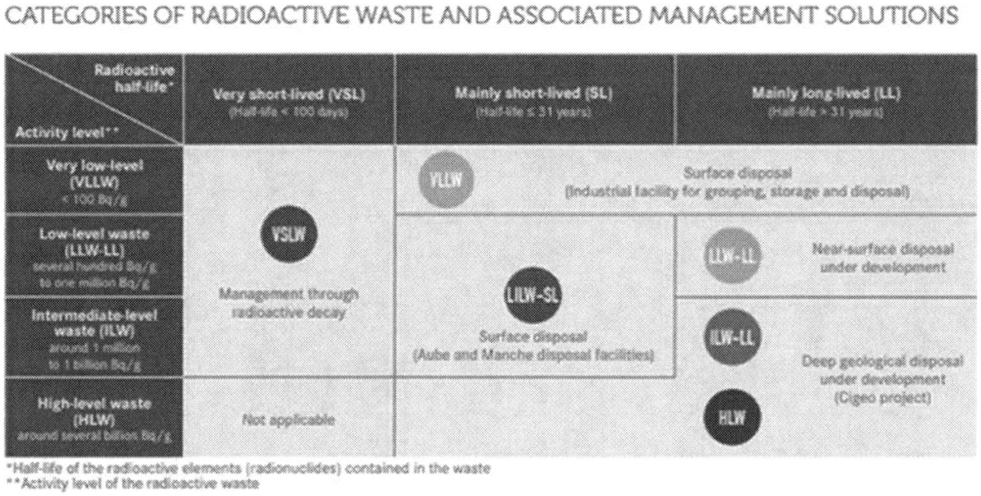

Fig. 14. – Categories of waste and management solutions. From [15].

3˙3. *Nuclear waste management.* – As explained above, there are different categories of waste and, depending on their activity level and the half-life of the isotopes present, different management options are considered. This is illustrated in fig. 14 from [17], which shows the management solutions adopted or under development in in France by the national agency in charge of nuclear waste, ANDRA. Nearly 90% of countries using nuclear energy have one or more long-term management solutions for radioactive waste in operation or under development [17]. Most of the storage in operation concerns low- and medium-level short-lived waste and is mainly on the surface. Clearly, the most critical question concerns long-lived high-level waste for which deep underground geological disposal in stable geological formations is foreseen. Plans for such sites are well advanced in Finland and in Sweden, for instance.

As explained above, the removal of plutonium from spent fuel, reused in MOX fuels, reduces the radiotoxicity of the waste to be stored. To go further, it has been envisaged to also separate the minor actinides and transmute ("burn") them, *i.e.* as far as possible make them fission, in suitable reactors. This would lead to a substantial reduction of the time needed for the radiotoxicity of the waste to reach the level of natural uranium ore from nearly 1 million years to a few hundreds of years, a time scale more manageable in terms of monitoring [18]. This is illustrated in fig. 15 from [19], which shows the evolution over time of the radiotoxicity of spent fuel without any recycling, with recycling of plutonium and after transmutation of minor actinides, relative to uranium ore.

Transmutation can be achieved more efficiently in fast neutron spectrum reactors. In fact, the introduction of minor actinides into fuels leads to a degradation of certain reactor safety parameters, such as the fraction of delayed neutrons, which is lower in MAs than in uranium 235. There are therefore two possible options to burn significant quantities of MAs: either adding small amounts of minor actinides in many fast reactors at a level acceptable for the safety, with the advantage that this might be done in

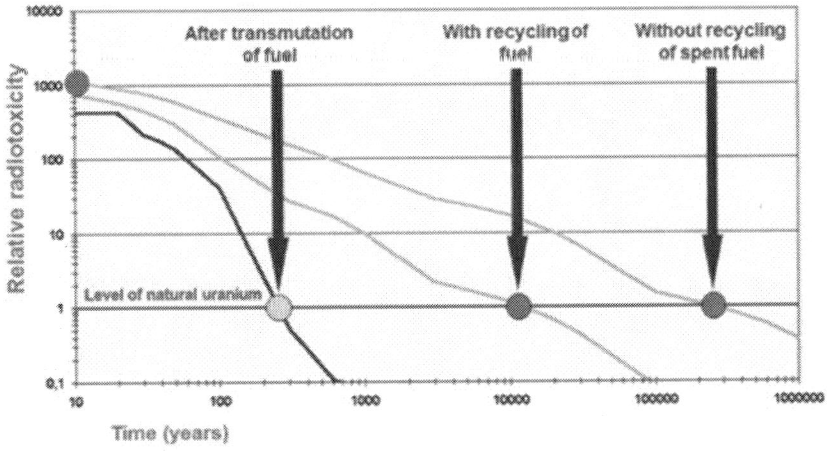

Fig. 15. – Radiotoxicity of UOX spent fuel relative to uranium ore, *vs.* time (years). From [19].

adapted commercial reactors; or burning large amount of minor actinides in dedicated accelerator-driven sub-critical systems, thus much less sensitive to the degradation of safety parameters.

Some studies about MA transmutation are conducted in Europe within the MYRRHA (Multi-purpose hYbrid Research Reactor for High-tech Applications) project in development at SCK•CEN, which will be able to work both in sub-critical (ADS) and in critical mode.

It has to be noticed however that to reach a high rate of transmutation of the produced MAs, multiple cycles of separation, fuel fabrication and transmutation have to be performed, which imply the handling and transportation of large quantities of highly radioactive materials, and therefore limit the interest of the process.

4. – Conclusion

Fission is a highly concentrated source of energy that does not generate CO_2 emissions. Fusion is even more concentrated and would have the advantage of not generating long-lived radioactive waste, but the viability of a commercial reactor remains to be demonstrated.

In this paper, it was shown that although nuclear power currently accounts for only a small portion of the world's energy production, with the growing need for electricity, the rise of emerging and developing countries, and the need to combine all possible solutions to combat global warming, it is quite clear that it will continue to play an important role. Most of the new nuclear power reactors are being built in emerging countries and the recent interest for Small Modular Reactors may facilitate the deployment in very different areas and use for non-energy applications.

It has also been shown that the global non-radioactive environmental impact of nuclear energy is generally lower than for other sources of energy and that long-term management solutions for radioactive waste are available.

REFERENCES

[1] Ripani M., these proceedings, p. 123.
[2] Lawson J. D., "Some Criteria for a Power Producing Thermonuclear Reactor", *Proc. Phys. Soc. Sect. B*, **70** (1955) 6.
[3] https://commons.wikimedia.org/wiki/File:Fusion_ntau.svg.
[4] Ongena J., these proceedings, p. 143.
[5] Romanelli F., these proceedings, p. 175.
[6] BP *Statistical review of world energy*, 2020.
[7] IEA, *Key world energy statistics*, 2020.
[8] Global Carbon Project, https://www.globalcarbonproject.org/carbonbudget/.
[9] Friedlingstein P. *et al.*, *Earth Syst. Sci. Data*, **12** (2020) 3269.
[10] IEA, *Net Zero by 2050*, report (2021).
[11] The World Nuclear Association, https://www.world-nuclear.org/information-library/facts-and-figures/world-nuclear-power-reactors-and-uranium-requireme.aspx.
[12] Generation IV International Forum, https://www.gen-4.org.
[13] IAEA, *Advances in Small Modular Reactor Technology Developments*, https://aris.iaea.org/Publications/SMR_Book_2020.pdf.
[14] International Atomic Energy Agency, *Nuclear Fuel Cycle Simulation System (VISTA) IAEA-TECDOC-1535*, IAEA, Vienna (2007), https://aris.iaea.org/Publications/SMR_Book_2020.pdf.
[15] ANDRA, Synthesis report (2021).
[16] Poinssot Ch. *et al.*, "Assessment of the environmental footprint of nuclear energy systems. Comparison between closed and open fuel cycles", *Energy*, **69** (2014) 199.
[17] https://www.andra.fr/panorama-mondial-ou-en-sont-les-autres-pays.
[18] Nuclear Energy Agency (NEA), *Physics and Safety of Transmutation Systems*, Status Report No. 6090, Paris, France (2006).
[19] Abderrahim H. A. *et al.*, NEA/NSC/R(2015)2.

Beta decay studies and search for octupole deformation in the $A \sim 225$ Po-Fr nuclei

M. Polettini for the S460 experiment and HISPEC-DESPEC Collaboration

Università degli Studi di Milano e INFN, Sezione di Milano - Milano, Italy

Summary. — In the landscape of nuclear shapes, dominated by reflection-symmetric forms leading to either spherical or axially deformed arrangements, the occurrence of asymmetric pear-like nuclei has long been searched for. Evidence for static octupole deformation has only been found in selective areas of the nuclear chart, in the mass regions $A \sim 222$ and $A \sim 146$, the so-called "Islands of Octupole Deformation" (IOD). This paper is focused on β decay studies in the Po-Fr nuclei in the $220 \leq A \leq 230$ island of octupole deformation exploiting the FRS+DESPEC setup at GSI in Spring 2021. The experimental setup and the analysis techniques employed to perform this study are here shown and discussed, together with preliminary results from the on-going analysis.

1. – Introduction

The HISPEC-DESPEC Collaboration aims at investigating the nuclear structure of exotic nuclei formed in high-energy projectile-fragmentation reactions. The experiment here discussed was performed as part of the Phase-0 GSI-FAIR campaign, with the aim of investigating $A \sim 225$ Po-Fr nuclei in the south-east frontier of the $A \sim 222$ IOD populating the nuclei of interest using β decay.

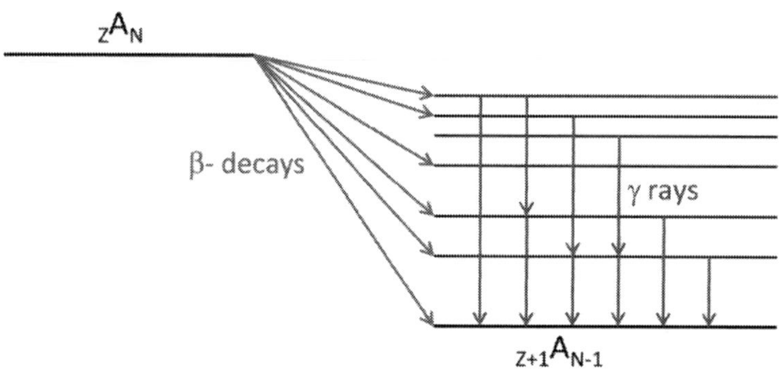

Fig. 1. – Example of a typical β-decay scheme of de-excitation.

Beta decay can serve as a tool to explore the inner structure of nuclei. The γ transitions caused by the de-excitation of the nucleus, and following the β decay, carries information on the de-excitation pattern of daughter nucleus, as shown in fig. 1. γ-γ coincidence studies allow us to to define the structure of the level scheme of the nuclei under investigation, while the relative intensities can give information on the probability of a particular level to be populated. One of the applications of β decay is the study of the decay heat of a reactor. Once the reactor is shut down, the energy released in radioactive decays provides the main source of heating. Estimation and control of the heat emitted by the decay of fission products has a key role in the safe operation of reactors. These calculations require extensive libraries of cross sections, fission yields, and decay data [1].

In our specific case, we use β decay to reach out to n-rich nuclei and extract details on the presence of permanent octupole deformation in our nuclei of interest. Octupole deformation is associated to octupole correlations, which result in a non-zero β_{30} term, resulting in reflection asymmetric shapes, also known as pear-like shapes [2]. Octupole correlations are the result of the long-range octupole-octupole interaction between nucleons occupying pairs of orbitals which differ by 3 units in both orbital- and total-angular momentum. These states approach each other at the Fermi surface when either Z or $N \sim 34, 56, 88, 134$, at values just greater than the magic numbers, where nuclei are nearly spherical [3].

These exotically deformed nuclei can be found only in selective areas of the nuclear chart, in particular in the mass region $A = 146$ (Xe-Sm region) and $A = 222$ (Rn-Th region), called Islands of Octupole Deformation (IOD). To find direct experimental evidences of these elusive shapes proves to be very challenging, and it is worth exploring several spectroscopic properties in a wide range of nuclei. The main experimental signatures are a large $B(E3)$ transition strength, an enhanced $B(E1)$ transition strength, due to the separation of the centre of charge and the centre of mass of the nucleus, and interleaving negative and positive parity states that are nearly degenerate in energy. This was confirmed by theoretical calculations performed for nuclei in the light actinide region, as shown in refs. [4, 5].

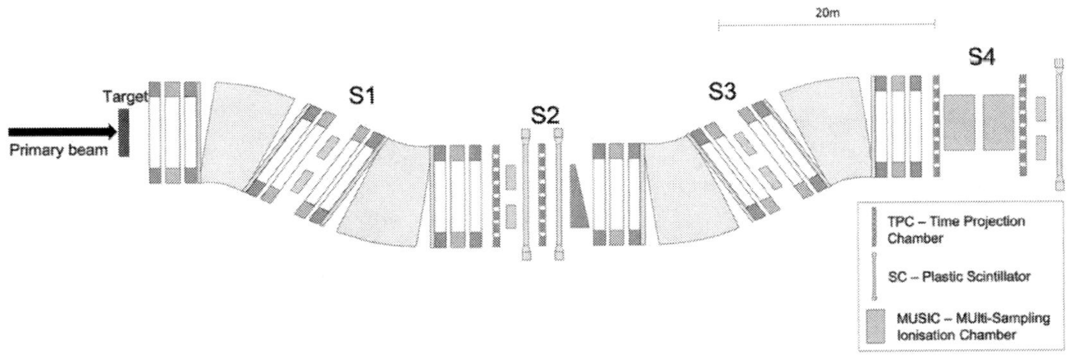

Fig. 2. – Schematic of the FRS magnetic spectrometer: the main detectors for ion identification are labeled in the box on the bottom-right.

Very few direct measurements of octupole correlations were performed in recent years, and $E3$ transitions were measured directly in selected cases, in particular in ^{220}Rn, ^{224}Ra [6] and ^{228}Th [7], highlighting a typical de-excitation pattern.

The experiment here described was performed at GSI Darmstadt, Germany in April 2021 with the following main aims: finding evidence of octupole deformation in the nuclei of interest, provide new β-decay data to test models for r-process, to measure super-deformed bands in $^{220-222}$Po. We plan to measure directly octupole correlations by using β decay and fast timing to locate 1^- and 3^- states and their correspondent transition strength, and compare them with predicted values [4]. In order to study shape isomers in $^{220-222}$Po, we plan to use delayed isomer spectroscopy to locate the super-deformed 2^+ states at low energies [8].

2. – Experiment

The experiment was performed at the GSI-FAIR laboratories within the DESPEC Phase-0 campaign. The ions of interest were formed in a fragmentation reaction of a ^{238}U beam impinging on a ^{9}Be target. The primary beam was accelerated using the UNIversal Linear ACcelerator (UNILAC) and the SIS-18 synchrotron, producing a beam with $\sim 10^9$ pps intensity and spill length on 4 s. The fragments' secondary beam here produced forms a so-called cocktail beam composed of many nuclear species. Therefore, there is a need to select and identify each different ion species in order to perform focused studies on their properties and decays. In order to do this, the fragments are selected with the $B\rho$-ΔE-$B\rho$ method using magnetic elements and energy loss in materials. The ions can also be identified with the TOF-$B\rho$-ΔE method [9] in the FRS FRagment Separator where several detectors collect information on the ions' time of flight, position in the focal planes, and atomic number Z (see fig. 2). This powerful magnetic spectrometer allowed us to produce and transport ions in the Po-Fr region with mass $A \sim 225$ towards our DESPEC setup.

The DEcay SPECtroscopy station [10] was commissioned in 2019 and used in the following campaigns. It comprises several detector systems used for decay experiments.

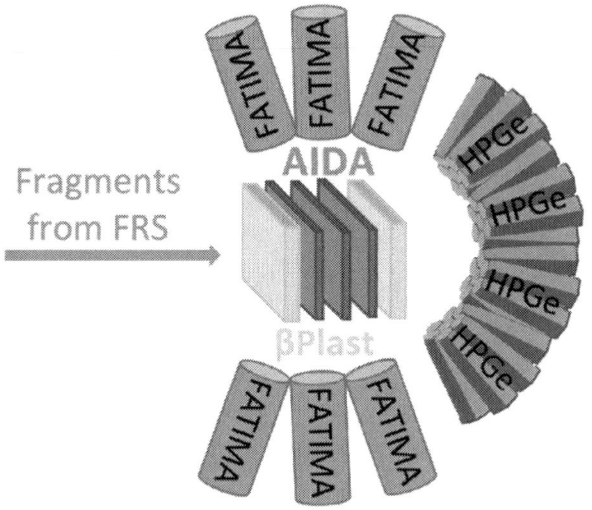

Fig. 3. – Schematic of the DESPEC decay station.

The ions are implanted in a stack of 3 Double-Sided Silicon-strip Detectors (DSSD) segmented in 128×128 strips (16384 pixels) with 0.6 mm strip pitch, the AIDA (Advanced Implantation Detector Array) [11]. This active stopper is also used to detect α and β particles emitted by the ion after its decay. The three silicon detectors are sandwitched between two plastic scintillators, the bPlast detector, which are made of tiles of BC-400 scintillator material, readout by a series of SiPM glued to each side, used for timing measurements of the β particle emitted in the decay. The γ rays emitted in the de-excitation of the daughter nuclei are detected using a hybrid array for γ-ray detection. The FATIMA (FAst TIMing Array), comprising 36 $LaBr_3(Ce)$ detectors, arranged in three concentric rings [12], is used for timing measurements. Each detector is composed of a crystal with a diameter size of $1.5 \times 2\,\text{inch}^2$ length coupled to Hamamatsu R9779 photomultiplier tubes. The detectors are equipped with a lead shield of 4 mm thickness around the crystal to minimise scattering among neighbouring detectors [13]. The FATIMA detector system has a full-energy peak efficiency of 2.9% at $\sim 1\,\text{MeV}$. Precision γ energy measurements are performed with four 7-fold HPGe clusters, in forward position as depicted in fig. 3. This germanium array has an efficiency of 2% at $\sim 1\,\text{MeV}$. All the subsystems involved have independent data acquisition systems that are synchronised using a time sorter. Time stamping correlations are established between the systems using a distributed clock: the so-called White Rabbit, and a 2 Hz pulser.

This setup allows to perform β-decay measurements by correlating the implanted ion and the β particle in the silicon detectors, using position and timing conditions. The γ rays detected in the FATIMA and HPGe detectors can then be correlated to the ion and β to probe the inner structure of the daughter nuclei, establishing β-γ-γ correlations. Alpha decay studies can be performed in a similar fashion by correlating the implanted ion with an alpha detected in the silicon detectors.

https://doi.org/10.1051/epjconf/202226800009

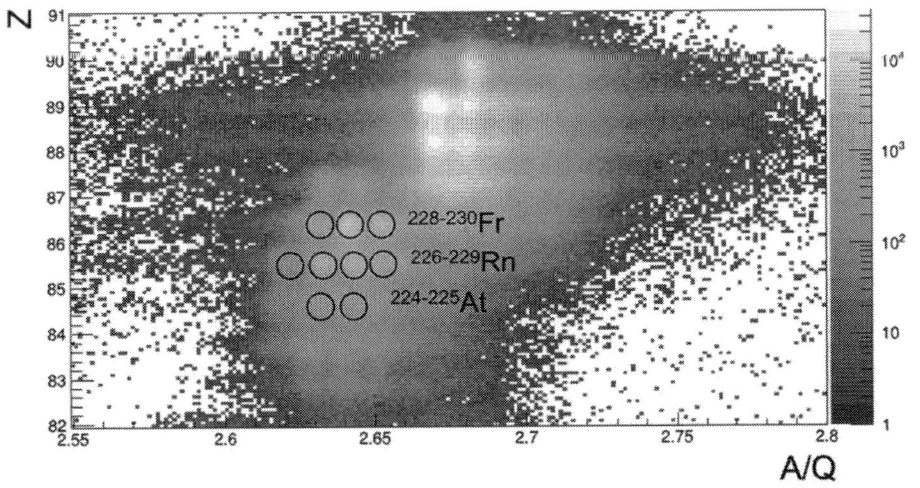

Fig. 4. – Particle identification plot obtained for our experimental run.

3. – Initial stages of analysis

The ions of interest were measured in the FRagment Separator as explained in sect. **2**. The identification was performed after accurate calibrations and corrections of all the detectors used. We therefore obtain an identification plot (also called ID plot) in which the atomic number Z of each ion produced is plotted against the mass over ionic charge ratio (A/Q). As shown in fig. 4, we were able to produce all the ions of interest: $^{228-230}$Fr, $^{226-229}$Rn, $^{224-225}$At, together with several other ion species that can therefore be investigated with this measurement. Some of the ions were produced in a different charge state, and form a structure on the right-hand side of fig. 4. The ions are well separated, and with software conditions it will be possible to focus on each single species to perform β decay and γ spectroscopy.

The data-processing is on-going and will be now focused on the ion-β coincidence reconstruction developing *ad hoc* codes for ion-β and β-γ-γ correlations. This analysis will be exploited to study of the inner structure of the daughter nuclei to identify new decay transitions and evaluate the presence of octupole deformation. In parallel, α decay analysis is being performed in Rn isotopes in order to confirm the ion-decay correlations and to enrich the knowledge on the α-decay mode of these nuclei.

4. – Conclusions

This paper reports an experimental study on octupole deformation exploiting beta decay in the Po-Fr region with mass $220 \leq A \leq 230$. The experiment was performed at GSI-FAIR in April 2021 within the DESPEC phase-0 campaign using the FRS-DESPEC setup to select and identify the ions of interest produced in fragmentation reactions of a ^{238}U beam. Preliminary results from the on-going analysis were presented in this work.

* * *

The authors would like to thank the FRS group and the GSI accelerator staff. Support by INFN is acknowledged.

REFERENCES

[1] Algora A. *et al.*, *Eur. Phys. J. A*, **57** (2021) 85.
[2] Butler P. A., *Phys. G: Nucl. Part. Phys.*, **43** (2016) 073002.
[3] Butler P. A., *Proc. R. Soc. A*, **476** (2020) 0202.
[4] Robledo L. M. and Butler P. A., *Phys. Rev. C*, **88** (2013) 051302(R).
[5] Xia S. Y. *et al.*, *Phys. Rev. C*, **96** (2017) 054303.
[6] Gaffney L. P. *et al.*, *Nature*, **497** (2013) 199.
[7] Chishti M. M. R. *et al.*, *Nat. Phys.*, **16** (2020) 853.
[8] Nerlo-Pomorska B. *et al.*, *Acta Phys. Pol. B*, **47** (2016) 943.
[9] Geissel H. *et al.*, *Nucl. Instrum. Methods Phys. Res. B*, **70** (1992) 286.
[10] Mistry A. K. *et al.*, *Nucl. Instrum. Methods Phys. Res. A*, **1033** (2022) 166662.
[11] AIDA technical report (2008).
[12] FATIMA technical report (2015).
[13] Rudigier M. *et al.*, *Nucl. Instrum. Methods Phys. Res. A*, **969** (2020) 163967.

Energy from nuclear fission

M. Ripani

INFN, Sezione di Genova - Genova, Italy

Summary. — The physics of nuclear fission will be briefly illustrated, from the basic mechanism behind this phenomenon to the relevant physical quantities like nuclear cross sections, neutron flux and reaction products, together with the accompanying phenomenon of neutron capture and its role in determining how the fuel transforms in a nuclear reactor. The basic concepts underlying the operation of different types of nuclear reactors will be illustrated, along with the concept of fuel cycle. The aspects of radioactive waste, fuel resources and safety will also be briefly illustrated.

1. – Physics of fission

What is fission and why can it produce energy? The basic reason is the behaviour of the nuclear mass and nuclear binding energy for different nuclear species. As is known from relativity, the mass of a bound system is given by the sum of the masses of its components, plus the binding energy of the system. The binding energy is by definition a negative quantity, *i.e.* the mass of the composite system is always less than the sum of the masses of its components. When applied to nuclear masses, this concept can be expressed through the equation

$$M(Z,A) = ZM_p + (A-Z)M_n + B(Z,A), \tag{1}$$

where $M(Z,A)$ is the mass of a nucleus with Z protons and A protons and neutrons (*i.e.* $A-Z$ neutrons), M_p is the proton mass, M_n is the neutron mass and $B(Z,A)$ is the

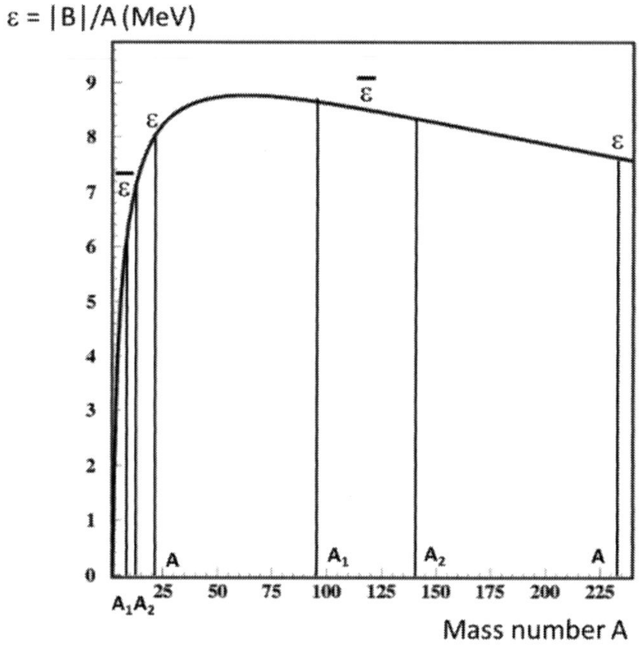

Fig. 1. – Absolute value of binding energy per nucleon ϵ, as a function of the mass number A. The figure also illustrates the case of the nuclei of mass number A_1 and A_2 undergoing fusion and merging into a nucleus of mass number A, as well as the case of a nucleus of mass number A undergoing fission and splitting into two fragments of mass number A_1 and A_2. $\bar{\epsilon}$ is the mass-number-weighted absolute binding energy per nucleon in both the fusion and fission cases (see sect. **1**˙1).

binding energy of the system, with $B(Z,A) < 0$, *i.e.* $M(Z,A) < ZM_p + (A-Z)M_n$. Z is called the atomic number, while A is called the mass number. The absolute value of the binding energy represents the energy that must be given to the system from outside to overcome the binding nuclear force and set all nucleons free. We can also define the binding energy per nucleon, *i.e.* the average quantity of energy that must be supplied per single nucleon when separating all nucleons from one another. Figure 1 shows the absolute value of the binding energy per nucleon, $\epsilon = \frac{|B|}{A}$. It is seen that such value increases from the lightest nucleus, the deuteron, to the region of iron, then it decreases towards the heavier nuclei. In other words, iron is more bound than the light nuclei, while, *e.g.*, uranium is less bound than iron.

1˙1. *Energy balance.* – Based on the above considerations, it is easy to see that a nuclear reaction where nuclei move from smaller to bigger binding energy will release energy. Indeed, in going from the initial to the final state in the reaction, the total mass will decrease, so that the mass defect will be transformed in energy. Looking at fig. 1, it is clear that two light nuclei undergoing nuclear fusion, *i.e.* a nuclear system moving from the left-hand side to the right-hand side of the graph, will move towards higher binding, thereby releasing nuclear energy, *i.e.* giving rise to an *exoenergetic* nuclear reaction, a reaction which yields a net amount of energy. This is the mechanism at the basis of nuclear fusion [1], *e.g.* in the reaction $^2\mathrm{H} + {}^3\mathrm{H} \rightarrow {}^4\mathrm{He} + n$, where about 17.5 MeV are released in the process. On the other hand, heavy nuclei above iron can undergo fission, where the nuclear system splits into two nuclei, moving from the right-hand side to the left-hand side of the graph in fig. 1, again moving towards higher binding. This is called the process of nuclear fission and can also be exoenergetic. Let us consider in detail the energy balance of a fission process, where a nucleus with Z protons and $(A-Z)$ neutrons splits into two fragments

$$(Z, A) \rightarrow (Z_1, A_1) + (Z_2, A_2), \tag{2}$$

where obviously $Z = Z_1 + Z_2$ and $A = A_1 + A_2$ and the masses of the nuclei are $M(Z, A)$, $M(Z_1, A_1)$ and $M(Z_2, A_2)$, respectively. The energy released in the reaction is called the Q-value and it is given by definition by

$$Q = M(Z, A) - M(Z_1, A_1) - M(Z_2, A_2). \tag{3}$$

By using eq. (1) we obtain (remember that ϵ is the *absolute* binding energy per nucleon, *i.e.* $\epsilon > 0$)

$$\begin{aligned} Q &= B(Z, A) - B(Z_1, A_1) - B(Z_2, A_2) \\ &= -\epsilon A + \epsilon_1 A_1 + \epsilon_2 A_2 = -\epsilon A + \bar{\epsilon} A = (\bar{\epsilon} - \epsilon) A, \end{aligned} \tag{4}$$

where

$$\bar{\epsilon} = \frac{\epsilon_1 A_1 + \epsilon_2 A_2}{A_1 + A_2}. \tag{5}$$

Figure 1 tells us that if the initial nucleus is a heavy one (as indicated at the right of the picture), its fission into two medium-mass nuclei is such that $Q > 0$. By repeating a similar calculation for two light nuclei undergoing fusion, it is easy to see that also in that case one can get $Q > 0$.

The fact that fission of heavy nuclei is possible in terms of energy balance does not mean that fission will easily happen spontaneously. To understand that, imagine the two daughter nuclei (called *fission fragments*) being bound together in a potential well (the basketball in fig. 2). In order to be separated from each other and transform into the smaller basketball and the tennis ball on the right of fig. 2, they will have to pass through a Coulomb repulsion barrier. This would be impossible in classical physics but it is possible in quantum mechanics, thanks to the quantum tunneling effect, which leads to the so-called *spontaneous fission*. The lifetime for the nucleus to decay via spontaneous fission will vary depending on the probability for the two nuclei to pass through the barrier. If energy is supplied from the outside to the initial heavy nucleus, the system will jump to a more excited quantum state, *i.e.* a higher energy state of the two fragments in the potential well, accompanied by a deformation of the nucleus schematically illustrated by

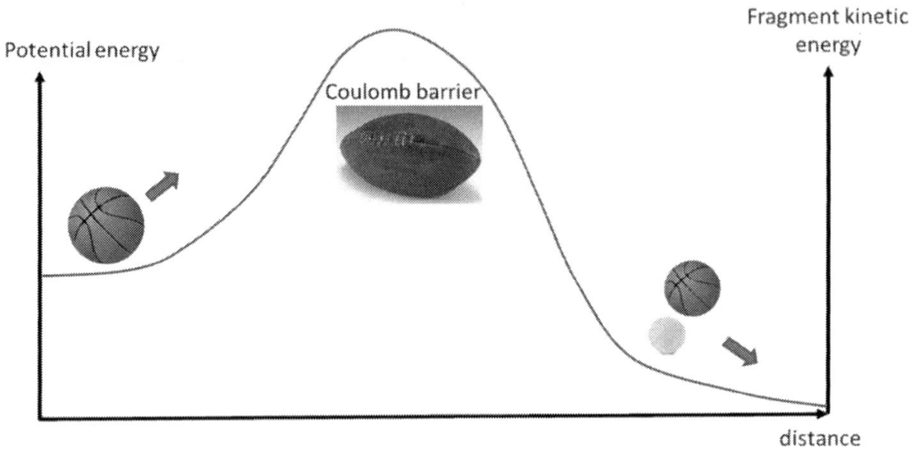

Fig. 2. – Nuclear potential energy as a function of the distance between the two fragments.

the rugby ball in fig. 2. Therefore the probability for the two fragments to penetrate the barrier will increase. This is what we call *induced fission*, which can actually happen with significant probability for nuclei with $A >\simeq 230$.

How can we supply energy to the nucleus from the outside? For example, suppose that an incoming neutron is captured in the potential well of a heavy nucleus (Z, A). Due to the generally significant binding energy (a few MeV) in the newly formed $(Z, A+1)$ nucleus, a few MeV of energy will be released and made available for the nucleus to approach the top of the barrier, thereby increasing by orders of magnitude the probability of the $(Z, A+1)$ nucleus to undergo fission. However, given the specific dependence of the binding energy on Z and A, typically for odd-A nuclei the binding energy of the extra neutron captured is higher than for even-A nuclei. The former nuclei are called *fissile* and they can undergo fission with high probability even when capturing a very slow neutron (even a neutron in thermal equilibrium with the surrounding medium). The latter nuclei are called *fissionable* and they can undergo fission with significant probability only when capturing a neutron with kinetic energy of the order of 1 MeV or more. Two famous examples of the above are the $^{235}_{92}$U and $^{238}_{92}$U Uranium isotopes. Indeed, uranium is found in nature as two isotopes, $^{238}_{92}$U and $^{235}_{92}$U, whose abundances are 0.7% and 99.3%, respectively. While ^{238}U is fissionable, ^{235}U is fissile. Both uranium isotopes are radioactive and undergo α decay, with emission of an α particle, *i.e.* a helium nucleus. Radioactive decays follow an exponential law, which means that, given an initial number of radioactive nuclei N_0, their number at a time t is given by

$$N(t) = N_0 e^{-t/\tau}, \tag{6}$$

where τ is the mean decay time. Often used is the *half-life*, $t_{1/2} = \tau\ \ln(2)$, which is the time it takes for the number of radioactive nuclei to drop by 50%. ^{238}U has a half-life of 4.47 billion years, about the age of the Earth, while ^{235}U has a half-life of about

704 million years. This means that when the Earth was formed, ^{238}U was twice more abundant than today, while ^{235}U was about 84 times more abundant.

Fission is not the only exoenergetic nuclear reaction induced by neutrons. Just to make two examples, the reaction $n + {}^{3}\mathrm{He} \rightarrow {}^{3}\mathrm{H} + p$, when initiated by a very slow neutron (whose kinetic energy we can neglect), can release about 0.76 MeV in form of the kinetic energy of the two final nuclei. The reaction $n + {}^{6}\mathrm{Li} \rightarrow {}^{3}\mathrm{H} + {}^{4}\mathrm{He}$ can release about 4.78 MeV, again taken by the motion of the final nuclei. However, fission has a distinctive feature that makes it different from all others. Indeed, the fission process is accompanied by the emission of a few *prompt neutrons*, *i.e.* in addition to the splitting of the initial nucleus into two fragments, a few neutrons are emitted by the fragments in a very short time after the splitting. Such neutrons are fast, *i.e.* they have high speed and high kinetic energy (about 2 MeV on average) and, as we will see shortly, if properly slowed down, can be very effective in inducing another fission. Due to the additional fission processes that can be initiated by these other neutrons, fission, under certain conditions, can give rise to a *chain reaction*, where neutrons emitted by one fission can produce additional fissions, and so on and so forth, resulting in a very significant release of energy by the fissile material.

Another important process is the *radiative neutron capture*, where the energy released during capture of a neutron excites the resulting nucleus to a quantum state that then decays by emission of electromagnetic energy in form of energetic photons (gamma rays). Neutron capture often leads to the formation of radioactive nuclei. An example is the formation of a radioactive Cobalt isotope, ^{60}Co by exposure of steel (which contains Cobalt) to neutrons, via the process

$$n + {}^{59}\mathrm{Co} \rightarrow {}^{60}\mathrm{Co} + \gamma \rightarrow {}^{60}\mathrm{Ni} + e^{-} + \bar{\nu}. \tag{7}$$

We see that radiative capture is accompanied by "prompt" gamma emission, *i.e.* the high-energy photons are emitted immediately upon capture, but in the above case the resulting nucleus ^{60}Co is radioactive and undergoes β decay, with emission of an electron and an antineutrino. ^{60}Co has a half-life of about 5.3 years, however β decay half-lives in different nuclei can vary by several orders of magnitude, depending on the involved quantum nuclear transitions. The radiative capture process can also occur on uranium nuclei and it can be a competitor to fission. The case of radiative capture on ^{238}U is a special one, as it leads to the formation of Plutonium, ^{239}Pu, a fissile element not existing in nature, via the chain of reactions

$$n + {}^{238}\mathrm{U} \rightarrow {}^{239}\mathrm{U} + \gamma \rightarrow {}^{239}\mathrm{Np} + e^{-} + \bar{\nu} \rightarrow {}^{239}\mathrm{Pu} + e^{-} + \bar{\nu} \tag{8}$$

Let us discuss the energy balance of the capture process in more detail, by considering the process $n + A \rightarrow A+1$. In particular, we want to calculate the energy excess ϵ created by the capture process.

We can write the energy conservation as

$$m_n + T_n + M(Z, A) = M(Z, A+1) + T_{A+1} + \epsilon\,, \tag{9}$$

with m_n, $M(Z,A)$, $M(Z,A+1)$ being the neutron, initial nucleus and final nucleus masses, respectively, T_n the neutron's kinetic energy, T_{A+1} the recoil nucleus' kinetic energy and ϵ the excess energy. Therefore we obtain

$$m_n + T_n + M_A = M_{A+1} + \frac{p_{A+1}^2}{2M_{A+1}} + \epsilon = M_{A+1} + \frac{p_n^2}{2m_n}\frac{m_n}{M_{A+1}} + \epsilon, \tag{10}$$

where we used momentum conservation, $|\vec{p_n}| = |\vec{p_{A+1}}|$, assuming the initial nucleus to be at rest. Approximating $M_{A+1} \sim (A+1)m_n$ in the ratio m_n/M_{A+1} (by neglecting the proton-neutron mass difference and the binding energy, which amounts to a small correction to the final result), we get

$$T_n + m_n + M_A - M_{A+1} = T_n \frac{1}{A+1} + \epsilon\,, \tag{11}$$

i.e.

$$T_n\left(1 - \frac{1}{A+1}\right) - B_n = \epsilon\,, \tag{12}$$

where $B_n = M_{A+1} - m_n - M_A < 0$ is the captured neutron's binding energy in the nucleus $A+1$. Finally, we can write

$$\epsilon = T_n \frac{A}{A+1} + |B_n|. \tag{13}$$

We can see that even in the case of capture of slow neutrons ($T_n \to 0$) there is an energy excess $\epsilon \simeq |B_n|$. This energy excess necessarily goes into internal nuclear degrees of freedom, which means that the compound nucleus $A+1$ is excited and must eventually release the excess energy in form of γ rays, nuclear particles, or the fission process.

Let's now calculate the actual value of $|B_n|$ in the two cases of interest: capture by ^{235}U and by ^{238}U. By applying the definition of binding energy and by using tabulated values of the Uranium isotope masses, we find that in the case

$$n + {}^{235}\mathrm{U} \to {}^{236}\mathrm{U} \tag{14}$$

about 6.5 MeV are released by the capture process, while in the case

$$n + {}^{238}\mathrm{U} \to {}^{239}\mathrm{U} \tag{15}$$

about 4.8 MeV are released. This difference is due to the fact that the captured neutron is more bound in the ^{236}U nucleus, with an even number of protons (92) and neutrons (144), than in the ^{239}U nucleus, with the same even number of protons (92) but an odd number of neutrons (147). Indeed, in a semiempirical description of nuclear masses, even-even nuclei have one more negative term in the expression for the mass as a function of Z

and A, which namely means that the binding is stronger. The opposite happens for odd-odd nuclei, where that term is positive, which means less binding. For even-odd or odd-even nuclei, such term is absent, so that the captured neutron is more bound (*i.e.*, releases more energy upon capture) in ^{236}U (even-even) than in ^{239}U (even-odd). Such additional term in the mass is related to the so-called *pairing* part of the nuclear force, a term essentially telling us that a nucleus is more bound when both protons and neutrons can be coupled in pairs.

We discussed above that we can see a heavy nucleus as being formed by two fragments that have to cross a barrier to get separated from one another in the fission process. Therefore, from a quantum mechanics point of view, if we excite the nucleus to a state which is energetically closer to the barrier, its fission probability will increase. At variance with that, the decay of the excited nucleus via gamma radiation does not depend on any barrier and so it is not so much dependent on the excitation energy obtained via neutron capture. As a result, radiative neutron capture can happen in both ^{235}U and ^{238}U no matter what the initial neutron energy is. On the contrary, while in ^{235}U the energy excess from capture of a slow neutron (which we saw above to be just the absolute value of the binding energy of that neutron in the final nucleus) is enough to cross the barrier and make fission happen, in the ^{238}U case the capture of a slow neutron does not supply enough energy and the neutron has to possess a kinetic energy of the order of 1 MeV for the fragments to cross the barrier and make fission happen.

1·2. *Energy released and reaction products.* – By calculating the specific mass balance given by eq. (3), it is easy to see that when a uranium nucleus fissions into two nuclear fragments, about 0.1% of uranium mass appears as a fission energy of ~ 200 MeV. Such an energy release is bigger than any other exoenergetic nuclear reaction by more than an order of magnitude (see for instance the above example of neutron capture on ^{6}Li) and bigger than chemical reactions by several orders of magnitude. Just for comparison, in the chemical reaction where a methane molecule is burnt, $CH_4 + 2O_2 \rightarrow CO_2 + 2H_2O$, about 9.2 eV of energy are released, more than seven orders of magnitude less than nuclear fission. In terms of specific energy, *i.e.* the amount of energy that can be obtained from a certain mass of material, measured in MJ/kg (MegaJoule per kilogram), Uranium produces about 8×10^7 MJ/kg, while for example methane produces about 56 MJ/kg. In terms of energy density, *i.e.* the amount of energy that can be obtained from a certain volume of material, measured in MJ/L (MegaJoule per liter), uranium produces about 1.6×10^9 MJ/L, while for example methane (at 15 °C and atmospheric pressure) produces 0.0378 MJ/L.

As anticipated above, the chain reaction is possible because along with the fission fragments an average of 2.5 prompt neutrons are emitted, with a mean kinetic energy per neutron of about 2 MeV. The average number of neutrons emitted is called ν and in general is of the order of 2-3, depending on the nucleus undergoing fission and on the incident neutron energy. Out of the 200 MeV released per fission, about 180 MeV appear as kinetic energy of the emerging two nuclei, plus kinetic energy of the prompt neutrons, plus energy in form of prompt gamma rays (photons) emitted when nuclear fragments are produced in a nuclear excited state. Due to the longer time scale of β decay of the fission fragments, additional energy is released later after fission and is carried by the electron, the associated antineutrino and possibly additional gammas. Antineutrinos escape from the reactor due to their negligible interaction with matter and therefore do not contribute to the process thermal budget, while electrons and gammas do, contributing with about 13 MeV per fission. When radioactive fragment build-up reaches a steady state, these 13 MeV (6.5% of fission energy) contribute to the instant reactor thermal power. At the exact moment that the reactor is shutdown by stopping the chain reaction, the thermal power due to these radioactive decays, called *residual heat*, remains and decays slowly with time (see later).

1·3. *Cross sections and flux.* – The cross section is a physical quantity and an observable that characterizes a nuclear reaction (elastic, inelastic scattering, etc.). The cross section is connected to the range and strength of the involved forces and represents the effective area of a nuclear target in a certain type of reaction or collision. Here we will only consider the so-called total cross section, defined as follows. Given a flux of particles incident on a single nucleus (target) per unit area dS per unit time

$$\frac{\mathrm{d}N_{\mathrm{in}}}{\mathrm{d}S\mathrm{d}t} \tag{16}$$

and given a reaction rate, which represents the number of interacting particles (scattered or absorbed projectiles) per unit time,

$$R = \frac{\mathrm{d}N_{\mathrm{reac}}}{\mathrm{d}t} \tag{17}$$

then the total reaction cross section σ is

$$\sigma = \frac{\frac{\mathrm{d}N_{\mathrm{reac}}}{\mathrm{d}t}}{\frac{\mathrm{d}N_{\in in}}{\mathrm{d}S\mathrm{d}t}}\,. \tag{18}$$

As is evident from the above equation, σ has the physical dimensions of a surface and can be interpreted as an effective target area.

Given a macroscopic target comprising several nuclei with mass density ρ and small thickness x, struck by a particle beam of intensity I (particles/sec), we have

$$R = I\frac{\rho x}{A}N_A\sigma, \tag{19}$$

where A is the target atomic weight and N_A is the Avogadro number. For a target of arbitrary thickness, by dividing it in thin slices of thickness dx, one obtains that

$$I(x) = I(0)\exp\left(-\frac{\rho}{A}N_A\sigma x\right). \tag{20}$$

In nuclear physics, the cross section unit is the *barn*, corresponding to $10^{-24}\,\mathrm{cm}^2$.

The quantity

$$\Sigma = \frac{\rho}{A} N_A \sigma \tag{21}$$

is called the *macroscopic cross section* and represents the probability of interaction per unit length of material crossed (and therefore has dimensions of the inverse of a length). It is easy to show that $1/\Sigma$ is the mean free path of an incident particle in the material and Σv is the frequency at which a projectile with speed v interacts with the crossed material, *i.e.* it represents the number of interactions of the projectile with the target nuclei per unit time.

Since the nuclear radius is roughly 10^{-12} cm, the geometrical cross sectional area of the nucleus is roughly 10^{-24} cm = 1 barn. However, the combination of quantum mechanical and specific dynamical effects from the nuclear forces can make nuclear cross sections extremely variable over a few orders of magnitude. As an example, in fig. 3 are reported the cross sections for the processes of fission and radiative capture for the two uranium isotopes found in nature and used in nuclear reactors, ^{238}U and ^{235}U, as a function of the neutron energy [2]. It is clearly seen that in ^{238}U, only fast neutrons with kinetic energy above 1 MeV can cause fission, with a cross section of the order of the barn, while for lower energies only radiative capture is significant. On the contrary, in ^{235}U, both fission and radiative capture cross sections are relevant at all neutron energies, with the fission cross section being much bigger than the one of radiative capture. Radiative capture for both nuclei and fission for ^{235}U follow a $1/v$ law, which makes fission in ^{235}U much more likely to occur at low neutron energies or, in other words, for "slow" neutrons. This is the main characteristic of fissile nuclei: the fission process can occur at all neutron energies and the corresponding cross section falls down with a $1/v$ law. As we saw in sect. **1**˙1, such a difference in the fission process is due to the nuclear pairing interaction, which makes the capture neutron more bound in ^{236}U than in ^{239}U, thereby making ^{236}U crossing the barrier and undergoing fission even when the captured neutron possesses a negligible kinetic energy, while ^{239}U needs additional energy besides that obtained via the capture process, which is supplied by a fast neutron possessing at least 1 MeV kinetic energy.

1˙4. *Neutron density and flux.* – We define as neutron density the expected number of neutrons with energy between E and $E + \mathrm{d}E$, in the volume $d^3\vec{r}$ at the space position $\vec{r}$, at a time t in a nuclear assembly. This quantity is indicated by $n(\vec{r}, E, t)$ and can be measured for instance in neutrons per cm^3 (or in cm^{-3}). The reaction density $R(\vec{r}, E, t)$ is defined as the number of reactions in the volume $d^3\vec{r}$ about $\vec{r}$, at a time t, initiated by neutrons with energy between E and $E + \mathrm{d}E$ and is given by the equation

$$R(\vec{r}, E, t) = n(\vec{r}, E, t)\, \Sigma\, v, \tag{22}$$

where again v is the neutron speed. The reaction density is measured in reactions per cm^3 per second. We give a special name to the quantity $n(\vec{r}, E, t)v$: it is called the

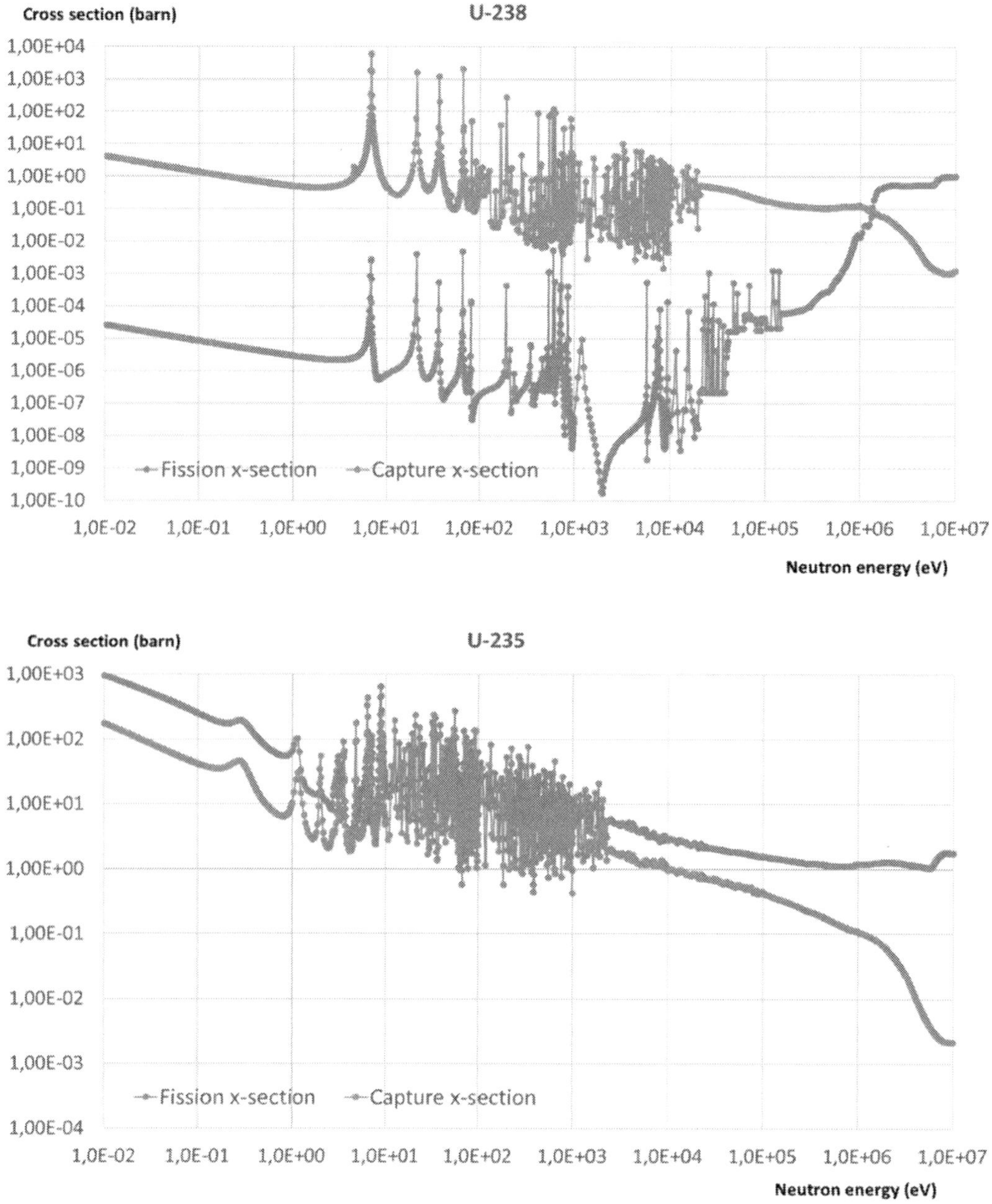

Fig. 3. – Fission (blue) and radiative capture (red) cross sections for the two isotopes of uranium.

neutron "flux" $\phi(\vec{r}, E, t)$ and is measured in neutrons per cm^2 per second. This quantity is essentially the number of neutrons arriving from all directions and crossing a unit surface in unit time (it may differ from other definitions of flux in physics but we will not discuss these differences here). Therefore, given a neutron flux ϕ and a macroscopic cross section Σ, the reaction density, number of reactions per unit volume and per unit time, R is simply given by $\Sigma\,\phi$

$$R(\vec{r}, E, t) = \Sigma\phi. \tag{23}$$

As an immediate example, let us consider a thermal reactor (see later) with 3 GW thermal power = $3 \cdot 10^9$ J/s (indicated by the symbol 3 GWth, with a 30% conversion efficiency it will produce about 1 GW electric power, 1 GWe). Considering that each fission releases order of 180 MeV energy = $3 \cdot 10^{-11}$ J at the moment of fission, it follows immediately that in the whole reactor about 10^{20} fissions/s occur, which corresponds to 2–$3{\cdot}10^{20}$ neutrons/s emitted and about $6{\cdot}10^{20}$ neutrinos/s emitted from fragment β decay. Assuming the reactor core to contain 27 tons of uranium oxide (UO_2, with density about 11 gr/cm^3), we get a total volume of the core of about 2.5 cubic meters, or $2.5 \cdot 10^6$ cm^3, which implies that the reaction density must be around $4 \cdot 10^{13}$ fissions/cm^3/sec. By recalling eq. (21) and assuming an enrichment of 3.5% in fissile uranium (*i.e.* 3.5% of the uranium in the core is fissile ^{235}U, which means that the effective fissile uranium oxide density will be 0.39 gr/cm^3, see sect. **1**˙5 below) and by using the known value of the fission cross section for thermal neutrons(¹) on ^{235}U (from fig. 3 we see that at 0.025 eV, corrisponding to room temperature, we have about 500 barn), we can calculate that $\Sigma \simeq 0.5\,\text{cm}^{-1}$. By recalling eq. (23), we can see that the neutron flux ϕ inside the reactor must be in the order of $8 \cdot 10^{13}$ neutrons cm^{-2} s^{-1}. This is a lot of neutrons incident on structural materials and since a good percentage of them are fast, energetic ones, possible damage to the materials has to be taken into consideration(²). This is even more the case in fast reactors, where most of the neutrons are fast, as well as in fusion reactors, where neutrons are essentially only fast ones (see [1]).

1˙5. *Fuel.* – We already encountered the definition of *fissile* (*e.g.* ^{233}U, ^{235}U, ^{239}Pu) and *fissionable* (*e.g.* ^{238}U) nuclei. There are also nuclei that can produce a fissile isotope via neutron radiative capture and subsequent β decay. These nuclei are called *fertile*, meaning that they can be used to produce fuel and essentially comprise ^{238}U and ^{232}Th. As an example, ^{238}U can produce a fissile nucleus via the chain of reactions in eq. (8).

Natural uranium comprises 0.7% ^{235}U and 99.3% ^{238}U, however reactors using mixed ^{235}U–^{238}U fuel typically need 3–5% ^{235}U, so that a procedure called uranium *enrichment* is needed to fabricate reactor fuel. The process of plutonium production, where actually a mixture of several Pu isotopes is produced, is called Plutonium *breeding*. Pu breeding is important in cases where the used fuel is reprocessed to extract the Plutonium isotopes. Under certain conditions, a reactor can produce more Pu than it consumes, in which case it is called *breeder*. From the cross sections shown in fig. 3 it follows that, in a reactor using a mixed ^{235}U–^{238}U fuel and for neutron energies < 1 MeV, ^{235}U provides

(¹) Thermal neutrons at room temperature have T_n around 0.025 eV or 25 meV, *i.e.* the value of kT, where k is the Boltzmann constant and T is the temperature of the system; see sect. **1**˙8.
(²) A fast neutron can collide with material nuclei and transfer a significant amount of energy, so that *e.g.* atoms in a metal can be displaced from their position in the lattice.

the dominant contribution to the fission rate. It is worth mentioning that thorium, an abundant element in nature, can be used to produce a different type of fuel. Indeed, in the reaction

$$n + {}^{232}\mathrm{Th} \rightarrow {}^{233}\mathrm{Th} + \gamma \rightarrow {}^{233}\mathrm{Pa} + e^- + \bar{\nu} \rightarrow {}^{233}\mathrm{U} + e^- + \bar{\nu} \qquad (24)$$

the final product is ^{233}U, which is fissile. Use of thorium as a breeder of ^{233}U is studied as a possible way to increase the available stock of fissile elements. The whole process of mining, extraction of the uranium mineral from the ore, purification, fuel fabrication through enrichment, irradiation and final storage or disposal is called *fuel cycle*. In the *open* or *once through* fuel cycle the spent fuel is put in temporary storage waiting for final disposal. In the (partially) *closed* fuel cycle the spent fuel is reprocessed to extract the Pu, which is then used to fabricate new fuel typically in the form of Mixed Oxide of Uranium and Plutonium (MOX).

1·6. *Nuclear waste.* – As discussed above, many of the fission fragments are radioactive, typically undergoing β decay very often accompanied by gamma emission. Such radioactive elements increase the radioactivity of the irradiated fuel, so that the fuel extracted from the reactor is a highly radioactive material that needs to be protected and shielded. While most of the radioactive fragments have relatively short half-lives, the longest being in the order of a few tens of years, a small group of fragments has very long half-lives in the order of 10^5 years, the so-called Long Lived Fission Products, LLFP. The aforementioned plutonium production is one example of process leading to the appearance of *Transuranics* in the fuel, *i.e.* chemical elements beyond uranium that do not exist in nature. Reaction (8) leads to production of the ^{239}Pu isotope, but further neutron captures produce other isotopes $^{240-242}$Pu, of which ^{241}Pu β-decays to americium, ^{241}Am and so on. Many of the Transuranics decay by α radioactivity and have long half-lives, from a few hundred years to a few hundred thousand years. Reprocessing of the irradiated fuel gives the possibility to recycle the Pu to fabricate new fuel, but whenever reprocessing is not applied or stopped, the resulting spent fuel is considered to be a *nuclear waste* or *radioactive waste* which needs to be properly stored and eventually disposed of. Indeed, such radioactive nuclei can be dangerous for the environment and for human health, due to direct exposure, especially in the case of the penetrating γ radiation, but also in case of skin/eyes exposure to β radiation, and due to ingestion or inhalation in the case of α and β decay. Additional radioactive waste is produced in the reactor by neutron capture in the surrounding materials, producing *activated* materials with a varying degree of radioactivity. The International Atomic Energy Agency has come up with a classification scheme that is used as a template by each country to define categories of waste depending on its radioactivity, isotope content and possible heat production due to the radioactivity itself [3].

1·7. *Fuel consumption.* – Let us go back to our ideal 1 GWe reactor. About 10^{20} fissions/s take place in the reactor, meaning that 10^{20} (^{235}U nuclei)/s disappear (actually somewhat more because of the concurrent radiative capture process). Therefore roughly

40 mg/s of ^{235}U are "burnt" in the reactor. For 1 year of operation at 80% load factor (the actual working time), this means a consumption of about 1 ton of ^{235}U: in volume of pure metallic ^{235}U, this would be a cube of about 36 cm side. For comparison, the same amount of electric power is obtained by burning 1.6 Mtoe, million tons of oil equivalent (considering a slightly higher conversion efficiency of oil-fired power plants), or about 2 billion cubic meters of natural gas, or 2.2 million tons of coal. In a real fission reactor ^{235}U consumption is partly compensated by plutonium (^{239}Pu) breeding according to eq. (8). Indeed, since ^{239}Pu is also a fissile nucleus, it is both produced and burnt in the fission core. In practice, for an initial load of about 27 tons of uranium enriched at 3.5% fissile ^{235}U (of which about 950 kg are ^{235}U), after 1 year of operation 280 kg of ^{235}U remain: about 560 kg of ^{235}U and 380 kg of other isotopes, mostly ^{239}Pu, have been burnt, which is approximately the 1 ton we have been estimating above.

1'8. *Fast and slow neutrons.* – We have seen that the cross sections for nuclear reactions induced by neutrons depend a lot on the energies involved, therefore it is important to introduce some classifications of the neutrons according to their energies. It is customary to adopt the following classification: *slow* neutrons are those with kinetic energy $T_n < 1\,\text{eV}$, in particular thermal neutrons at room temperature have T_n around 0.025 eV or 25 meV (the value of kT, where k is the Boltzmann constant and T is the temperature of the system); *epithermal* neutrons are those with $1\,\text{eV} < T_n < 100\,\text{keV}$ (0.1 MeV); finally, fast neutrons have $0.1\,\text{MeV} < T_n < 20\,\text{MeV}$. Obviously neutrons in general can have energies above 20 MeV but this is an extreme limit in reactor physics (*e.g.*, neutrons from fusion of a deuteron and a tritium nucleus have 14 MeV fixed energy).

It is easy to show in non-relativistic kinematics that after a scattering off a nucleus with mass number A, *on average* the kinetic energy of the neutron changes according to the ratio

$$\frac{T'_n}{T_n} \simeq \frac{1+A^2}{(1+A)^2} \tag{25}$$

(where T_n and T'_n are the kinetic energies before and after the scattering, respectively). For a heavy nucleus $A \gg 1$, one has that $T'_n \simeq T_n$ or in other words, the neutron has to undergo many collisions in order to significantly lose energy. Consider instead the case $A = 1$ (this is relevant for any medium containing hydrogen, where hydrogen nuclei, *i.e.* protons, will be among the targets). In this case $T'_n = \frac{T_n}{2}$, *i.e.* on average a neutron will lose half of its energy at each collision and therefore few collisions are sufficient to rapidly decrease its energy. For collisions on a light nucleus like carbon ($A = 12$), the situation will be similar to that of hydrogen, although the average loss of energy per collision will be lower. Light materials containing hydrogen (water, paraffin, plastic materials in general etc.) or containing light nuclei (graphite, etc.) are called *moderators*, for their effectiveness in slowing down neutrons. In a thermal reactor, the fuel is immersed in water, which is used both as coolant and as moderator: the fast neutrons from fission will be slowed down very quickly via collisions with protons and oxygen nuclei in water and therefore they will have a very high fission cross section for ^{235}U (see fig. 3).

2. – Chain reaction and simple reactor kinetics

For the reactor to work effectively and safely, the chain reaction must not diverge and must not die away. This means that, amongst the 2-3 neutrons emitted on average in the fission process, precisely one has to induce another fission event, not more, not less. The remaining fission neutrons will then either be absorbed by radiative capture or will leak out from the system. The character of the chain reaction depends essentially on the reactor composition and on its geometry and is defined through the quantity called k, also called multiplication coefficient

$$k = \frac{\text{number of neutrons in one fission generation}}{\text{number of neutrons in the preceding fission generation}} . \tag{26}$$

The condition $k = 1$ corresponds to a *critical* reactor, *i.e.* a reactor actually working and producing thermal power from fission. For $k > 1$ we have a supercritical reactor (the number of fission reaction diverges) and for $k < 1$ we have a subcritical reactor (fission reactions die away). In a "simple minded" reactor time evolution, called *kinetics*, it is relatively easy to show that the number of neutrons in the system evolves as

$$n(t) = n(0) \exp\left(\frac{k-1}{\tau}t\right), \tag{27}$$

where $n(0)$ is the initial number of neutrons and τ represents the average lifetime of a neutron in the system before it disappears due to fission, capture or escape (leakage). Here we clearly see that $k = 1$ corresponds to a steady state, while $k > 1$ implies that the system is diverging and $k < 1$ corresponds to a decreasing population. Based on this simple kinetics, we see that the time constant governing the response of the reactor would be $\frac{\tau}{k-1}$. Since the typical neutron lifetime in a thermal power reactor is of the order of 10^{-4} s, we see that even for $k = 1.001$, the neutron population and therefore the reactor power would increase by a factor 2.7 in 0.1 s, which would not be a manageable situation from a practical point of view. But there is one missing piece in the picture of fission discussed above. Actually, there is a small fraction (less than a percent) of neutrons originating from the fission process that are not emitted immediately, but some time later. These neutrons are emitted by highly excited fission fragments on a time scale from milliseconds to seconds and for this reason are called *delayed neutrons*. In simple terms, a nuclear reactor is made critical thanks to such small fraction of delayed neutrons and therefore they are the ones that dominate the reactor response time making it much longer, thereby allowing reactor control by neutron-absorbing control rods.

2·1. *A bit more on neutron multiplication and transport.* – By considering in more detail what processes can intervene along the route of the next-generation neutrons emitted after a single fission, one can come up with the following definition of multiplication coefficient k:

$$k = \eta f P_{NL}, \tag{28}$$

where

- f is the conditional probability that, if a neutron is absorbed, it will be absorbed in the fuel instead of structural elements;
- η is the average number of neutrons produced per neutron absorbed in the fuel, which is given by the average number of neutrons per fission ν multiplied by the probability that the absorption process leads to fission instead of radiative capture $\eta = \nu \frac{\sigma_f^F}{\sigma_a^F}$, where σ_a^F is the absorption cross section in the fuel (absorption=fission+radiative capture) and σ_f^F is the fission cross section in the fuel;
- P_{NL} is the probability that a neutron does not escape from the reactor (No-Leakage).

For an infinite reactor clearly $P_{NL} = 1$ and we get $k_\infty = \eta f$. In practice, an effective multiplication factor, k_{eff} is defined, which takes into account neutron leakages and the reaction dependences on neutron energy. It is interesting to look at the behavior of the quantity η as a function of neutron energy, shown in fig. 4 for the ^{235}U and the ^{239}Pu fissile nuclei. It is clear that for thermal neutrons, this number is around 2, which means that effectively there are two neutrons available for further processes, one of which can trigger another fission, while another can be lost in captures or leak out of the system. In case of capture on a fertile nucleus like ^{238}U, plutonium breeding can occur, but overall this will happen with relatively low probability. On the contrary, for high kinetic energies about 10–100 keV (fast neutrons), η becomes larger than 2 and so more neutrons will be available for breeding new fuel. Therefore, a so-called *fast reactor*, *i.e.* one in which neutrons are not moderated (see sect. **3**·1: "Reactor types"), may produce more plutonium than it consumes, in which case we will have a *fast breeder*.

3. – Main features of a nuclear fission reactor

We saw that neutrons slow down when undergoing collisions with nuclei (in particular with light nuclei), therefore their energies in the core can go from about 10 MeV (usually the maximum energy of fission neutrons), down to as low as 10^{-3} eV. We also saw that neutron cross sections have a strong dependence on neutron energy, in particular fission cross sections decrease very strongly with energy for fissile nuclei like ^{235}U. Therefore, when using fuels containing these nuclei, it is easiest to maintain a fission chain reaction by using slow neutrons. Hence most nuclear reactors until now (the so-called Generation I to III+) use low mass number materials such as water or graphite to moderate the fast fission neutrons, thereby slowing them down to energies comparable to the thermal energies of the nuclei in the reactor core. By definition, a *thermal reactor* is one where the

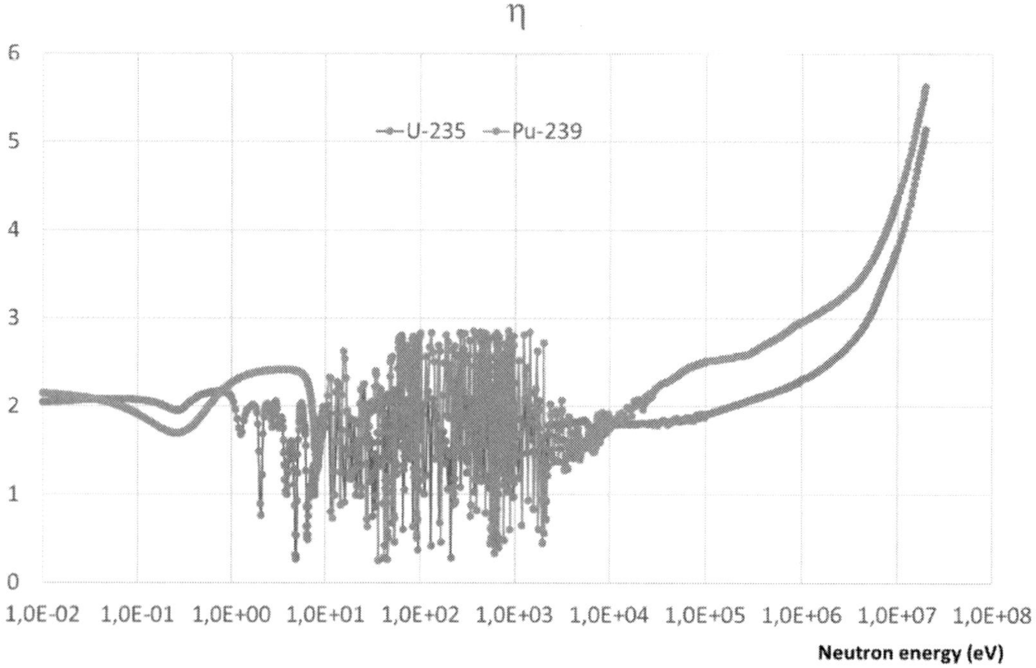

Fig. 4. – The parameter η as a function of neutron kinetic energy for the two main fissile nuclei.

average neutron energy is comparable to thermal energies (order of several tens of meV). Because of the very high-fission cross section at thermal energies (see fig. 3), this type of reactor requires the minimum amount of fissile material for fueling. As an example, a Light Water Reactor (LWR) can start with 3.5% ^{235}U + 96.5% ^{238}U (a mixture enriched with respect to natural uranium). Burn-up (consumption) of ^{235}U is partly compensated by breeding of ^{239}Pu via neutron radiative capture and subsequent β decays and after 1 year operation, the core may contain something like 1% ^{235}U + 1% plutonium (all Pu isotopes, including ^{239}Pu). At the contrary, in a *fast reactor* the average neutron energies in the core are typically above 100 keV. As previously discussed, this can be accomplished if the core contains mostly high mass-number materials. This means that for core cooling the choice will be a liquid metal like sodium, lead or a lead-bismuth mixture, or alternatively the coolant should be a low-density material like Helium gas. On the other hand, σ_f is smaller for fast neutrons, which means that a much higher percentage of fissile material is needed to sustain the chain reaction. At the same time, based on similar considerations to those in sect. **1**˙4, it is easy to see that the neutron flux must be a couple of orders of magnitude higher than the thermal reactor, therefore more around 10^{16} neutrons cm^{-2} s^{-1}. Therefore, in a fast reactor the absolute neutron flux is much higher than in a thermal reactor and it is also composed by a majority of fast, energetic neutrons. This is why the long-term damage to the structural elements by neutron bombardment is a more important issue in a fast reactor than in a thermal one. In summary, the typical conceptual schemes of a thermal and a fast reactor are illustrated in fig. 5.

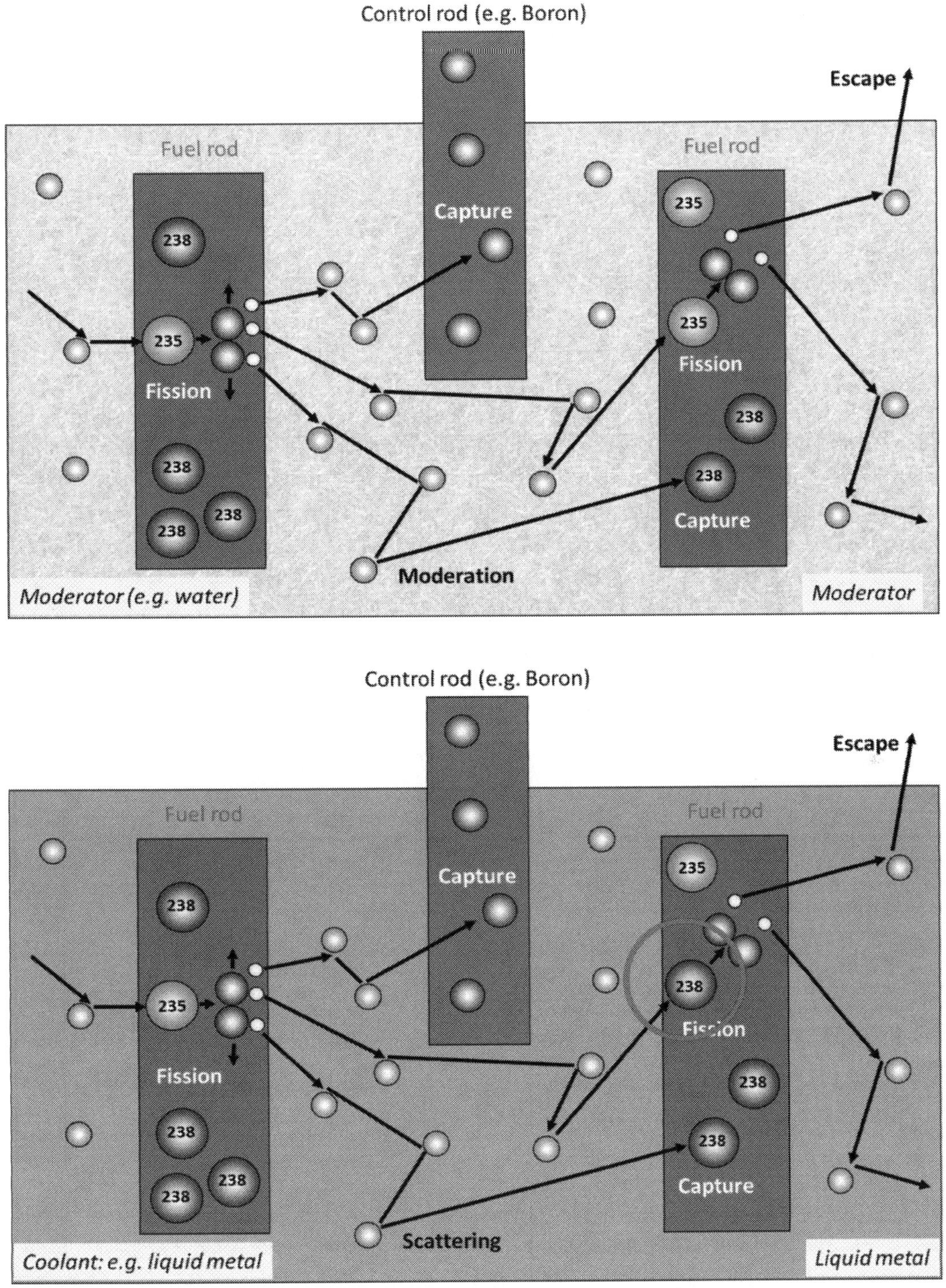

Fig. 5. – Conceptual schemes of thermal (top) and fast reactor (bottom). In the thermal reactor (top), the neutron energy spectrum exhibits a strong slow component from neutrons in thermal equilibrium with the moderator, with a kinetic energy of 25 meV at 290 K, corresponding to the most probable speed of the Maxwellian distribution. In the fast reactor (bottom), the neutron energy spectrum is much harder and most neutrons can have kinetic energies up to a few MeV.

3˙1. *Decay heat.* – Decay heat is the heat produced in the reactor core as a result of radioactive decay: the energy of the alpha, beta or gamma radiation is converted into atomic motion. Decay of the short-lived radioisotopes created by fission continues for some time after shut down. A practical approximation is given by the formula

$$\frac{P}{P_0} = 6.6 \cdot 10^{-2} \left[\frac{1}{(\tau - \tau_s)^{0.2}} - \frac{1}{\tau^{0.2}} \right], \qquad (29)$$

where P is the decay power, P_0 is the reactor power before shutdown, τ is the time since reactor startup and τ_s is the time of reactor shutdown measured from the time of startup (in seconds). At shutdown, the heat power is about 6.5% of the previous core power ($\simeq$ 200 MWth for a 1 GWe reactor), which is sufficient to melt the core. About 1 hour after shutdown, the decay heat will be about 1.5% of the previous core power, after 1 day it will fall to 0.4% and after a week it will be 0.2%, much smaller but still significant. For this reason, spent fuel rods discharged from a reactor core after a period of operation are kept for long time in special cooling water pools, before being further processed. Removal of decay heat is therefore a very important aspect in designing a reactor and implementing safety measures, as the Fukushima accident has dramatically stressed.

3˙2. *Safety.* – Safety is a very important aspect in nuclear power production. The core of a nuclear reactor produces radiation both while in operation and after shutdown, because of the high content of radioactive materials in the fuel and because of the neutron irradiation of materials in the core and surrounding the core, which produces radioactive nuclei starting from stable ones. Therefore, the safety culture is an essential aspect of modern nuclear industry. Attention to safety is present from the design stage down to operation, shutdown and decommissioning of the plants and particular care is devoted to the analysis of possible accidental scenarios. A fundamental principle is the *defence-in-depth*, which means that a series of barriers, starting from the solid fuel form itself up to the reactor containment building, are put in place to contain radioactivity as much as possible at all times and in all instances. Obviously, in this respect, the role of independent regulatory authorities is crucial in overseeing all stages of a plant lifetime.

3˙3. *Radioactive waste.* – We saw in the preceding sections that a nuclear reactor is producing a significant amount of radioactive substances, due to fission itself, as well as due to neutron capture in the fuel and in structural elements. Fission products with a half-life until several days are responsible for the large amount of decay heat that is produced within the core immediately after reactor shutdown, *i.e.* after stopping the chain reaction. However, their radioactivity dies away in a few days or weeks, which means that in case of accidental release of radioactive products they pose a threat only for such a limited amount of time and when the spent fuel is stored in the cooling pools they do not produce anymore heat after a relatively short amount of time. Other Short-Lived Fission Products with half-lives until a few years can instead be a safety concern also after an accordingly longer period of time, as are Medium-Lived Fission Products with half-lives until about 100 years (*e.g.*, ^{137}Cs and ^{90}Sr, with half-life of about 30 years). Such isotopes contribute to heat production in the spent fuel and can be a safety concern for decades if accidentally released into the environment. Finally, Long-Lived Fission Products (LLFP) have half-lives over about 200000 years (no fission product has half-life between about a 100 years and 200000 years). Transuranics, *i.e.* plutonium together with the so-called *Minor Actinides, MA*, neptunium, americium, curium, etc., decay predominantly by α decay and have half-lives ranging from a few hundreds of years up to above 2 million years (^{237}Np). The typical inventory from a 1 GWe PWR fuelled with about 27 tons of U (3.5% ^{235}U) at discharge (when the spent fuel is removed from the core) is approximately 280 kg of ^{235}U, 266 kg of Pu of which 156 kg of fissile ^{239}Pu, 20 kg of MA, 13 kg of ^{90}Sr and 30 kg of ^{137}Cs (both with 30 years half-life), and 63 kg of LLFP [4]. The very long decay time of both LLFP and MA is the reason why disposal of spent fuel and similar waste requires very long-term storage, for which one possible solution envisaged is storage deep underground in the so-called geological repositories. Currently, Finland, Sweden and France have advanced projects for building such repositories for final disposal. Some countries apply reprocessing of the spent fuel, where plutonium is extracted from the fuel rods and used to produce new fuel containing both uranium and plutonium, so that the spent fuel is directed to the underground repository only after partial recycling.

3·4. *Waste production and incineration.* – In the effort to improve the safety, security and efficiency of nuclear plants, new concepts of reactors have been developed with goals including the minimization of the production of MAs, a better and more efficient use of the fuel, a better thermodynamic efficiency, the possible production of hydrogen at high temperatures and finally improved safety features to minimize the risk of accidents. All these various concepts are considered within the so-called Generation-IV reactors, which are the subject of an international initiative [5]. As an example, we saw that in a fast reactor, by using *e.g.* a liquid metal or gas as coolant, the energy spectrum of the neutrons is harder. In such a reactor core, it becomes possible to burn not only fissile elements like ^{235}U and ^{239}Pu, but also fissionable elements like ^{238}U for which fission occurs significantly only above a certain energy threshold around 0.5 MeV, as can be seen in fig. 3. Although given the much lower fission cross section typically the fuel has to be richer in fissile content, this means that also ^{238}U can to some extent be considered a fuel as well, which has obvious implications in terms of how long uranium resources will last. Another important advantage in terms of fuel cycle in fast neutron systems is that, for energies, say, above 0.5 MeV, fission becomes dominant over capture, as can again be seen in fig. 3. This means that the MA production, which occurs namely via capture, is suppressed. Yet another important aspect is that in a fast reactor the MA produced can be partly destroyed in the reactor itself by fission. This is because many of the MA are fissionable, *i.e.* they can undergo fission only for neutron energies above around 0.5 MeV. An important physics aspect when considering fission of MA is that the amount of delayed neutrons produced upon their fission is significantly less than for ^{235}U and ^{239}Pu. This means that for instance fueling a reactor purely with MA is not possible, as one would lose the crucial safety feature of the reactor response times being dominated by the relatively long delayed neutron emission times. However, for Accelerator Driven Systems (ADS), which are based on a subcritical reactor core that requires an external neutron source, the role of delayed neutrons is not so crucial and therefore there is the possibility to mix a larger quantity of MA into the fuel, thereby offering the possibility to use the apparatus to perform *transmutation* of the MA, *i.e.* incinerate the MA into fission products, thereby reducing the radiotoxicity of the final materials in the fuel. This is called *P&T, partitioning and transmutation* approach, meaning that first some nuclides are separated from the spent fuel (partitioning) and then they are irradiated in a special core to be either destroyed, transformed into stable elements or into very short-lived isotopes reaching fast the stability (transmutation [4,6]). Currently, the MYRRHA project in Belgium [7] started a staged approach to the construction of a multipurpose ADS that will also demonstrate the concept of waste incineration.

REFERENCES

[1] Ongena J., "Fusion: a true challenge for an enormous reward", these proceedings , p. 143; Romanelli F., "Fusion energy: technological challenges", these proceedings, p. 175.
[2] Java-based Nuclear Data Information System, `http://www.oecd-nea.org/janis/`.
[3] *Classification of Radioactive Waste*, No. GSG-1, General Safety Guide (IAEA) 2004, `https://www-pub.iaea.org/MTCD/Publications/PDF/Pub1419_web.pdf`.
[4] Nifenecker H., Meplan O. and David S., *Accelerator Driven Subcritical Reactors* (Institute of Physics Publishing, Bristol and Philadelphia) 2003.
[5] *Generation IV International Forum*, `https://www.gen-4.org/`.
[6] Achuthan P. V. *et al.*, *Implications of partitioning and transmutation in radioactive waste management*, Technical Reports Series No. 435 (IAEA) 2004, (Physics Publishing, Bristol and Philadelphia) 2003, `https://www.iaea.org/publications/7112/implications-of-partitioning-and-transmutation-in-radioactive-waste-management`.
[7] Abderrahim H. Aït *et al.*, "Partitioning and transmutation contribution of MYRRHA to an EU strategy for HLW management and main achievements of MYRRHA related FP7 and H2020 projects: MYRTE, MARISA, MAXSIMA, SEARCH, MAX, FREYA, ARCAS", *EPJ Nucl. Sci. Technol.*, **6** (2020) 33.

Fusion: a true challenge for an enormous reward

J. Ongena

Plasma Physics Laboratory, Royal Military Academy
Renaissancelaan 30, 1000, Brussels, Belgium

Summary. — A source of energy which would be inexhaustible, inherently safe and environmentally friendly, is this not a marvellous prospect? Nuclear fusion is a possible candidate for this role. It has been the energy source of our Sun and the stars in the universe for billions of years. The process requires temperatures of tens of millions of degrees, so extremely high and foreign to our daily experience that it seems out of reach. Nevertheless, these extremely high temperatures are routinely realised in several laboratories all over the world. Since the early 1990s, tens of MW of fusion power have been released from fusion reactions. Progress in the last years shows that fusion holds the promise to be a clean and safe solution for mankind's long-term energy needs. We are witnessing the birth of a new technology destined to meet the gigantic future energy needs of mankind with minimal impact on the environment.

1. – Fusion reactions in our Sun and on Earth

The dominant fusion reaction in our Sun is the p-p reaction, converting hydrogen into helium. Other fusion reactions occur in stars, *e.g.* the carbon-nitrogen-oxygen or CNO cycle. It is dominant in stars that are about 1.3 times heavier than our Sun. In our Sun only about 1% of the ^{4}He nuclei originate from the CNO cycle [1].

At first glance it seems easy to overcome the repulsion of the positively charged nuclei involved in fusion reactions. Why not simply use a particle accelerator to provide the nuclei with sufficient energy and make them collide? Unfortunately, the probability for a fusion reaction is extremely small. Rather than fuse, the nuclei are mostly scattered in all directions and only a few fusion reactions take place.

Our Sun offers a more promising solution: in this gaseous body, the particles that are confined in the core of the Sun undergo a much larger number of collisions and, hence, fusion reactions per unit of time. Gravity is responsible for the confinement and the heating of the fusing protons. An estimate of the temperature in the centre of the Sun can be found by remarking that the protons need to be sufficiently fast (*i.e.*, have sufficient energy and thus be sufficiently hot) to overcome the compressional forces from gravitation. This means: kinetic energy of a proton in the centre = potential energy from gravity, or

$$\frac{3}{2} k T_{\text{proton in centre}} = \frac{G m_{\text{proton}} M_{\text{Sun}}}{R_{\text{Sun}}}, \tag{1}$$

$$T_{\text{proton in centre}} = \frac{2 G m_{\text{proton}} M_{\text{Sun}}}{3 k R_{\text{Sun}}}, \tag{2}$$

with $k = 1.38 \times 10^{-23}$ J/K Boltzmann's constant, $G = 6.67 \times 10^{-11}\,\text{Nm}^2/\text{kg}^2$ the gravitational constant, $M_{\text{Sun}} = 1.989 \times 10^{30}$ kg and $R_{\text{Sun}} = 695\,500\,\text{km} = 6.955 \times 10^8$ m (more than 2 times the distance Earth-Moon...) we obtain

$$T_{\text{proton in centre}} = 1.56 \times 10^7\,\text{K} = 15.6\,\text{MK}. \tag{3}$$

From the formula above, it is immediately clear that a contraction of the star under its own gravity will further raise the central temperature, resulting in the production of heavier elements. Fusion in heavy stars will continue to keep the star alive as long as the reactions are exothermic and are thus able to help the star defend itself against a full collapse under its own gravity. The mass defect curve (fig. 1) shows that energy can be gained from fusion reactions up to ^{62}Ni, which is the tightest bound nucleus [2]. Once the star tries to fuse heavier nuclei, gravity finally wins, and a collapse of the star follows. This ends in a spectacular supernova, leading to a neutron star or for the heaviest stars, the formation of a black hole.

The p-p and CNO fusion reactions in our Sun have a far too low reaction rate to be useful on Earth. One of the underlying reasons is that from every four protons taking place in the reaction chain that leads to ^{4}He, two of them need to be converted into a neutron. In the p-p reaction, this occurs via the formation of a deuteron involving the slow beta-plus decay reaction:

$$\text{p} + \text{p} \rightarrow \text{D} + \text{e}^+ + \nu + 0.42\,\text{MeV}. \tag{4}$$

Once the neutron is created, a further set of reactions involving ^{3}He and the unstable nucleus ^{6}Be finally leads to the formation of ^{4}He. Despite this complex and slow chain

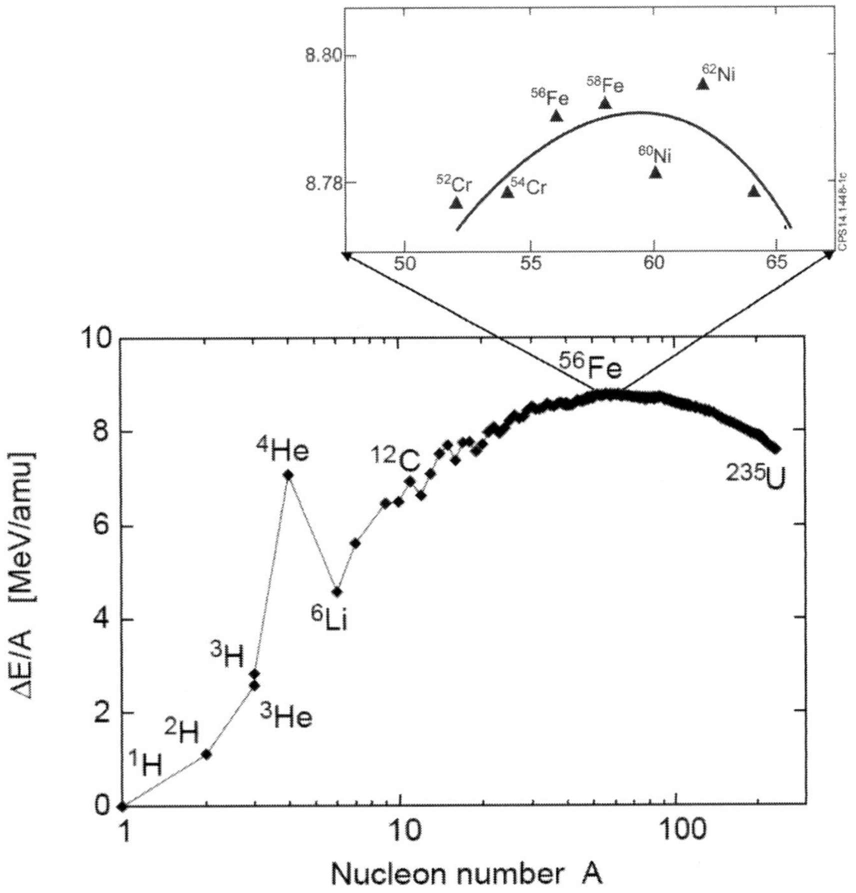

Fig. 1. – Binding energy per nucleon *vs.* atomic mass number showing the maximum in the curve reached for ^{62}Ni. Note the exceptionally large binding energy of the ^{4}He nucleus among the light nuclei. An approximate indication is given for the difference in binding energy per nucleon in fusion reactions of light nuclides with ^{4}He as one of the reaction products ($\sim 5\,\text{MeV/nucleon}$) and fission of ^{235}U ($\sim 0.9\,\text{MeV/nucleon}$). The exact numbers for the mass defects of selected nuclides with mass numbers up to 20 can be found in table II.

of reactions, every second about 700 million tonnes of H are converted in our Sun into 695 million tonnes of ^{4}He; it thus means that every second about 5 million tonnes of the Sun's mass disappear and are totally transformed into energy.

On Earth we have to use nuclear fusion reactions that avoid the difficult conversion process from protons to neutrons. The evident solution is to use reaction products that already contain neutrons, needed for the formation of ^{4}He. Instead of protons as reaction products, we use the nuclei of the isotopes D and T or also of the rare isotope ^{3}He. Here D symbolizes deuterium (the stable isotope of hydrogen with a nucleus consisting of one proton and one neutron) and T is the symbol for tritium (the radioactive hydrogen isotope with a nucleus of 2 neutrons and 1 proton). A number of fusion reactions [3] together with their cross-sections and energy released are listed in table I. Note that the

TABLE I. – *Fusion reactions.*

Reaction	σ at 10 keV (barn)	σ_{max} (barn)	Center-of-mass energy (keV) for σ_{max}	Energy released (MeV)
$D + T \to {}^4He + n$	2.72×10^{-2}	5.0	64	17.59
$D + D \to T + p$	2.81×10^{-4}	0.096	1250	4.04
$D + D \to {}^3He + n$	2.78×10^{-4}	0.11	1750	3.27
$T + T \to {}^4He + 2n$	7.90×10^{-4}	0.16	1000	11.33
$D + {}^3He \to {}^4He + p$	2.2×10^{-7}	0.9	250	18.35
$p + {}^6Li \to {}^4He + {}^3He$	6×10^{-10}	0.22	1500	4.02
$p + {}^{11}B \to 3\,{}^4He$	4.6×10^{-17}	1.2	550	8.68
$p + p \to D + e^+ + \nu$	3.6×10^{-26}	–	–	$1.44 + 0.27(\nu)$
$p + {}^{12}C \to {}^{13}N + \gamma$	1.9×10^{-26}	1.0×10^{-4}	400	1.94

energy released is the largest in reactions that lead to the formation of a ^{4}He nucleus because of its exceptionally large binding energy (fig. 1). The fusion reaction with the largest cross section and resulting in the formation of ^{4}He is the so-called D-T reaction:

$$D + T \to {}^4He\ (3.5\,\text{MeV}) + n\ (14.1\,\text{MeV}). \tag{5}$$

In this reaction, "only" a rearrangement of the nuclides is needed to form ^{4}He, which is much more easy to realise: the cross-section σ for this reaction at, *e.g.*, 10 keV is about 10^{24} times larger than that for the p-p reaction.

The reaction products are an α-particle (^{4}He nucleus) and a neutron. Twenty percent of the reaction energy is carried by the α-particles and 80% by the much lighter neutron. This follows directly from the conservation of momentum. In magnetic confinement systems the neutron does not feel the presence of the magnetic field (because it is not charged) and escapes immediately from the reactor volume, while the charged α-particle remains confined by the magnetic field. The kinetic energy of these escaping fast neutrons will be converted into heat in an appropriate blanket surrounding the future fusion reactor and then into electricity using conventional technology (steam cycle). About one million times more energy is released from a fusion reaction in comparison with a chemical one (MeV's instead of eV's for the latter). This is the reason why so little fuel can produce so much energy: about 15 g of a 50%/50% DT mix suffices to produce the electricity needed by one EU citizen for 80 years.

Other possible fusion reactions of interest between isotopes of hydrogen and helium are

$$D + D \to {}^3He\ (0.82\,\text{MeV}) + n\ (2.45\,\text{MeV}), \tag{6}$$

$$D + D \to T\ (1.01\,\text{MeV}) + H\ (3.02\,\text{MeV}), \tag{7}$$

$$D + {}^3He \to {}^4He\ (3.6\,\text{MeV}) + H\ (14.7\,\text{MeV}). \tag{8}$$

TABLE II. – *Atomic mass and binding energy per nucleon [42], stability and natural abundance for light nuclides. Note the remarkable absence of stable nuclides with mass number 5 and 8.*

Nucleus	Stable/unstable (half-life)	Natural abundance	Binding energy per nucleon (keV)	Atomic mass (u)
^{1}H	s	99.985%	n.a	1.0078250
n	u (12 min)	n.a.	n.a	1.0086649
^{2}H=D	s	0.015%	1112.283	2.0141018
^{3}H=T	u (12.32 yr)	n.a.	2827.265	3.0160493
^{3}He	s	0.000137%	2572.680	3.0160293
^{4}He	s	99.999863%	7073.915	4.0026033
^{6}Li	s	92.41%	5332.331	6.0151229
^{7}Li	s	7.59%	5606.439	7.0160034
^{9}Be	s	100%	6462.668	9.0121831
^{10}B	s	19.8%	6475.083	10.0129369
^{11}B	s	80.2%	6927.732	11.0093052
^{12}C	s	98.89%	7680.144	12.0000000
^{13}C	s	1.11%	7469.849	13.0033548
^{14}N	s	99.634%	7475.614	14.0030740
^{15}N	s	0.366%	7699.460	15.0001089
^{16}O	s	99.762%	7976.206	15.9949146
^{17}O	s	0.038%	7750.728	16.9991318
^{18}O	s	0.2%	7767.097	17.9991596
^{19}F	s	100%	7779.018	18.9984032
^{20}Ne	s	90.48%	8032.240	19.9924402
^{21}Ne	s	0.27%	7971.713	20.9938467
^{22}Ne	s	9.25%	8080.465	21.9913851

These are more difficult to realise and have a lower power density than the D-T reaction [4, 5] but show even more benign environmental features. The D-D reaction would eliminate the need for tritium and produce neutrons with lower energies that are easier to absorb and shield. A reactor based on the D-^{3}He reaction would proceed with low neutron production (some neutrons would be produced in competing but much less occurring D-D and secondary D-T reactions) and lead to much less induced radioactivity in the reactor structures. However, the prospects for these "advanced" fuels are still too speculative and only the D-T reaction has immediate future prospects. For the interested reader, table II provides the masses and binding energies per nucleon for a selection of lighter nuclides, including those involved in reactions (5) to (8).

In order to undergo fusion reactions, the reacting particles have to approach each other to within short distances. A first estimate of the energy necessary to realise fusion reactions can be obtained from the potential energy of one nucleus in the presence of the

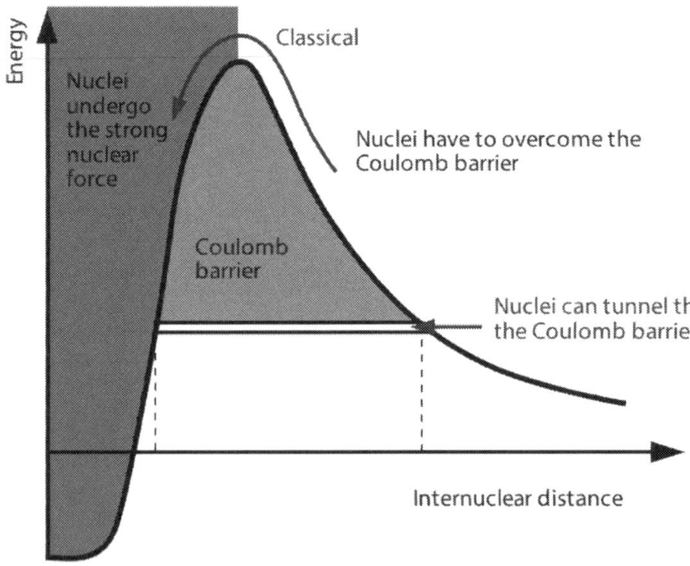

Fig. 2. – Atomic particles can overcome the Coulomb barrier (electrostatic repulsion) at much lower energies than the maximum by "tunnelling" through it. This increases the probability for fusion.

electric field of another one at a distance such that, in the classical representation, the nuclei just touch each other. The maximum of the Coulomb barrier between two nuclei with radius R_1 and R_2 and charges qZ_1 and qZ_2 can be found from

$$V_C[\mathrm{J}] = \frac{q^2 Z_1 Z_2}{4\pi\varepsilon_0(R_1 + R_2)}, \tag{9}$$

or

$$V_C[\mathrm{MeV}] = 1.02\,\frac{Z_1 Z_2}{A_1^{1/3} + A_2^{1/3}}, \tag{10}$$

where we made use of the formula for the radius R of the nuclei [6]:

$$R = 1.4\,A^{1/3} \times 10^{-15}\,\mathrm{m}, \tag{11}$$

with A the mass number of the nucleus. For the reaction between deuterium and tritium we find 0.38 MeV. In the classical picture, nuclei should therefore at least acquire this energy before any fusion reaction can take place. The equivalent temperature of a Maxwellian plasma with this value as average energy amounts to more than 4 billion degrees.

A reduction of this huge temperature is made possible by the tunnel effect. Owing to the wave character of matter, nuclei have a finite probability to "tunnel" through the Coulomb barrier and thus can fuse at energies much lower than those given by the

 https://doi.org/10.1051/epjconf/202226800011

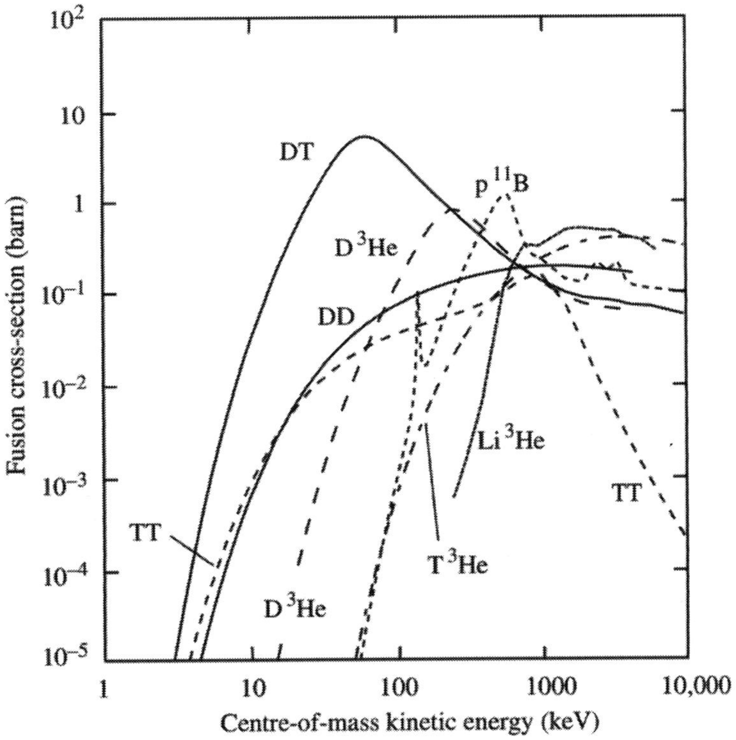

Fig. 3. – Cross-sections (in barn) as a function of the centre-of-mass energies of the reacting particles for various fusion reactions.

maximum of the barrier (fig. 2). As we shall see further, this reduction is substantial, and the optimum temperature for a magnetically confined D-T plasma is about 13 keV.

Cross-sections [3] as a function of the centre-of-mass energies of the reacting particles for the reactions (5) to (8) above are shown in fig. 3. The largest cross-section is found for the D-T reaction (about 5 barn) at the centre-of-mass energy of the colliding particles $1/2\ \mu\ (v_1 - v_2)^2$, where $\mu = 1/2(m_1 m_2/(m_1+m_2))$ of about 60 keV. However in a gas one has to take into account that the velocities of the colliding particles are described by a Maxwell-Boltzmann distribution and values for the reactivity $\langle\sigma v\rangle$ averaged over this distribution as a function of the temperature are shown in fig. 4 for various fusion reactions. The maximum is now much broader but still around energies of about 60 keV for the D-T reaction. Thus one is inclined to conclude that the plasma has to be heated to temperatures of over 600 million degrees to obtain the maximum fusion rate. However, one has to remember that in a magnetic fusion device, tokamak or stellarator, the operating pressure $p = nkT$ is limited due to various instabilities and turbulence in the plasma (see, *e.g.*, [7]; for recent experimental progress towards the reduction of plasma turbulence, see [8, 9]). The fusion reaction rate $R = \frac{n^2}{4}\langle\sigma v\rangle$ can thus be written as $R = \frac{1}{4}\frac{\langle\sigma v\rangle}{T^2}$ and the maximum reactivity is determined by the maximum of this function [10] as shown in fig. 5. It is clear that for the D-T reaction the peak is located at about 13 keV. For the D-D or D-^{3}He reactions, the temperature is higher (or the power density is smaller) at

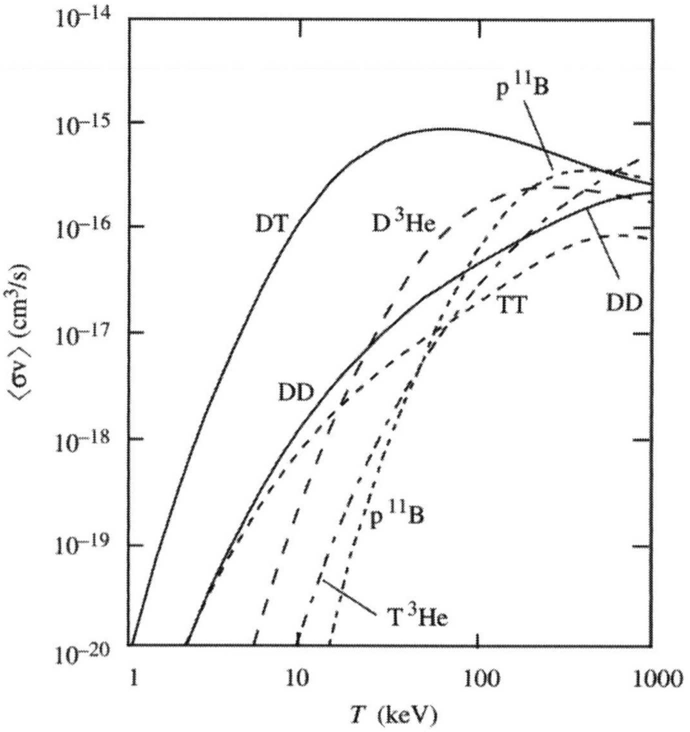

Fig. 4. – Reactivity averaged over a Maxwell-Boltzmann distribution as a function of temperature for the fusion reactions of fig. 3.

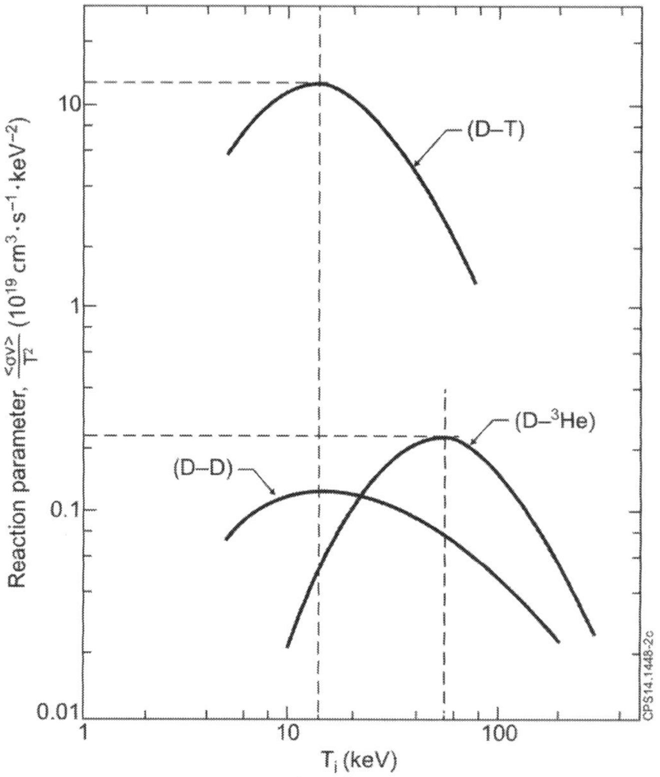

Fig. 5. – Maxwellian averaged values for $\frac{\langle \sigma v \rangle}{T^2}$ as a function of temperature for various fusion reactions.

the maximum of the $\langle \sigma v \rangle$ curve. For the D-T reaction we find that the reactivity is maximized at a temperature of about 150 million degrees (the temperature of a Maxwellian gas with an average energy of 1 eV corresponds to 11600 K), which is 10 times larger than in the Sun's core. In other words, the realisation of nuclear fusion on Earth requires the creation of a small but very hot Sun. Two questions then immediately arise: how does one realise such extreme temperatures and —equally important— how can one confine a gas at such high temperatures in a safe way?

2. – Plasma confinement

Two radically different approaches have been followed towards the realisation of nuclear fusion on Earth: inertial and magnetic confinement. In the first approach, lasers or particle beams are used to rapidly compress the nuclear fuel in order to achieve the conditions required for fusion. To this end, a small sphere with a diameter of a few hundred μm is filled with equal amounts of deuterium and tritium. Powerful laser or particle beams irradiate the surface of this little sphere as uniformly as possible, for about a billionth of a second. The outermost layers of the sphere vaporise nearly immediately, generating an inward-propagating spherical shock wave, which compresses the sphere's contents enormously. At the end of this short compression phase, a density of a thousand times that of water is reached at a temperature of some tens of millions of degrees centigrade. Under these conditions, the deuterium and tritium nuclei start to fuse. The energy released by these first fusion reactions further heats the rest of the strongly compressed fuel, allowing fusion reactions to spread throughout the sphere.

The second approach is magnetic confinement. Since this has proven to be the most successful method to date, magnetic fusion research is under development in laboratories all over Europe and throughout the world. The rest of this article is focussed on magnetic confinement. For a recent overview of the status of inertial confinement, we refer to [11]. Recent inertial fusion results with the achievement of burning plasmas, are announced in [12, 13] and a nice summary of these results is presented in [14].

At the high temperatures necessary for nuclear fusion, atoms break up into their constituents, electrons and nuclei, and the fuel becomes fully ionised: a plasma. To confine a hot plasma one can make use of magnetic fields, since charged particles will follow a helical path around the field lines owing to the Lorentz force. Possible movements perpendicular to the field are in this way highly restricted.

The most obvious way to realise this is a purely cylindrical configuration, which, however, suffers from plasma losses at both ends. Bending the magnetic field lines onto themselves, leading to a doughnut-shaped or toroidal magnetic field configuration can avoid these losses. A toroidal magnetic field, however, is not homogeneous. Its curvature destroys most of the nice confinement properties of a purely cylindrical configuration. As a result, ions and electrons move in opposite vertical directions (fig. 6), giving rise to an electric field which, when combined with the toroidal magnetic field, expels all particles to the outside, irrespective of their charge. Without special measures, it is thus impossible to confine a toroidal plasma in a stable way. However, if the magnetic field can be given

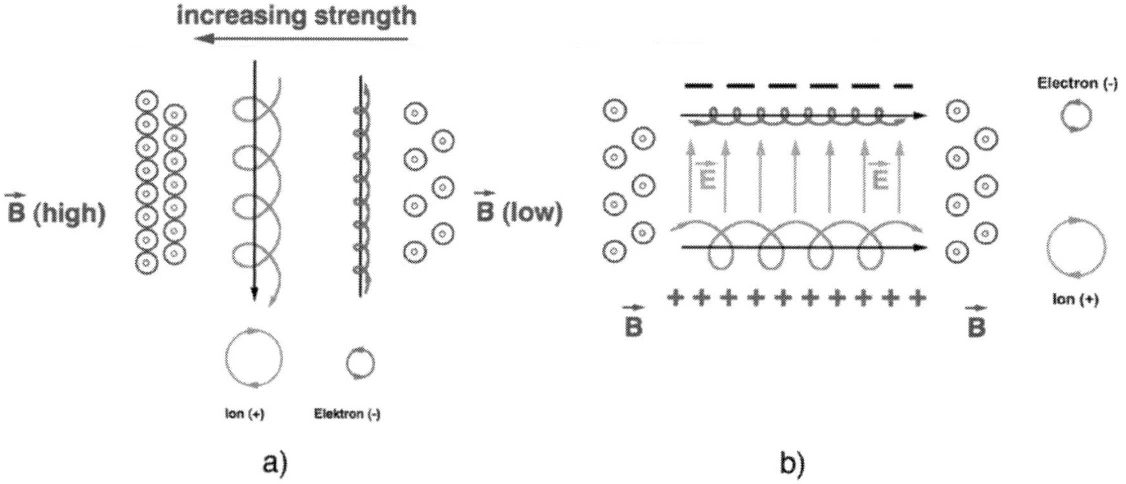

Fig. 6. – Illustration of the effect of the inhomogeneity of the toroidal magnetic field on plasma confinement. The effect can be easily explained in two steps considering the formula for the Larmor radius: $\rho = \frac{mv_\perp}{qB}$. For the first step see panel (a) on the left: formation of the electric field due to the inhomogeneity of the toroidal field. The figure shows a cross-section of the plasma; the orbits of ions and electrons are no longer perfectly circular, but the particles spiral vertically up and down: indeed from the formula follows that ρ decreases as B increases and increases as B decreases. As ions and electrons rotate in opposite directions, they move vertically in opposite directions, creating an electric field in the plasma. Second step, see panel (b) on the right: effect of the created electric field on the plasma particles. We consider a small section of the previous figure, such that B is to a good approximation homogeneous over the section, and thus the effect of E alone can be studied. As B is constant over this section, the only parameter influencing the movement of the particles is the velocity. Positive ions are accelerated towards the negative layer of charges and decelerated towards the positive layer of charges, leading to an increase, respectively, decrease of the velocity of the particle. From the formula follows that with increasing velocity the Larmor radius increases and decreases with decreasing velocity: thus the particle spirals vertically outwards. A similar reasoning for the electrons leads to exactly the same result. Both electrons and ions move thus horizontally outward: the plasma is unstable. To solve this problem, the formation of the electric field has thus to be avoided. Adding a poloidal component to the toroidal field is a way to do this.

an extra twist, vertical drifts can be cancelled and a suitable trap can be made for fusion plasmas. This extra twist can be produced by means of electrical currents flowing inside or outside of the plasma ring. The *tokamak* concept relies on currents produced inside the plasma, whereas the *stellarator* concept relies on currents external to the plasma.

In a tokamak, a set of coils placed around the doughnut-shaped plasma chamber produces the main toroidal magnetic field. The electrical current necessary for the stability of the plasma flows in the plasma ring, which serves as the (only) secondary winding of an enormous transformer. The induced current in the plasma ring generates a poloidal magnetic field. The combination of this poloidal field with the main toroidal field results in a helical magnetic field (fig. 7). The magnetic structure thus generated consists of an infinite set of nested toroidal magnetic surfaces, each with a slightly different twist,

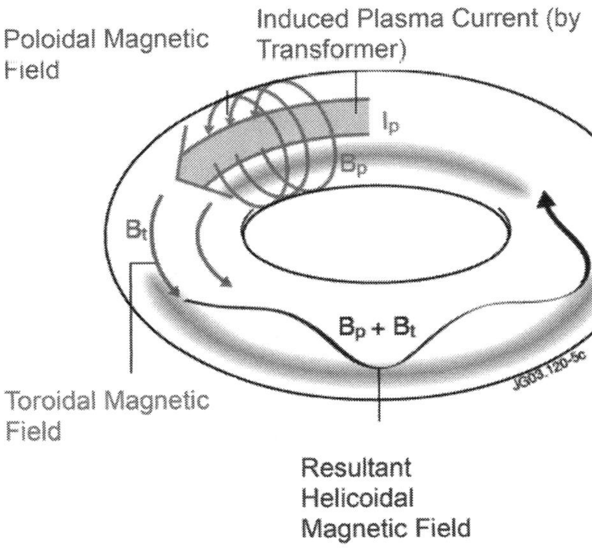

Fig. 7. – Principle of the tokamak: A large current induced in the plasma ring generates a poloidal magnetic field. The combination of this poloidal field with the main toroidal field results in a helical magnetic field, necessary for a stable confinement of the plasma in a toroidal configuration.

reducing further the leakage of particles and heat from the plasma. On each of these surfaces, the plasma pressure is constant. Tokamak research started in the Kurchatov Institute in Moscow under the leadership of Lev Andreyevitch Artsimovich. The success of this configuration became at once clear at the Novosibirsk IAEA conference in 1968 when much higher temperatures were announced, obtained in a tokamak, compared to all other magnetic configurations under study at that time.

The tokamak is however a pulsed device, since the transformer needs a steadily increasing current in its primary winding to induce the plasma current. Continuous operation of a tokamak fusion reactor could be realised in theory using the bootstrap current [7] or non-inductive current drive by the heating systems. However this still will require considerable efforts before becoming a practically viable option [15].

A fusion device that operates without a plasma current and thus allows continuous operation, is the stellarator. Confinement in this device relies on currents *external* to the plasma to create the needed helical magnetic configuration. In its basic configuration, extra helical coils around the toroidal plasma provide the necessary additional twist to the toroidal magnetic field generated by the main field coils (fig. 8). These helical windings around the plasma ring complicate the construction. In addition such "classical" stellarator configurations lack good confinement properties. Modern stellarators have optimised confinement properties, and are equipped with a set of coils with a complex shape that is determined numerically to very high precision (fig. 9). Several devices of the stellarator type are in operation or in construction at this moment all over the world. The largest stellarator in the world that started operations in 2016 is Wendelstein 7-X, located in Greifswald, Germany (fig. 10) with successful experimental campaigns [16, 17]

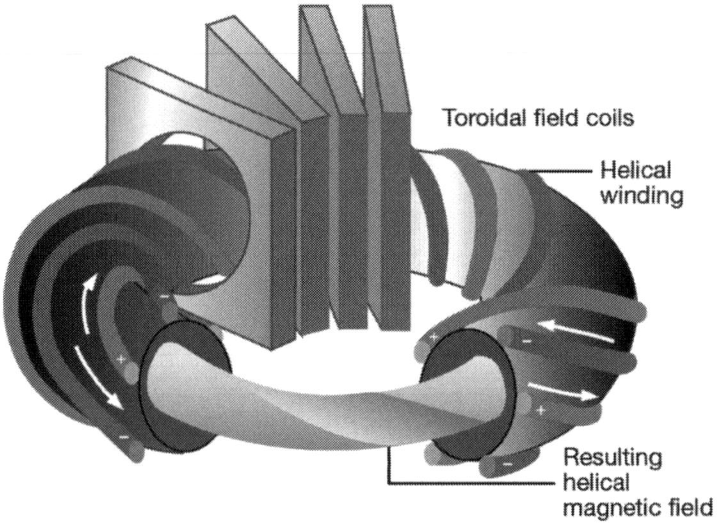

Fig. 8. – In a stellarator, helical coils external to the plasma ring create a supplementary magnetic field that, in combination with the main toroidal magnetic field, confines the plasma.

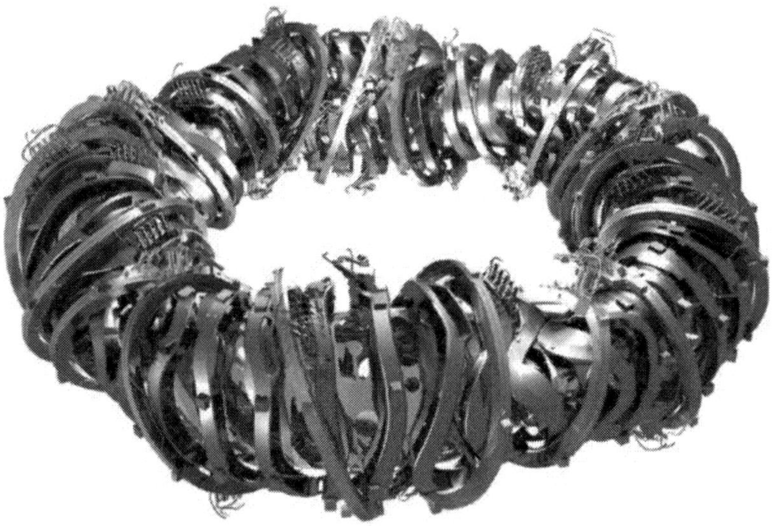

Fig. 9. – An illustration of the complex shapes of the coils of a modern stellarator, which combine the functions of the separate coil systems of fig. 8.

and further campaigns planned for the coming years. The device will be further upgraded in the coming years and in its final configuration will be able to run plasma pulses of 30 minutes. The largest stellarator-like device outside Europe is the LHD (Large Helical Device) of the National Institute for Fusion Science (NIFS, near Nagoya, Japan) that started operations in March 1998. Also this device has led to several rather important contributions [18].

Fig. 10. – A picture of the stellarator Wendelstein 7-X, at the time of construction in November 2011.

3. – Plasma heating

The plasma current in a tokamak, needed to create the helical magnetic field providing stable confinement is a first source of heat. Since the heating results from the finite resistance of the plasma, just as in an electrical heating element, this is referred to as Ohmic heating. The plasma resistance, however, decreases with increasing plasma temperature, and at some millions of degrees, the efficiency of Ohmic heating becomes too low to be useful for further heating. In fact, at these temperatures the plasma becomes a better conductor than copper at room temperature. That is why auxiliary heating methods have to be used to reach the temperatures required for fusion.

One additional heating method consists of injecting energetic particle beams of neutral hydrogen or deuterium into the plasma. First ions are created in a plasma source and accelerated with voltages up to 150000 volts. The accelerated particles need to be neutralized because only neutral particles can penetrate the magnetic fields needed to confine the plasma. They pass through a cloud of neutral gas where they "steal" electrons from atoms in a charge-exchange reaction, and become energetic neutral particles. These in turn enter the hot tokamak plasma without impediment from the confining magnetic fields of the fusion device and, once in the hot plasma, are immediately ionised again, thus confined, and start to deposit their energy via collisions to the rest of the plasma particles. Powers of up to several million watts per neutral injector can be delivered to the plasma using this technique. Note that for higher acceleration voltages negative ions have to be used, as is now considered for ITER and DEMO (see sects. **8**˙3 and **8**˙4). The reason is that the neutralization of positive ions becomes very inefficient at higher voltages.

A second additional heating method sends electromagnetic waves into the plasma, by antennas or wave guides at the plasma edge. The principle underlying this method is similar to that of a microwave oven: the energy from the waves is most easily absorbed if the emitted frequency is equal to a "natural" frequency of the particles that need to be heated. The cyclotron frequency with which the charged plasma particles gyrate around the magnetic field lines is an evident choice. Thus, two different heating systems exist: Ion Cyclotron Resonance Heating (ICRH) and Electron Cyclotron Resonance Heating (ECRH), depending on whether ions or electrons are heated. Ion cyclotron frequencies are in the MHz range (20 MHz and upwards), while electron cyclotron frequencies are in the range of 100–200 GHz, due to the smaller mass of the electrons. Heating powers for high-frequency systems range from 100 kW to several tens of MW.

4. – Characterizing the fusion reactivity of plasmas

The power amplification factor Q is defined as the ratio of the power produced by fusion reactions to the total heating power injected in the reaction chamber. Two important milestones for the value of Q are customarily used in fusion research. The first, *breakeven*, is reached when the heating power is equal to the power produced from fusion reactions, corresponding to a Q value of one. The second, *ignition*, is reached when the additional heating systems can be switched off and the heat of the fusion reactions alone is sufficient to maintain the high temperatures needed for fusion. This corresponds to an infinite value for Q.

5. – Conditions for a fusion reactor: Lawson criterion

The Lawson criterion determines conditions for an operating reactor. In its simplest form it can be defined as follows: the sum of the power densities from auxiliary heating P_{heat} and α particles has to be at least equal to the power densities lost by radiation $P_{\mathrm{radiation}}$ and transport, $P_{\mathrm{transport}}$ (convection, conduction). This can be written as

$$P_{\mathrm{heat}} + P_\alpha \geq P_{\mathrm{transport}} + P_{\mathrm{radiation}}. \tag{12}$$

To simplify the calculations, we consider in what follows a pure D-T plasma without impurities (*i.e.*, effective charge of the plasma $Z_{\mathrm{eff}} = 1$). It thus implies also that we neglect the presence of He ash from the reaction. We assume only Bremsstrahlung losses for $P_{\mathrm{radiation}}$ (given by the formula P_{Brems} (in $\mathrm{W/m^3}$) $= C_B T^{1/2} n^2$ if T is expressed in keV, and n in $\mathrm{m^{-3}}$ with $C_B = 5.35 \times 10^{-37}\,\mathrm{W\,m^3\,keV^{-1/2}}$ [7]) and we characterise all transport related energy losses, expressed by the term $P_{\mathrm{transport}}$, making use of the confinement time τ_E, defined by the relation $\mathrm{d}W_p/\mathrm{d}t = W_p/\tau_E$, where W_p is the energy density in the plasma. From this definition it follows that the energy confinement time can also be defined as the characteristic time for the plasma to cool down, once all heat sources have been switched off. We also take into account that

$$P_{\text{heat}} Q = P_{\text{fusion}} = 5\, P_\alpha \tag{13}$$

and that the total plasma energy density is the sum of the ion and electron energy densities

$$W_p = \frac{3}{2}(n_e k T_e + n_i k T_i)\,, \tag{14}$$

with T_i and T_e in K, n_e and n_i in m^{-3}, and $k = 1.38 \times 10^{-23}$ J/K the Boltzmann constant.

In a neutral D-T plasma electron and ion densities are the same $n_e = n_i = n$; we also assume that the plasma is in thermal equilibrium, *i.e.* there is no difference between electron and ion temperatures $T_e = T_i = T$. Therefore

$$W_p = 3\,\text{nkT}. \tag{15}$$

Assuming a pure 50%D–50%T plasma leads also to $n_\text{T} = n_\text{D} = {}^1\!/_2 n_e = {}^1\!/_2 n$. The expression for the power density delivered to the plasma by the α particles, can thus be written as

$$P_\alpha = n_\text{D} n_\text{T} E_\alpha \langle \sigma v \rangle = \frac{1}{4} n^2 E_\alpha \langle \sigma v \rangle. \tag{16}$$

Expressing E_α and T_k in keV, (12) can be written as

$$\frac{1}{4} n^2 E_\alpha \langle \sigma v \rangle \left(\frac{Q+5}{Q} \right) \geq \frac{3nT_k}{\tau_E} + C_B^* Z_{\text{eff}} T_k^{1/2} n^2 \tag{17}$$

and finally

$$n\tau_E \geq \frac{3T_k}{\frac{\langle \sigma v \rangle E_\alpha (Q+5)}{4Q} - C_B^* Z_{\text{eff}} T_k^{1/2}}\,, \tag{18}$$

where $C_B^* = C_B / 1.602 \times 10^{-16} = 3.34 \times 10^{-21}\,\text{W}\,\text{m}^3\,\text{keV}^{-1/2}$.

The right-hand side of the formula shows a minimum for about $T_k = 25\,\text{keV}$. Approximating $\langle \sigma v \rangle$ [in $\text{m}^{-3}\,\text{s}$] $= 1.1 \times 10^{-24}\ T_k^2$ (valid to within 15% in the interval $10\,\text{keV} < T_k < 25\,\text{keV}$) and setting $Z_{\text{eff}} = 1$ (pure D-T plasma) we find

$$\text{for breakeven } (Q = 1)\text{: } n\tau_E \geq 2.5 \times 10^{19}\,\text{m}^{-3}\text{s} \tag{19}$$

and

$$\text{for ignition } (Q = \infty)\text{: } n\tau_E \geq 1.5 \times 10^{20}\,\text{m}^{-3}\text{s}. \tag{20}$$

An alternative formulation of the Lawson criterion is in the form of the so-called fusion triple product. Considering again only the temperature range where the minimum of eq. (18) occurs ($kT_e \sim 10$–$25\,\mathrm{keV}$), we can neglect to a first approximation the corrections for the Bremsstrahlung losses. Thus we find, with the same conventions for the various physical quantities as in eq. (18):

$$n\tau_E T_k \geq \frac{12QT_k^2}{(Q+5)\langle\sigma v\rangle E_\alpha}\,. \tag{21}$$

The minimum for the right-hand side is obtained at the maximum for $\langle\sigma v\rangle/T_k^2$, as shown above, about $T_k = 13\,\mathrm{keV}$. We thus find

$$\text{for breakeven } (Q=1)\text{: } n\tau_E T_k \geq 5\times 10^{20}\,\mathrm{m^{-3}s\,keV} \tag{22}$$

and

$$\text{for ignition } (Q=\infty)\text{: } n\tau_E T_k \geq 3\times 10^{21}\,\mathrm{m^{-3}s\,keV}. \tag{23}$$

These simplified expressions should give a good idea of the physics behind the famous Lawson criterion. For a more detailed treatment, taking into account radiation, impurities etc. we refer to the interesting paper by Rebhan and Van Oost [19].

6. – Outline of a fusion reactor power plant

The concept of a future fusion power device based on a tokamak is outlined in fig. 11. Due to the high reactivity of the D-T fusion reaction, deuterium and tritium will most likely be selected as fuel for the first generation of fusion power stations. The electrically charged α-particles from fusion reactions are confined by the magnetic field. These high-energy helium nuclei collide with the background plasma particles, *i.e.* with the D and T ions in the plasma and maintain the temperature rendering additional heating systems quasi unnecessary as soon as ignition is reached.

In contrast to the α-particles, the neutrons generated by the D-T reaction have no electrical charge and escape immediately from the confining magnetic field in the plasma chamber. These neutrons are not lost, but serve a dual purpose in the reactor. First, they produce tritium in a blanket surrounding the reactor containing lithium or lithium compounds, according to the following reactions:

$$^{7}\mathrm{Li} + \mathrm{n} \rightarrow {}^{4}\mathrm{He} + \mathrm{T} + \mathrm{n} - 2.47\,\mathrm{MeV}, \tag{24}$$

$$^{6}\mathrm{Li} + \mathrm{n} \rightarrow {}^{4}\mathrm{He}\ (2.05\,\mathrm{MeV}) + \mathrm{T}\ (2.73\,\mathrm{MeV}). \tag{25}$$

This is necessary to refuel the plasma with tritium, consumed in fusion reactions. Second, the energy of the neutrons deposited in the blanket is converted into heat, which is in turn is transported to the exterior by the cooling system and used to generate steam and drive classical turbines for the production of electricity.

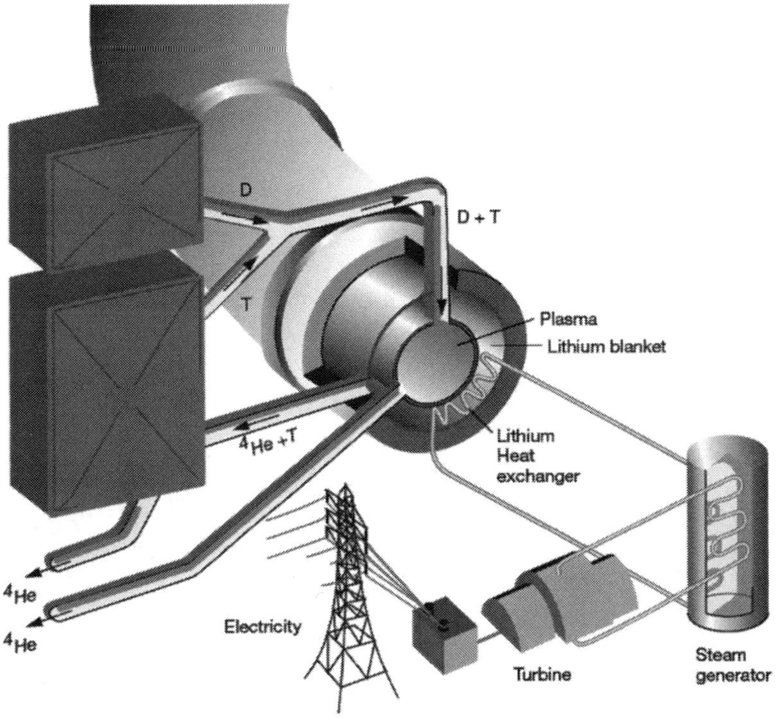

Fig. 11. – Many obstacles still have to be overcome before the first nuclear fusion power plant will be put into operation. Eventually, heat generated in fusion reactions will be used to generate steam to drive a turbine, with electrical power as final result.

7. – Progress in recent years

The progress obtained in fusion research in the past three decades is clearly visible from values obtained for the fusion product. Compared to the first fusion experiments, this value has increased by a factor of more than 10 000! The results obtained on the largest tokamak in the world, JET (Joint European Torus, located at the laboratories of the Culham Centre for Fusion Energy (CCFE), close to Oxford, UK), have contributed significantly to this progress, An important milestone was reached in 1991 with the first production of MWs of fusion power from controlled fusion reactions, as tritium was used for the first time as fuel in a tokamak [20]. These experiments, obtained with a mixture of 90%D and 10%T, generated fusion powers in the megawatt range for nearly 2 seconds, with a maximum of about 1.7 MW, corresponding to a Q value of about 0.15.

Further successes were obtained early in 1994 in the American tokamak TFTR (Tokamak Fusion Test Reactor) at the Plasma Physics Laboratory of Princeton University. In plasmas consisting of a mixture of 50% deuterium and 50% tritium, multi-megawatt level fusion powers were generated for about one second, with a maximum of 6.3 MW [21]. Plasma temperatures in excess of 300 million degrees were reached in the plasma centre, 20 to 30 times hotter than in the centre of the Sun! In November 1994, fusion powers of more than 10 MW were generated, corresponding to Q values of about 0.27 [22].

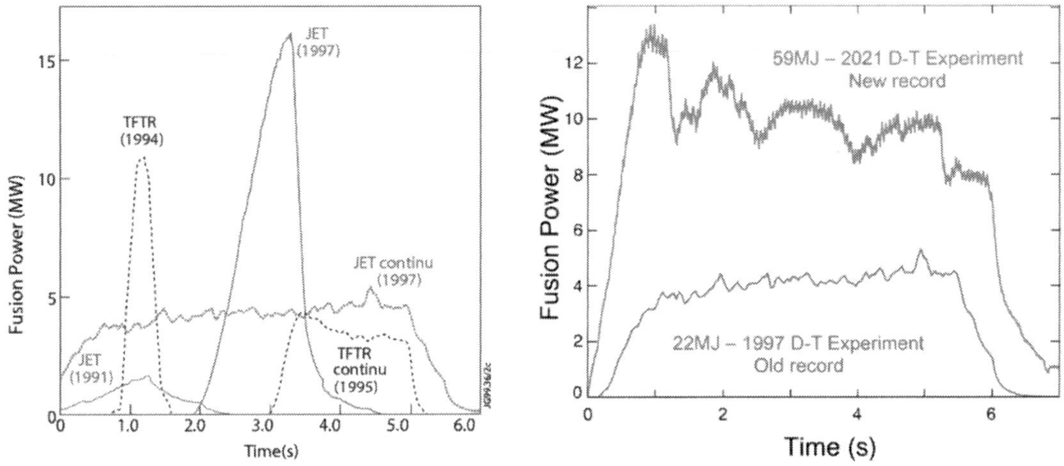

(a) (b)

Fig. 12. – (a) Time traces of the fusion power released in different high performance deuterium-tritium fusion experiments in TFTR and JET in the 1990s. (b) Time traces of the D-T experiments with record fusion energy production in JET in December 2021.

Other record values were reached in 1996 on the Japanese Tokamak JT-60U of the Naka Fusion Research Establishment, located 150 km north of Tokyo, a division of the Japanese Atomic Energy Research Institute (JAERI). This machine demonstrated temperatures in excess of 520 million degrees [23], the highest temperature ever realised by man on Earth. Even more important, a record value for the fusion triple product was obtained in pure deuterium plasmas. If the same conditions had been realised in deuterium-tritium plasmas, it would have resulted in a Q value of about 1.25, *i.e.* conditions better than breakeven [24].

The most impressive results in fusion research up to now were obtained in the JET. Experiments in 50%D–50%T plasmas resulted both in maximum fusion power and record fusion energy production values in JET. Over 16 MW of fusion power for about 1 second, with Q values in excess of 0.65 [25, 26] were reached in 1997. These are the highest fusion powers and Q values ever reached in D-T reactions, thereby effectively resulting in the first demonstration of breakeven in reactor grade D-T fusion plasmas. In December 2021, a world record amount of 59 MJ fusion energy has been produced in pulses of 5 seconds, a duration limited mainly by the use of copper magnetic field coils. This is an improvement by a factor of about 3 of the D-T results obtained in 1997. A summary of the different high performance D-T pulses obtained on JET and TFTR, is presented in figs. 12(a) and (b).

The success of the D-T experiments on JET and TFTR proves the scientific feasibility of controlled nuclear fusion. In addition, those experiments also allowed to get a first glimpse of the effects of fast fusion born alpha particles on the stability and the confinement of the plasma.

8. – Future prospects: JET, JT-60SA, ITER and DEMO

8·1. *JET as a testbed for ITER.* – JET is currently focused on providing experimental evidence to questions raised by the ITER team. ITER, short for International Thermonuclear Experimental Reactor, is the next step device in fusion research and currently under construction in Cadarache, France, as a combined effort between Europe, Japan, the Russian Federation, South Korea, India, China and the United States (see sect. **8**·3). Because of its worldwide unique capabilities (able to operate with both beryllium as wall material and tritium) and its large size, JET is ideally placed to do so. JET's "Programme in Support of ITER" has been elaborated along three main axes: i) experimentation with an ITER-like wall (combination of W and Be as a first wall), ii) development of plasma configurations and parameters towards the most ITER-relevant conditions achievable today and iii) integrated experimentation in deuterium-tritium. The JET auxiliary heating systems are upgraded with currently 45 MW of installed heating power, allowing access to ITER-relevant operational conditions. This will provide access to conditions of melt layer formation both on the beryllium first wall and the tungsten divertor and, help to make further progress in hybrid anded advanced scenarios for ITER, which require full or partial current profile control, thereby making use of new dedicated diagnostics. Results from recent operations with the ITER like wall in JET are summarized in [27].

8·2. *JT-60SA.* – In February 2007, EURATOM and Japan signed the so-called Broader Approach agreement. It aims to complement the ITER Project and to accelerate the realisation of fusion energy by carrying out R&D and developing some advanced technologies for future demonstration fusion power reactors (DEMO). Within the Broader Approach, three main projects are being implemented: i) the construction in Japan of the large superconducting tokamak JT-60SA aimed at developing operational scenarios relevant for ITER (International Thermonuclear Experimental Reactor) and DEMO; ii) the construction of facilities to test components to be adopted in the research machine IFMIF (International Fusion Material Irradiation Facility) aimed to study the material damage caused by high energy neutron flux generated in fusion reactions, the construction of a prototype of the liquid lithium IFMIF target; iii) the creation of the International Fusion Energy Research Centre (IFERC) including a centre for advanced materials development and a supercomputing centre [28].

Parameters for JT-60SA are summarized in table III. JT-60SA is a fusion experiment designed to support the operation of ITER and to investigate how best to optimise the operation of fusion power plants that are built after ITER. It is a joint international research and development project involving Japan and Europe, and is to be built in Naka, Japan using infrastructure of the existing JT-60 Upgrade experiment, for which we cited several important experimental results in sect. **7** above. SA stands for "super, advanced", since the experiment will have superconducting coils and study advanced modes of plasma operation. According to the current planning, first plasmas on JT-60SA are foreseen in the coming months. More information on this interesting tokamak can be found in [29].

TABLE III. – *Main parameters of important tokamaks.*

	T-10	DIII-D	JT-60SA	JET	ITER
Country/ Organisation	Russian Federation	USA	Japan	GB/ EURATOM	International
Plasma shape	circular	elliptical (D)	elliptical (D)	elliptical (D)	elliptical (D)
Minor radius (m)	0.3	0.67(hor) 1.74(vert)	1.18(hor) 2.30(vert)	1.25(hor) 2.1(vert)	2.0(hor) 3.7(vert)
Major radius (m)	1.5	1.67	2.96	2.96	6.2
Toroidal magnetic field (T)	2.5	2.2	2.25	3.5	5.3
Plasma current (MA)	0.7	3.0	5.5	5.0	15 (17)
Pulse length (s)	4	10	100	60	300–500
Injection of neutral particle beams (MW)	–	20	34	30	
					73 (130) in total
Injection of electromagnetic waves (MW)	1.5	8	7	38	

8·3. *ITER.* – The results summarized in sect. **7** provide crucial information for the design of a next large tokamak, ITER, aimed at demonstrating the technical feasibility for large-scale energy production. ITER will thus for the first time in history allow mankind to produce hundreds of MW of fusion power from nuclear fusion reactions at temperatures over 100 million degrees. ITER is expected to generate fusion powers of the order of 500 MW in pulses of 300–500 seconds. Specifications for ITER (and a few other major tokamaks) are summarised in table III. After ITER, the construction of a demonstration reactor, currently termed DEMO, is foreseen, which should show not only the technical, but also the economical feasibility of fusion.

8·4. *DEMO.* – Before DEMO can become a practical demonstration of electricity generation on a power-plant scale that satisfies various socio-economic goals, major challenges must be resolved. Although ITER will bring significant advances, a large gap

TABLE IV. – *Main differences between ITER and DEMO.*

ITER	DEMO
Experimental Device	Close to commercial plant
400s pulses Long interpulse time	Long pulses, high duty cycle or steady state
Many diagnostics	Minimum set of diagnostics only needed for operations
Many H&CD systems	Reduced set of H&CD systems
No T breeding required	Self sufficient T breeding
316 SS structural material	Reduced activation structural material
Low n-fluence, low dpa Low material damage	High n-fluence, high dpa Significant material damage

remains between ITER and DEMO. The main differences between ITER and DEMO are summarized in table IV.

DEMO is currently based on the tokamak configuration, as this is the most advanced fusion concept to date. In parallel, reactor studies are also being developed for Helical Devices (see, *e.g.*, [30-33]). However, a decision on a next step stellarator/helical device can only take place when the main results of the current large helical devices in operation or construction have been obtained.

To lead a coordinated effort in the EU towards DEMO, the Power Plant Physics and Technology Department (PPP&T) has been established in 2011 by the then European Fusion Development Agreement (EFDA) [34]. The aims of the DEMO studies in Europe are: i) to quantify key physics and technology prerequisites for DEMO; ii) to identify the most urgent technical issues that need to be solved in physics and technology and iii) to plan and implement supporting physics and technology R&D.

Two DEMO design options are currently being investigated by PPP&T (see table V for main characteristics):

i) DEMO Model 1: A "conservative baseline design" that could be delivered in the short to medium term, based on the expected performance of ITER with reasonable improvements in science and technology, *i.e.* a large, modest power density, long-pulse inductively supported plasma in a conventional plasma scenario.

ii) DEMO Model 2: an "optimistic design" based upon more advanced assumptions which are at the upper limit of what may be achieved, leading to a steady state plasma scenario where a large fraction of the plasma current is induced non-inductively, *i.e.* without making use of the transformer. This is currently a rather speculative option.

From table IV follows immediately a minimum set of research topics that need to be (much) further developed: heating systems capable of heating the large plasmas foreseen for DEMO, diagnostic systems that can sustain the heavy neutron loading, tritium generating modules (so-called tritium blankets) to foresee tritium self-sufficiency, and new materials that will be compatible with the high heat and neutron loads. This

TABLE V. – *Main parameters of the early DEMO 1 and more advanced DEMO 2 model currently under investigation by the PPP&T Department of EFDA. Shown are the thermal output power (P_{th}), the net electrical power to the grid (P_{net}), the recirculating power (P_{rec}), the auxiliary heating power (P_{aux}), major radius (R_0) and minor radius (a) of the device, plasma current (I_p) and toroidal magnetic field on axis (B_t).*

Device operation mode	DEMO 1 pulsed	DEMO 2 steady state
P_{th} (MW)	2200	2700
P_{net} (MW)	500	500
P_{rec} (MW)	594	600
P_{aux} (MW)	50	350
R_0 (m)	9.0	8.15
a (m)	2.25	3.0
I_p (MA)	14.1	19.8
B_t (T) on axis	6.8	5.0

implies also that new laboratories will have to be built to be able to explore possibilities and further develop existing systems. Among others, a material research facility to study the behaviour of various candidate materials under a flux of 14.1 MeV neutrons (a prototype facility, LIPAc [35], exists already in Japan and preparations for a fusion materials research facility in Europe are underway [36]), a facility to develop, *e.g.*, Neutral Beam Heating systems (if these are viable at all for DEMO) with acceleration energies between 1 and 2 MeV, based on negative ion sources (in contrast to the positive ion sources used currently, as mentioned before) etc. This research is done in the PRIMA facility in Padova, Italy, for the development of the ITER NBI systems [37]. But more efforts in various other fields will have to follow, and this will unavoidably take time.

The European Fusion Development Agreement (EFDA) released in November 2012 a roadmap for the realization of fusion electricity to the grid by 2050; the EUROfusion consortium recently published an updated document [38]. ITER is the key facility of the roadmap as it is expected to achieve most of the important milestones on the path to fusion power. The second period is focused on maximizing ITER exploitation and on preparing the construction of DEMO. Building and operating DEMO is the subject of the last roadmap phase (time horizon about 2050). For more technical details on the challenges in the design of a fusion reactor, we refer to the paper by Romanelli "Fusion energy: technological challenges", in this volume.

9. – Will fusion be a safe, clean and inexhaustible energy source?

The main advantage of fusion is the absence of any long-lived radioactive waste and its inherent safety. The quantity of fuel in the reaction chamber at any moment is only a few grams, in contrast to the presence of several tons of fuel in the core of a fission reactor. Furthermore, the fusion process is not susceptible to an uncontrolled release of energy, as it is in essence a "gas burner" and thus can be halted in a very short time by cutting the supply of fusion fuel. There is thus no risk of a runaway reaction, and Chernobyl-like accidents are absolutely excluded.

As previously mentioned, the fuels for the most accessible fusion reactions are deuterium and tritium. Deuterium can be extracted in large quantities from ordinary water found in rivers and oceans, rendering each country less dependent on fuel supply from other parts of the planet. Moreover, a minimal amount of fuel is necessary to cover the lifetime electricity needs of a European citizen: about 10 g of D and 15 g of T (and about twice this amount for a citizen of the United States). As explained above, tritium will be produced *in situ* from the irradiation of Li with the neutrons from the fusion reaction. Thus the real combustibles are in fact D and Li, both of which are non-radioactive. World reserves of Li are sufficient for many thousand years of operation of D-T fusion power plants. Studies are underway to also extract it from seawater [39, 40]. If at a later stage the D-D reaction could be used (requiring, however, even much higher plasma temperatures), we could have at our disposal a virtually unlimited energy source, as reserves of D are sufficient for several million years of energy production. In both cases, the reaction product is helium, which is chemically inert and non-radioactive. With nuclear fusion there is thus no release of gases which contribute to the greenhouse effect, acid rain or the depletion of the ozone layer. Moreover, helium is the only gas still liquid at temperatures down to absolute zero, thus offering the only possibility for cooling low-temperature superconductors (as, *e.g.*, in modern MRI devices). It is thus not a "waste product" but in fact is essential for all applications involving ultra-low temperatures. This is especially important since helium is rather scarce on Earth.

The high-energy neutrons will activate the metallic reactor structure and hence make it radioactive. Low activation materials have to be selected to minimise this effect. In this way, the total radioactive inventory of a fusion reactor could be made one million times smaller than that of existing fission reactors. The activation problem could be drastically reduced in a distant future by the use of other fusion reactions, which liberate fewer neutrons. The radioactivity induced by the D-^{3}He reaction for example, would be about 50 times lower.

The various steps that remain to be taken by thermonuclear fusion research indicate that some decades will still be needed to realise a practically useable energy source. The most recent results support the attractiveness of the fusion concept for energy production. A sustained effort —in laboratories worldwide— brings the vision of an energy supply by thermonuclear fusion ever closer to realisation. We all know that fusion is a challenging undertaking and that patience will be needed, but it is more than worth the effort given the difficulties we are facing in the future with our current energy supply and its suspected influence on climate. It will evidently be up to the young researchers to tackle these interesting problems. If successful, this will be an important contribution to the benefit of all people on Earth. A true challenge for an immense reward!

Further reading

A recent overview of the status of fusion research (magnetic and inertial) can be found in the "Insight Section" (66 pages) of the May 2016 volume of *Nature Physics* (`http://www.nature.com/nphys/insight/nuclear-fusion/index.html`) [41].

REFERENCES

[1] Agostini M., Altenmüller K., Appel S. *et al.*, *Nature*, **587** (2020) 577.
[2] Fewell M., *Am. J. Phys.*, **63** (1995) 653.
[3] NACRE (Nuclear Astrophysics Compilation of Reaction Rates) `http://pntpm.ulb.ac.be/nacre.htm`; Angulo C. *et al.*, *Nucl. Phys. A*, **656** (1999) 3.
[4] Raeder J. *et al.*, in *Controlled Nuclear Fusion: Fundamentals of Its Utilization for Energy Supply* (John Wiley & Sons, New York) 1986, Chapt. 1, ISBN-13: 978-047110312.
[5] Rebhan E., in *Heisser als das Sonnenfeuer* (Piper Verlag, München and Zürich) 1992, Chapt. 5, ISBN-13: 978-3492031097.
[6] Atzeni S. and Meyer-ter-Vehn J., in *The Physics of Inertial Fusion: Beam Plasma Interaction, Hydrodynamics, Hot Dense Matter* (Oxford University Press, Oxford), 2004, Chapt. 1, ISBN-13: 978-0198562641.
[7] Wesson J., *Tokamaks*, fourth edition (Oxford University Press, Oxford) 2011, Chapts. 5 and 7, ISBN 978-199592234.
[8] Mazzi S., Garcia J., Zarzoso D., Kazakov Ye. O., Ongena J. *et al.*, *Nat. Phys.*, **18** (2022) 776.
[9] Nespoli F., Masuzaki S., Tanaka N., Shoji M., Gilson E. P. *et al.*, *Nat. Phys.*, **18** (2022) 350.
[10] Emmert G. A., El-Guebaly L., Kulcinski G. L. *et al.*, *Nucl. Fusion*, **29** (1989) 1427.
[11] Betti R. and Hurricane O. A., *Nat. Phys.*, **12** (2016) 435.
[12] Zylstra A. B., Hurricane O. A., Callahan D. A. *et al.*, *Nature*, **601** (2022) 542.
[13] Kritcher A. L., Young C. V., Robey H. F. *et al.*, *Nat. Phys.*, **18** (2022) 251.
[14] Atzeni S., Batani D., Danson C. N. *et al.*, *Europhys. News*, **53** (2022) 18.
[15] Giruzzi G., Artaud G. F., Baruzzo M. *et al.*, *Nucl. Fusion*, **55** (2015) 073002; Garcia J., Giruzzi G., Artaud G. F. *et al.*, *Nucl. Fusion*, **48** (2008) 075007.
[16] Wolf R., Alonso A., Äkäslompolo S. *et al.*, *Phys. Plasmas*, **26** (2019) 082504.
[17] Beidler C. D., Smith H. M., Alonso A. *et al.*, *Nature*, **596** (2021) 221.
[18] Osakabe M., Takahashi H., Yamada H. *et al.*, *Nucl. Fusion*, **62** (2022) 042019.
[19] Rebhan E. and Van Oost G., in *Proceedings of the Sixth Carolus Magnus Euro-Summer School On Plasma And Fusion Energy Physics, Brussels, Belgium*, Fusion Sci. Technol., **45** (2004) 15.
[20] The JET team, *Nucl. Fusion*, **32** (1992) 187.
[21] Strachan J., Adler H., Alling P. *et al.*, *Phys. Rev. Lett.*, **72** (1994) 3526; Hawryluk R. J., Adler H., Alling P. *et al.*, *Phys. Rev. Lett.*, **72** (1994) 3530.
[22] McGuire K. M., Adler H., Alling P. *et al.*, *Phys. Plasmas*, **2** (1995) 2176.
[23] Ishida S., Neyatani Y, Kamada Y. *et al.*, in *Proceedings of the 16th IAEA Conference on Fusion Energy, Montreal*, Vol. **1** (IAEA, Vienna) 1997, p. 315.
[24] Fujita T., Kamada Y., Ishida S. *et al.*, *Nucl. Fusion*, **39** (1999) 1637.
[25] Keilhacker M., Gibson P., Gormezano C. *et al.*, *Nucl. Fusion*, **39** (1999) 209.
[26] Jacquinot J. and the JET Team, *Plasma Phys. Control. Fusion*, **41** (1999) A13.
[27] Mailloux J., Abid N., Abraham K. *et al.*, *Nucl. Fusion*, **62** (2022) 042026.

[28] Tsunematsu T., *Fusion Eng. Des.*, **84** (2009) 122.
[29] Barabaschi P., Kamada Y. and Ishida S., in *Proceedings of the 25th IAEA Conference on Fusion Energy, St-Petersburg, Russian Federation, 2014*, paper OV/3 (IAEA, Vienna) 2015.
[30] Beidler C. D. Harmeyer E., Hernegger F. *et al.*, "Stellarator Fusion Reactors – An Overview", in *Proceedings of the 12th International Toki Conference on Plasma Physics and Controlled Nuclear Fusion (ITC-12), Toki-city, Japan* (2001) Paper XI-3.
[31] Wobig H., Andreeva T. *et al.*, in *Proceedings of the 36th EPS Plasma Physics Conference Sofia, Bulgaria* (2009) Paper P4.192.
[32] Sagara A., Mitarai O., Imagawa S. *et al.*, *Fusion Eng. Des.*, **81** (2006) 2703.
[33] Sagara A., Imagawa S., Mitarai O. *et al.*, *Nucl. Fusion*, **45** (2005) 258.
[34] Federici G., Biel W., Gilbert M. R. *et al.*, in *Proceedings of the 25th Symposium on Fusion Engineering (SOFE), 10–14 June 2013, San Francisco, USA* (2013) paper WO1-1.
[35] Kondo K., Akagi T., Arranz F. *et al.*, *Fus. Eng. Des.*, **153** (2020) 111503.
[36] Królas W., Ibarra A., Arbeiter F. *et al.*, *Nucl. Fusion*, **61** (2021) 125002.
[37] Sonato P., Antoni V., Bigi M. *et al.*, in *Proceedings of the 3rd International Symposium on Negative Ions, Beams and Sources (NIBS), Jyväskylä, Finland, AIP Conf. Proc.*, **1515** (2013) 549.
[38] Donne A. J. H., *Philos. Trans. R. Soc. A*, **377** (2019) 20170432.
[39] Hoshino T., *Desalination*, **359** (2015) 59.
[40] Swain B., *Sep. Purif. Technol.*, **172** (2017) 388.
[41] Ongena J., Koch R., Wolf R. and Zohm H., *Nat. Phys.*, **16** (2016) 398.
[42] Wang M., Huang W. J., Kondev F. G. *et al.*, *Chin. Phys. C*, **41** (2017) 030003.

EPJ Web of Conferences **268**, 00012 (2022) https://doi.org/10.1051/epjconf/202226800012
LNES 2021

Conceptual design of a water-cooled first wall component for a tokamak machine

GABRIELE DE SANO

Department of Industrial Engineering, University of Rome "Tor Vergata"
Via del Politecnico 1, 00133 Rome, Italy

Summary. — In a tokamak, the first wall is a barrier protecting the internal parts of the machine from the intense fluxes of heat and particles coming from the plasma. This contribution presents the critical issues of a water-cooled first wall component based on steel as structural material, that is a relevant case study for the design of future fusion power plants.

1. – Layout of an actively cooled first wall component

The exploitation of nuclear fusion for commercial electricity production is a promising solution to ensure the future energy needs, given its important advantages: the fuel is virtually unlimited and widely available, no greenhouse gases are produced, the process is intrinsically safe, and radioactive waste is recyclable after a relatively short time. Fusion reactions powering our Sun and the stars can be performed inside a device on Earth by creating a suitable environment for the fuel. The fusion process needs extremely high temperatures, on the order of 10 keV, thus the fuel exists in a plasma state enabling the positively charged nuclei to collide with each other and to react at a sufficiently high rate [1]. A potential architecture for a fusion reactor relies on the tokamak layout, namely a toroidally shaped device which is able to confine the hot plasma in a finite volume by

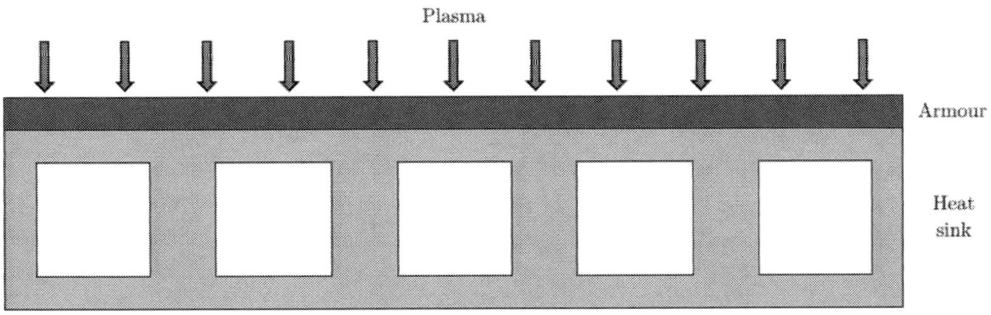

Fig. 1. – Cartoon schematic illustrating a possible layout of an actively cooled first wall component, with multiple rectangular cooling channels in the heat sink and a flat tile armour.

means of strong magnetic fields [2]. During the operation of a tokamak, heating power is continuously supplied to the plasma in order to maintain it at the necessary high temperatures and sustain the reactions. At steady state, the same amount of heat is emitted by the plasma and carried by fluxes of photons and charged particles, which interact with the surrounding elements of the tokamak. For this reason, the walls of the reaction chamber are covered by the so-called plasma facing components (PFCs), that exhaust the plasma heating power and shield the internal parts of the machine. Since a tokamak is characterized by a pulsed operation, PFCs are loaded only for periods of time. If the loads are applied for a few seconds, the effects of thermal inertia prevent materials from reaching their technological limits. However, all upcoming devices will operate with pulse lengths of hundreds or even thousands of seconds [3], thus it becomes mandatory for all PFCs to actively exhaust the thermal power by means of a cooling system. Water is frequently adopted as coolant.

The first wall, which is the main focus of the present contribution, consists of PFCs covering most of the plasma facing surface. The layout of an actively cooled first wall component consists of two main parts, namely the *armour* and the *heat sink*. The first is a layer directly facing the plasma, while the second is the structural material in which the cooling channels are manufactured (see fig. 1). For all PFCs, the need for an armour results from the interaction with charged particles, which collide with the surface layer of the walls causing sputtering, that is the ejection of atoms which then become impurities in the plasma. To date, tungsten (W) is the most promising candidate as plasma facing material because it shows a high energy threshold to sputtering and a resulting low erosion yield [4]. At the same time, neutrons produced by the reactions escape the plasma and penetrate deeply inside the whole reactor, damaging both functional and structural materials. Nuclear reactions with the nuclei can occur, leading to radioactivity. Moreover, high-energy neutrons can displace nuclei from their lattice positions and trigger a dislocation cascade. These phenomena are responsible for embrittlement and loss of strength in materials. The unit "displacement per atom" (dpa) is typically used to express lattice damage by neutrons and is defined as the number of times that an atom is displaced, on average, for a given neutron fluence.

TABLE I. – *Properties (at 20 °C) of some materials employed in PFCs* [5,6].

	Coeff. of th. expansion α 10^{-6} [K^{-1}]	Young's modulus E [GPa]	Th. conductivity λ [$W\,m^{-1}\,K^{-1}$]	$\frac{\alpha E}{\lambda}$ 10^3 [$s\,m^{-2}$]
CuCrZr	16.7	128	318	6.7
Eurofer (Steel)	10.4	217	28	80.6
Tungsten	4.5	398	173	10.4

2. – Structural materials for the heat sink

The thermal power deposited on a PFC generates temperature gradients within the volume. The presence of temperature differences between two regions results in internal stress, since those regions elongate in a different way and undergo compression or traction. Considering a one-dimensional heat flow, a temperature difference across a material of thickness t depends both on the thermal conductivity λ of the material and the heat flux density q, according to Fourier's law of heat conduction:

$$\Delta T = \frac{tq}{\lambda}. \tag{1}$$

As a first approximation, the internal stress depends on the coefficient of thermal expansion α and Young's modulus E of the material, and increases for increasing ΔT:

$$\sigma \sim \alpha E \Delta T = \frac{\alpha E}{\lambda} tq. \tag{2}$$

Resulting from eq. (2), q has a direct impact on the internal stress, but it is linked to the applied heat load, which is fixed by tokamak operations and the precise location on the chamber walls. On the other hand, stress can be limited using materials with a reduced value of the factor $\frac{\alpha E}{\lambda}$. Table I compares the properties of some materials of interest, showing that the CuCrZr alloy is an excellent solution thanks to the high thermal conductivity λ of copper (Cu). In fact, CuCrZr is considered as the state-of-the-art material for the heat sink of a PFC. Unfortunately, its mechanical properties degrade even for a few dpa [7], after which the component shall be replaced, leading to high-radioactivity waste. The effects of neutron damage are negligible for current and upcoming experimental tokamaks, but they will be critical in the next step fusion device, the demonstration power plant DEMO [3], where a high neutron fluence is expected. DEMO will be the first fusion device to deliver electricity to the grid. Among other aims, DEMO should prove the economic feasibility of a fusion power plant, hence it is mandatory to limit the amount of waste [8] and minimize the substitution of components. Neutron resilient materials shall be developed in order to increase the lifetime of PFCs, particularly regarding the first wall that will be intensively bombarded by the high-energy fusion neutrons emitted by the plasma [9]. Reduced Activation Ferritic/Martensitic (RAFM) steels represent a reliable candidate since they exhibit a high resistance to neutron damage and are characterized by advantageous low-activation properties. Eurofer, that is the selected material for DEMO, has proved to maintain acceptable mechanical properties up to a 80 dpa [10]. However, the use of steel leads to higher internal stress compared to Cu-based materials because of the lower λ, under the same geometry and heat flux. This fact could compromise the structural integrity of components, limiting the manageable power deposition.

Moreover, one must also compare the coefficients of thermal expansion α of the two materials used in the heat sink and the armour, since a difference in terms of thermal strain results in additional stress in the structure. As reported in table I, the α of bulk W is much smaller than that of the two heat sink materials. The resulting interface stress could lead to a failure of the structure-armour bond. In water-cooled PFCs, the use of W in both the heat sink and the armour is not viable because W behaves like a brittle material in the operating temperature range of water, thus it cannot perform a structural function as heat sink material.

3. – Preliminarily assessment of the internal stress in a first wall component

In this final section, the critical issues of a steel-based PFC are discussed on the basis of steady-state thermo-mechanical simulations in ANSYS Workbench Mechanical 2021 R1. A simple 2D geometry of the first wall component is considered, in order to arrive at a preliminarily and didactic analysis. It consists of a heat sink region surrounding a single water channel, with a deposited flat tile of armour facing the plasma. Eurofer and bulk W are employed in the two domains respectively, and their thermo-mechanical properties are taken from [5, 6]. As previously discussed, the internal stress in a PFC depends on the heat load. According to recent studies for DEMO [11], the heat flux deposited on the first wall at steady state remains everywhere below the value of $1\,\mathrm{MW\,m^{-2}}$, with conservative assumptions. The following parameters are assumed in the calculations:

- Operative water pressure: $p = 4\,\mathrm{MPa}$;
- Operative water bulk temperature: $T_{\mathrm{bulk}} = 70\,°\mathrm{C}$;
- Water velocity: $u = 1\,\mathrm{m\,s^{-1}}$;
- Heat load deposited on the top surface of W: $q = 0.5\,\mathrm{MW\,m^{-2}}$.

The heat transfer between the Eurofer domain and the coolant is modelled by imposing T_{bulk} and the heat transfer coefficient. The latter can be evaluated through the Dittus-Boelter equation [12], that is a widely used correlation valid for turbulent forced convection. Once q and the heat transfer conditions are applied in the simulation, the temperature contours in the solid domains are obtained (see fig. 2). The temperature is maximum on the plasma facing surface and decreases as it approaches the fluid domain. The temperature gradient in Eurofer is higher than in W, since the λ of W is an order of magnitude larger.

Internal stresses are calculated after imposing the following loadings: operative water pressure in the cooling channel and temperature distribution in the solid domains. In

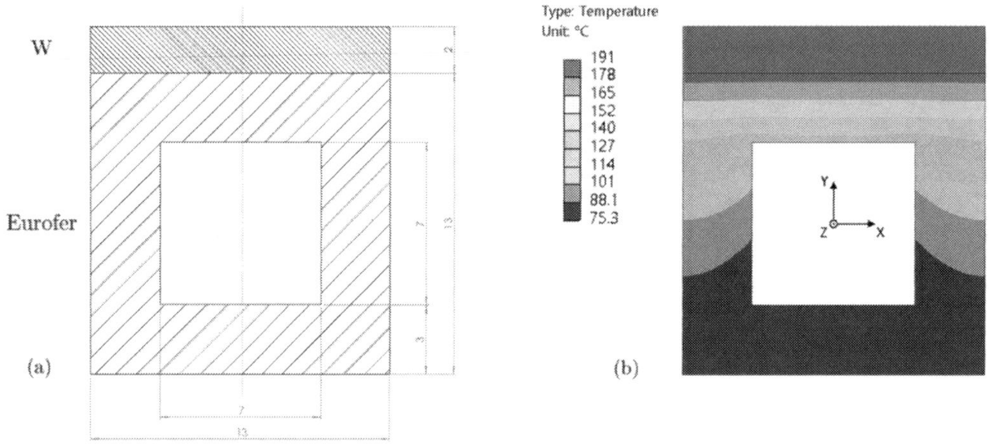

Fig. 2. – (a) 2D geometry of the model. (b) Temperature contours in the solid domains.

particular, water pressure is a mechanical load leading to primary stress P, while temperature gradients are a source of secondary stress Q. The thickness of the body along z is equal to 7 mm. The bottom face of the body is fixed in position along y ($U_y = 0$). On the left and right faces a coupling (U_x = constant) is imposed in order to simulate the presence of other channels in the first wall. Similarly, a coupling is imposed on the front and rear faces (U_z = constant) to model a longer component. The contours of von Mises stress [13] on the x-y plane are depicted in fig. 3, first considering the Eurofer domain only and then the whole geometry. As expected, stresses are higher in the second case because of the α mismatch between Eurofer and W, and the maximum values are concentrated in the region of the interface. The results are then verified to comply with the safety rules defined in the ITER Structural Design Criteria for In-vessel Components [5], that is a specific document developed for experimental tokamaks. The prescribed pro-

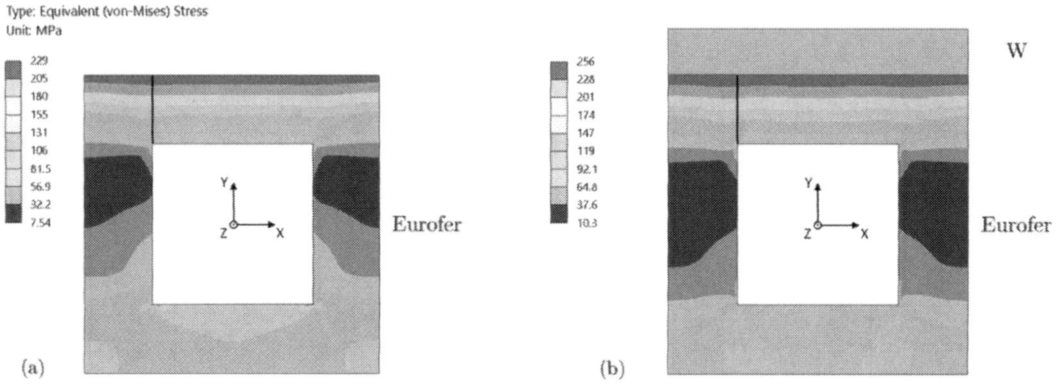

$\overline{P_L + Q_L} = 149.4$ MPa $\qquad \overline{P_L + Q_L} = 175.8$ MPa

Fig. 3. – Contours of von Mises stress without (a) and with (b) the W domain. The average value used for the criterion is calculated on the black path.

cedure is to carry out the linearization of the von Mises stress along some critical paths through thicknesses of the component and to compare the calculated values with the allowable limits defined by the code. Below, the primary plus secondary stress criterion is discussed, being the most severe for this case study:

$$\overline{P_L + Q_L} \leq S_e \, , \tag{3}$$

$\overline{P_L + Q_L}$ is the average von Mises stress on a path resulting from both primary and secondary loadings. S_e is the allowable primary plus secondary stress intensity, strictly related to the material tensile strength but also dependent on the dpa level. $S_e \sim 200\,\text{MPa}$ is considered in [14] to be a conservative value, given the embrittlement due to neutron damage. The criterion is tested on the black path shown in fig. 3 since it exhibits the greater average stress intensity both without and with the W domain. The calculated value of $\overline{P_L + Q_L}$ is reported in fig. 3 for both cases and it is smaller than 200 MPa even with the W domain, thus the criterion is satisfied.

However, the magnitude of the internal stress will increase for a greater heat load at steady state, or during transient events when a larger power deposition is expected [11]. A possible solution to ensure structural integrity is to reduce the contribution of secondary stress by decreasing the material thickness t, according to eq. (2), but the primary stress due to the pressure loading increases for decreasing t, requiring a proper compromise. Moreover, the internal stress at the interface can be minimised realizing an interlayer to couple heat sink and armour. Functionally graded materials are currently under investigation for this application [15, 16], combining Eurofer and W with a properly engineered composition to compensate the α mismatch at the interface.

REFERENCES

[1] Romanelli F., *EPJ Web of Conferences*, **246** (2020) 00013.
[2] Wesson J., *Tokamaks*, third edition (Clarendon Press, Oxford) 2004.
[3] Donné T. *et al.*, *European Research Roadmap to the Realisation of Fusion Energy* (EUROfusion) 2018, ISBN: 978-3-00-061152-0.
[4] Neu R., *Tungsten as a Plasma Facing Material in Fusion Devices* (IPP 10/25) 2003, p. 12.
[5] *ITER Structural Design Criteria for In-vessel Components (SDC-IC)*, (ITER) 2012.
[6] Tavassoli F., *Fusion Demo Interim Structural Design Criteria (DISDC)/Appendix A: Material Design Limit Data/A3. S18E Eurofer Steel*, (CEA) 2004.
[7] Li M. *et al.*, *J. Nucl. Mater.*, **393** (2009) 36.
[8] Dose G. and Noce S., *EPJ Web of Conferences*, **246** (2020) 00014.
[9] Federici G. *et al.*, *Nucl. Fusion*, **57** (2017) 092002.
[10] van der Schaaf B. *et al.*, *J. Nucl. Mater.*, **386-388** (2009) 236.
[11] Maviglia F. *et al.*, *Nucl. Mater. Energy*, **26** (2021) 100897.
[12] Dittus F. W. and Boelter L. M. K., *Univ. Calif. Berkeley Publ. Eng.*, **2** (1930) 443.
[13] von Mises R., *Nachr. Akad. Wiss. Göttingen Math. Phys. Kl.*, **1** (1913) 582.
[14] Aiello G. *et al.*, *J. Nucl. Mater.*, **414** (2011) 61.
[15] Qu D. D. *et al.*, *Fusion Eng. Des.*, **98-99** (2015) 1389.
[16] Emmerich T. *et al.*, *Nucl. Fusion*, **60** (2020) 126004.

Fusion energy: technological challenges

Francesco Romanelli

University of Rome "Tor Vergata"
Via del Politecnico 1, 00133 Rome, Italy

Summary. — This paper presents an overview of the main technological challenges of magnetic confinement fusion. Many of the challenges are being addressed in the context of the ITER construction and exploitation. Specifically, the demonstration of high-fusion gain regimes of operation will also provide a test of the technological solutions presently foreseen for the management of high heat and particle loads and the integration of the main technologies of a fusion power plant. In preparation of DEMO, reliable solutions for the breeding blanket and neutron resistant materials have to be developed.

1. – Introduction: fusion in the context of energy technologies

The world is vigorously pursuing a net zero CO2 emission strategy [1]. Fusion can play an important role in the future mix of energy technologies since it has a number of advantages:

- It is virtually unlimited.
- It does not produce greenhouse gases.
- It is intrinsically safe.
- The primary reaction does not produce radioactive materials. It does produce neutrons that activate the reaction chamber. However, with a proper choice of materials, radioactivity decays in a few tens of years and no geological repository is necessary.

TABLE I. – *Fusion reactions.*

$D + T \rightarrow {}^4He\ (3.5\,MeV) + n\ (14.1\,MeV)$	
$D + D \rightarrow T\ (1.01\,MeV) + p\ (3.02\,MeV)$	50%
$\rightarrow {}^3He\ (0.82\,MeV) + n\ (2.45\,MeV)$	50%
$D + {}^3He \rightarrow {}^4He\ (3.6\,MeV) + p\ (14.7\,MeV)$	
$T + T \rightarrow {}^4He + 2n + 11.3\,MeV$	
${}^3He + T \rightarrow {}^4He + p + n + 12.1\,MeV$	51%
$\rightarrow {}^4He\ (4.8\,MeV) + D\ (9.5\,MeV)$	43%
$\rightarrow {}^5He\ (2.4\,MeV) + p\ (11.9\,MeV)$	6%[(a)]
$p + {}^6Li \rightarrow {}^4He\ (1.7\,MeV) + {}^3He\ (2.3\,MeV)$	
$p + {}^7Li \rightarrow 2\,{}^4He + 17.3\,MeV$	20%
$\rightarrow {}^7Be + n - 1.6\,MeV$	80%
$D + {}^6Li \rightarrow 2\,{}^4He + 22.4\,MeV$	
$p + {}^{11}B \rightarrow 3\,{}^4He + 8.7\,MeV$	

(a) The ${}^5He$ isotope decays in 7×10^{-22} s in a neutron plus an alpha particle.

Several fusion reactions can in principle be exploited (table I). However, the reaction with the largest cross section is the reaction between deuterium and tritium that produces a 3.5 MeV alpha particle and a 14 MeV neutron. The amount of deuterium in sea-water (about 35 mg per litre [2, 3]) is virtually unlimited (making fusion with deuterium alone would supply the Earth, at the present consumption rate, for several billions years!). Tritium instead is radioactive and undergoes beta decay with a half-life of 12.3 years. Therefore, it has to be produced inside the reactor. This is accomplished by covering the inner part of the reaction chamber with a lithium blanket. The reaction between a neutron and a lithium nucleus produces a He nucleus together with a T nucleus that can be extracted and re-circulated in the reactor.

The cross section σ and the reactivity $\langle \sigma v \rangle$ (the product of the cross section and the particles relative velocity averaged over the distribution functions of the reacting species) are shown in fig. 1 for the DT, D^3He and DD reactions. In order to fuse D and T their mutual Coulomb repulsion must be overcome. This can be achieved by heating the DT gas to sufficiently high temperature (~ 20 keV or 200 M°C) in such a way that the ions in the tail of the Maxwellian distribution function provide a significant number of reactions. Matter at these temperatures is in the *plasma* state. The electrons are no longer bound to the nuclei and the system becomes the superposition of two gases of negatively charged electrons and of positively charged ions. To confine plasmas at these temperatures magnetic or inertial confinement [4] can be employed. We will consider here only magnetic confinement. In this case we can take advantage of the charged nature of the plasma constituents to confine them using intense magnetic fields (~ 100000 times larger than the average Earth magnetic field). In the presence of a magnetic field particles move along field lines as a train on a rail and are therefore confined in the plane

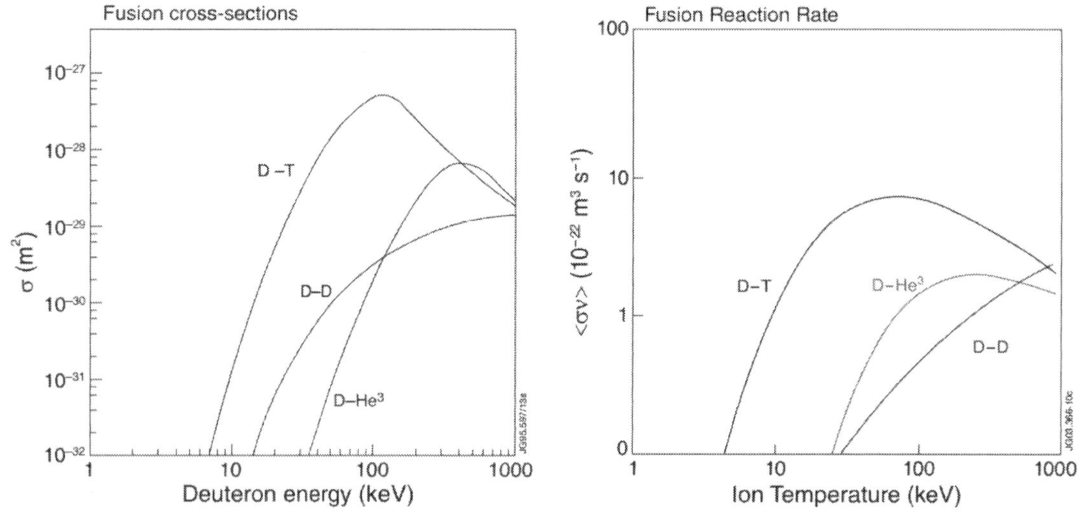

Fig. 1. – Fusion cross sections and fusion reactivity for the three main fusion reactions. The cross section is plotted as a function of the energy of the deuteron impinging on the target at rest.

perpendicular to the magnetic field. In order to confine plasmas also in the third direction the magnetic field lines are wound in such a way as to form a set of nested toroidal surfaces called magnetic surfaces, with each line lying on a magnetic surface. The combination of intense magnetic fields ($\sim 5\,\mathrm{T}$) and toroidal geometry enables plasma confinement.

It is important to stress that with both magnetic and inertial confinement plasmas of thermonuclear interest can be already routinely produced. In addition, a significant amount of controlled fusion reactions have been already obtained. In a dedicated campaign on the JET machine in 1997 [5] up to 16 MW of fusion power has been obtained in transient conditions to be compared with about 25 MW of power injected in the reaction chamber. Moreover, the National Ignition Facility, the largest inertial confinement fusion facility, has recently reported the production of 1.3 MJ of fusion energy which is about 70% of the energy delivered to the target [6]. Nevertheless, making a fusion power plant requires the solution to a number of other challenges that will be discussed in the following.

In 2012 the European Commission requested a Roadmap to fusion Energy in order to understand if fusion can start playing a role in the electricity production on the time scale of 2050 used for the Energy Roadmap of the European Union. The Fusion Roadmap [7] (an update has been recently produced by EUROfusion [8]) was based on the following assumptions:

- A program focussed around the priorities and organized in eight missions.
- A pragmatic approach: build on the shortest possible time scale a device capable of breeding tritium and produce electricity.
- Early involvement of industry.

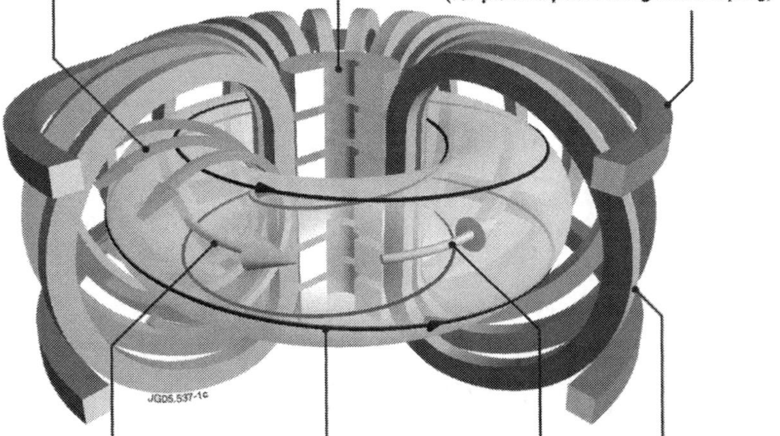

Fig. 2. – Magnetic confinement geometry illustrated for the tokamak configuration, see text (courtesy EUROfusion).

– Full exploitation of the opportunities arising from international collaborations.

The Roadmap has been the reference document for the EURATOM activities in fusion for Horizon 2020.

The Roadmap foresees two main facilities, ITER and DEMO:

– ITER will demonstrate burning plasma operation, *i.e.* operation with dominant plasma heating produced by the fusion generated alpha particles. ITER is also developing a large part of the technologies needed in a fusion power plant. However, ITER will neither produce electricity nor will breed the tritium it uses.

– DEMO will be the first demonstration fusion power plant. In addition to what ITER will accomplish, it will produce a net electricity output and will be self-sufficient for the production of tritium.

Both ITER and DEMO are machines based on a specific magnetic confinement configuration called *tokamak* [9] and schematically shown in fig. 2. The plasma is characterized by a doughnut shape (*torus*) with major radius R_0 and minor radius a. The magnetic field is the superposition of a toroidal component B_ϕ (mostly generated by a set of external coils) and a poloidal component B_p (mostly produced by a current flowing in the plasma itself in the toroidal direction). The combination of toroidal and poloidal magnetic field produces magnetic field lines that lay on toroidal surfaces called magnetic surfaces. The winding pitch of the magnetic field line on each magnetic surface is described by the *rotational transform* (the number of turns made in the poloidal direction for a single turn in the toroidal direction). The inverse of the rotational transform is the *safety factor*. The existence of a rotational transform is essential for plasma confinement. Without a rotational transform ions and electrons, under the effect of magnetic field curvature and spatial inhomogeneity, would drift vertically in opposite directions producing an electric field that, combined with the toroidal magnetic field, would push the plasma outwards in the radial direction in a very short time.

The safety factor is constant on a magnetic surface. For the sake of discussing the technological challenges we can use a simple approximation to the safety factor at the plasma edge [9]

$$q_{\text{edge}} = 2\pi B_\phi a^2 S/(R_0 \mu_0 I_p), \tag{1}$$

where I_p is the plasma current and S is a numerical factor that depends on the plasma shape. In order to have stable plasma operations [9] a necessary condition is $q_{\text{edge}} \geq 3$, which produces an upper bound to the plasma current that can circulate in a tokamak

$$I_{p,\max} = 2\pi B_\phi a^2 S/(3\mu_0 R_0). \tag{2}$$

Since, as we will see later, the plasma current is the main control parameter to achieve high fusion gain, the limit expressed by eq. (1) has to be taken into account to design a tokamak reactor. It is apparent that at fixed plasma shape and aspect ratio R_0/a, the maximum plasma current is proportional to $B_\phi a$. Thus, $I_{p,\max}$ can be increased either by increasing the reactor dimensions or by increasing the toroidal field.

After a confined plasma is established, it is heated up to temperatures of 10–20 keV by auxiliary heating systems until the fusion generated alpha particles become the dominant heating source. The heat generated in this way is transported from the center of the plasma to the edge (see sect. **2**) and removed from the reaction chamber. Similarly, the He ashes produced after the fusion alphas have transferred all their energy to the plasma need to be continuously removed to avoid poisoning the fuel (dilution). Plasma exhaust takes place at a special location, called *divertor*, in the reaction chamber (usually a niche at the bottom) sufficiently remote from the hot plasma such that the heat and particle removal can take place without disturbing the dynamics of the plasma inside the *magnetic separatrix*. The separatrix defines a sharp boundary between the hot plasma region (inside) and the plasma edge region in contact with the chamber wall and the divertor and therefore at much lower temperature.

In the following we review the main challenges of fusion research. The focus will be on the technological challenges of magnetic confinement although a number of challenges are common also to inertial confinement. In another paper of this school [10] the physical challenges will be reviewed in detail.

2. – Demonstration of regimes with a high energy gain

The first challenge is the demonstration of regimes in which the amount of fusion power absorbed in the plasma is much larger than the external heating power (burning plasma conditions). This challenge is briefly reviewed in this section in order to make contact with the technological challenges. A detailed discussion can be found in ref. [10]. The alpha particles produced in fusion reactions are charged and therefore can be confined by the same magnetic field that confines the plasma. They release their energy to the plasma via collisions. In burning plasma conditions the alpha particle heating can almost entirely maintain the high plasma temperatures. In ITER the target is a plasma in which alpha heating (100 MW) is twice the external heating (50 MW). Specifically, the power generated by fusion reactions is given by

$$P_{\text{fus}} = 17.6\,\text{MeV}\, n_{\text{D}}\, n_{\text{T}} \langle \sigma v \rangle V, \tag{3}$$

where n_{D} is the particle density of the deuterium nuclei (number of deuterium nuclei per unit volume), n_{T} is the particle density of the tritium nuclei, $\langle \sigma v \rangle$ is the Maxwellian fusion reactivity and V is the plasma volume. Since the 14 MeV neutrons carry four times more energy than the 3.5 MeV alpha particles, 20% of the power given by eq. (3) is in the form of alpha particles and 80% in the form of neutrons. The fusion power produced in the form of neutrons in ITER is 400 MW and the total fusion power is 100 MW + 400 MW = 500 MW and the fusion gain Q (given by the ratio between the fusion power and the external power injected in the reaction chamber) in ITER is Q = 500 MW/50 MW = 10. For comparison, the JET achievement quoted above (16 MW of fusion power with 25 MW of externally injected power [5]) corresponds to $Q \sim 0.67$.

Why high gain has not been achieved so far? The problem is associated with the conduction losses due to the small-scale turbulence destabilized by the free-energy sources always present in a confined plasma (the gradients of temperature and density). Radiation losses are also present although usually much smaller. Losses tend to cool down the plasma and must be balanced by the heating power. What we know from theory and experiments is that, at fixed density, temperature and magnetic field, the power lost via conduction is at most linearly increasing with the machine radius R [11], whereas the fusion power increases as R^3. Therefore the solution that has been pursued has been to make the machine size larger.

The power P_{cond} lost by the plasma through conduction is due to small-scale turbulence and can be quantified in terms of the energy confinement time τ_E

$$P_{\text{cond}} = W/\tau_E, \tag{4}$$

where $W = (3/2)\,(n_e T_e + \Sigma_j n_j T_j)\, V$ is the internal energy of the plasma with the sum extended to all the ion species and $T_e(T_i)$ the electron (ion) temperature. The energy confinement time is the characteristic time for the plasma to cool down once the heating sources are switched off. It has nothing to do with the time plasma is confined. In ITER τ_E is a few seconds whereas the plasma can be confined for hundreds of seconds.

The power balance equation in stationary conditions dictates that the fusion power released in the form of alpha particles ($3.5\,\mathrm{MeV}/17.6\,\mathrm{MeV} \sim 20\%$ of the total) plus the power P_{aux} injected from external sources must be equal to the power lost by the plasma. Neglecting the power lost by radiation we have

$$(3.5/17.6)P_{\mathrm{fus}} + P_{\mathrm{aux}} = P_{\mathrm{cond}}. \tag{5}$$

The fusion gain $Q = P_{\mathrm{fus}}/P_{\mathrm{aux}}$ becomes infinite for $P_{\mathrm{aux}} = 0$, meaning that the fusion reactions are self-sustained. This is the so-called ignition condition.

For a pure deuterium and tritium plasma ($n_{\mathrm{D}} = n_{\mathrm{T}} = n_e/2$) with equal electron and ion temperatures, the ignition condition can be written in terms of the so-called triple product(1)

$$n_e \tau_E T \geq 3.42 \times 10^{21}\,\mathrm{m^{-3}s\,keV}. \tag{6}$$

A careful analysis of the experimental data has produced the following scaling law for the energy confinement time in H-mode [11]

$$\tau_{\mathrm{ITER98}(y,2)}(\mathrm{s}) = \tag{7}$$

$$0.0562\, I_p(\mathrm{MA})^{0.93}\, B(\mathrm{T})^{0.15}\, P(\mathrm{MW})^{-0.69}\, n(10^{19}\,\mathrm{m^{-3}})^{0.41}\, M^{0.19}\, R_0(\mathrm{m})^{1.97} \epsilon^{0.58} \kappa^{0.78},$$

where I_p is the plasma current, B the toroidal magnetic field, P the heating power, n the line averaged plasma density, M the average ion mass, $\epsilon \equiv a/R$ the inverse plasma aspect ratio and κ the plasma elongation. The extrapolation to ITER gives a value of 4.3 s, with an estimated value of the triple product (assuming $n = 10^{20}\,\mathrm{m^{-3}}$ and $T = 20\,\mathrm{keV}$) $n_e\tau_E T \sim 8.6 \times 10^{21}\,\mathrm{m^{-3}s\,keV}$.

We refer to ref. [10] for a discussion of the ignition condition. In the context of the technological challenges we want to stress that upon combining the equations above it is possible to obtain an expression of the triple product in terms of the reactor parameters

$$n_e \tau_E T \propto n^{0.8} R^{0.7} P^{-0.4} {I_p}^{2.2}.$$

This expression shows that the triple product is mainly determined by the plasma current. Since, as noted above, the maximum plasma current is proportional to the product $B_\phi a$, high fusion gain conditions can be obtained either by small-dimensions high magnetic field devices or large-dimensions low magnetic field devices. The use of high magnetic fields leads to a more compact reactor design, but this path has been so far limited by the availability of superconducting materials capable of operating above 12 T for typical

(1) In deriving eq. (6) we have approximated the fusion reactivity as $\langle \sigma v \rangle \sim 10^{-24} T\,(\mathrm{keV})^2 \mathrm{m^3/s}$, a good approximation in the range 10–20 keV.

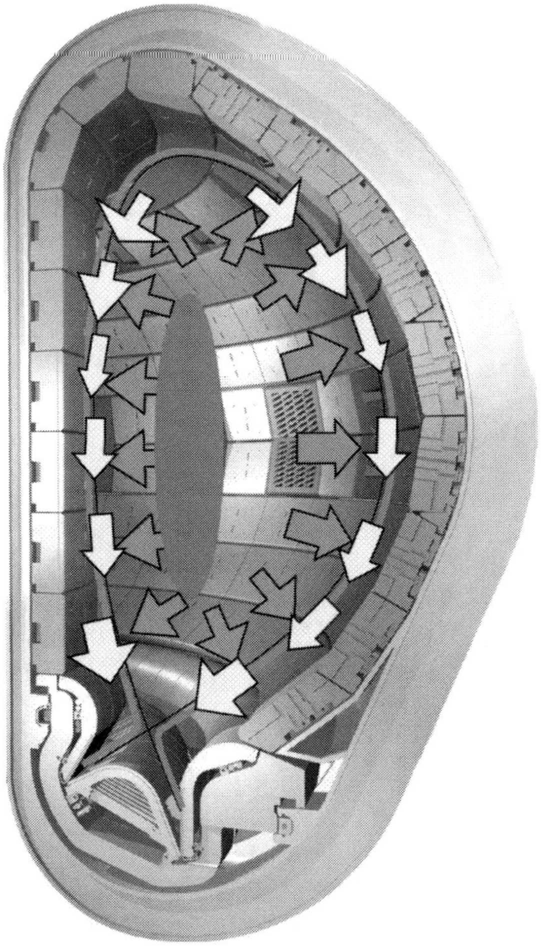

Fig. 3. – Divertor geometry. The orange line is the magnetic separatrix.

values of the current density in the magnet [12]. Recent developments in this field [13] have shown that coils capable of operating up to 20 T are indeed possible using high-temperature superconductors operated at liquid He temperatures.

Although the results of ref. [13] are indeed remarkable, it should be stressed that the design of compact fusion reactors must take into account all the constraints arising from the increase in B_ϕ.

3. – The challenge of the heat and particle load

The heat that crosses the separatrix due to conduction losses flows along the magnetic field lines (fig. 3) in a layer a few mm thick and is eventually deposited on the divertor [14]. Since all the heat is localized in a narrow layer, the heat load on the divertor in DEMO can reach values up to 60 MW/m^2, comparable with the heat load at the surface of the Sun!

In addition, the continuous flow of particles impinging on the divertor surface may produce large erosion. A simple estimate of erosion can be done as follows. In ITER with a volume of $800\,\mathrm{m}^3$ and a density of $10^{20}\,\mathrm{m}^{-3}$ there are about 8×10^{22} ions. If the rate at which they are lost through the separatrix is similar to the estimated energy confinement time ($\sim 4\,\mathrm{s}$) the number of particle arriving on the divertor per unit time is $2 \times 10^{22}\,\mathrm{s}^{-1}$. The exposed surface is of the order of $2\,\mathrm{m}^2$. If the probability of extracting an atom from the divertor surface (the so-called sputtering yield) is 10^{-4}, the number of atoms extracted in a year of operation is 3×10^{25} atoms per square meter, or taking into account that solid tungsten has a density of $6.4 \times 10^{28}\,\mathrm{m}^{-3}$, about 0.5 mm would be eroded every year.

To keep the sputtering yield at low values requires that the temperature in front of the divertor plates must be in the range of few eV. This can be understood as follows. Since electrons have a larger mobility than ions they tend to be more easily lost to the divertor plates which become negatively charged with a potential that is a fraction of the electron temperature at the plate divided by the electron charge. This negative potential accelerates the plasma ions towards the plate up to velocities of the order of the local plasma sound velocity. Such a flow of ions is beneficial in opposing the diffusion out of the divertor region of the impurities released from the plates and of the He ashes (thus avoiding plasma contamination and allowing efficient He pumping), however it also increases the probability of impurity extraction. An ion is indeed accelerated to energies proportional to the electron temperature at the plate and the probability of extracting an impurity atom from the plate becomes significant unless the temperature is low.

The solution to this challenge is made of two recipes. The first recipe is the development of plasma facing components that can withstand high heat and particle fluxes. The solution foreseen for ITER is the so-called tungsten monoblock made by a tungsten dice with a cooling channel made of CuCrZr attached to tungsten through a Cu interlayer. The W-monoblock has been shown to withstand more than 1000 cycles up to $20\,\mathrm{MW/m}^2$. This however would not be enough even for ITER. The second recipe is to produce semi-detached divertor conditions through the formation of a cloud of neutral gas that absorbs the energy and momentum of the incoming particles.

In this way it is possible to achieve temperatures in front of the plate of the order of few eV, sufficiently low to avoid substantial erosion in stationary conditions (or also under the effect of slow transients) in ITER (see fig. 4 (right)) where the sputtering yields for C and W under the bombardment of different hydrogen isotopes are plotted using the fits of ref. [14]).

A specific issue is the effect of transient heat loads such as those generated by ELMs and disruptions. Transient loads may drastically reduce the lifetime of plasma facing components (see *e.g.* fig. 7 of ref. [15]). Therefore appropriate disruption mitigation systems must be in place and the ELM amplitude must be kept small such that the local deposition is below $1\,\mathrm{MJ/m}^2$ per ELM.

The standard divertor solution is expected to be sufficient to cope with the ITER heat loads. However it is unclear whether it can work also for the much larger heat loads of DEMO. In principle, it can work provided a large fraction of the heating power is radiated before crossing the separatrix. In this case the heat would be exhausted on the

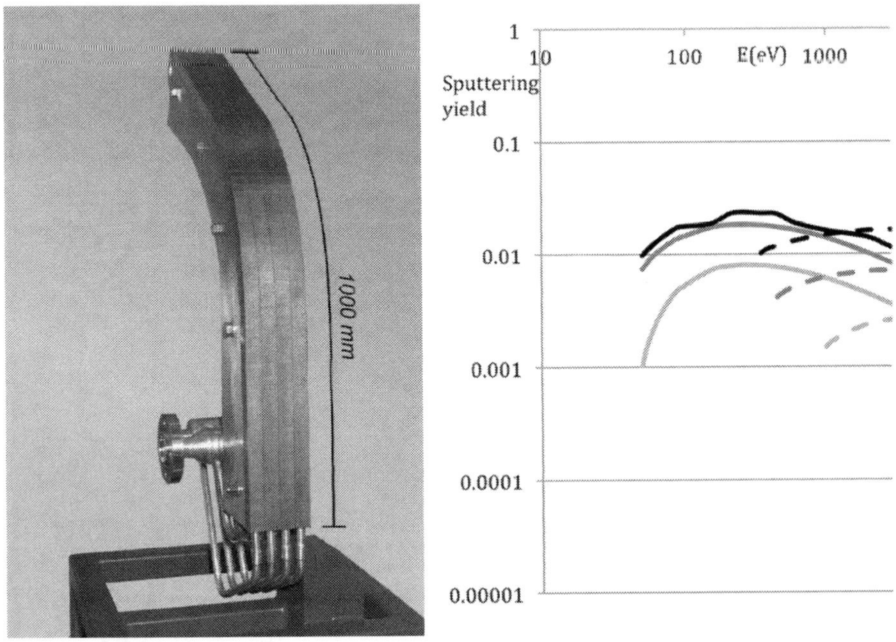

Fig. 4. – Tungsten monoblock mock up (courtesy EUROfusion) and physical sputtering yield *vs.* incident ion energy for tungsten (dashed lines) and carbon (continuous lines) under the bombardment by tritium (black), deuterium (blue) and hydrogen (orange).

large area of the main wall whereas the divertor would continue to play its role for the plasma/impurity density control. However, plasma regimes with high radiation usually exhibit a lower energy confinement time. The development of regimes that simultaneously radiate a large fraction of the heating power from the region inside the separatrix and maintain high confinement is one of the most important research lines for ITER and DEMO.

What if the predicted DEMO heat load will be too large to be managed with the conventional divertor solution? There is some room of maneuver with the present divertor configuration if the bottom divertor is paired by a divertor at the top or if the magnetic separatrix contact points are slowly swept up and down on the target plate in order to spread the heat load. However, if this will be not enough radically new solutions will be needed. These can be divided in two classes:

- The use of advanced magnetic field configuration with a larger divertor wetted surface. These configurations ("snowflake", super-X, etc.) are presently tested at the proof of principle level [16, 17] (see fig. 5).
- The use of liquid metals. In this case the surface exposed to the heat flux is continuously re-formed and its damage is no longer an issue.

The debate about the compatibility of these alternative solutions with the constraints of a reactor is still open. Nevertheless it is clear that the challenge of the exhaust problem

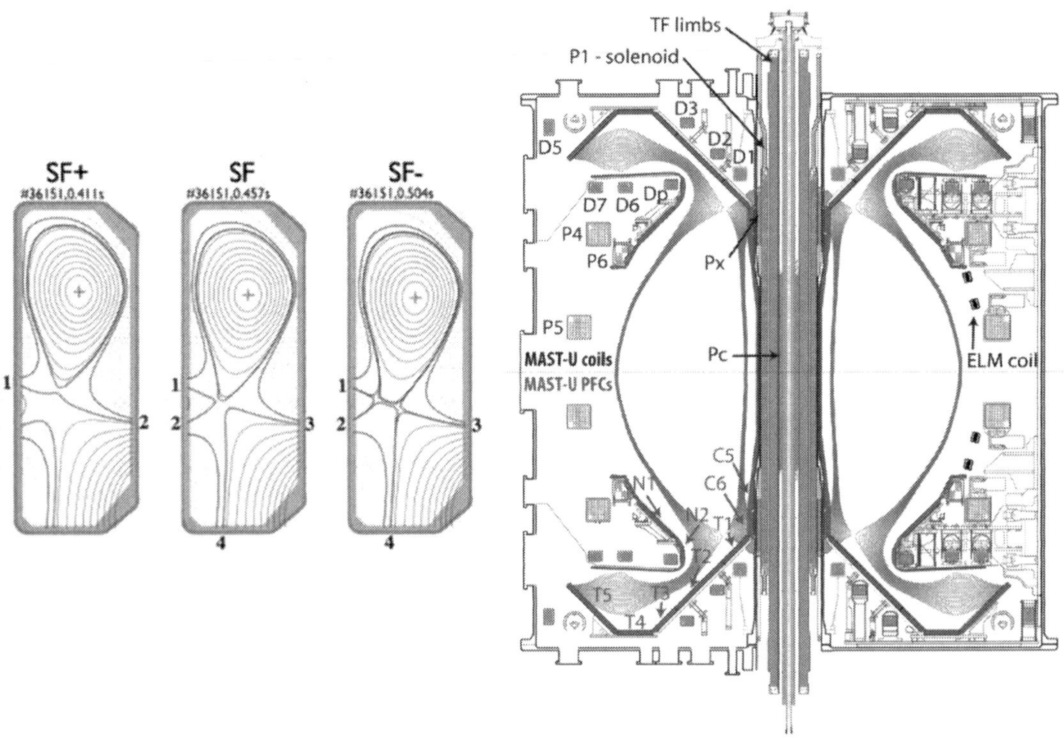

Fig. 5. – (a) Snowflake divertor configuration tested on the TCV tokamak (from ref. [16]). (b) Super-X configuration to be tested on the MAST-U tokamak (from ref. [17]).

deserves a wide investigation. To this goal, the possibility of a Divertor Tokamak Test facility was advocated in the Roadmap [6] and the realization of a dedicated experiment is presently under way [18].

DTT is a superconducting tokamak capable of producing plasmas with a current up to 5.5 MA and heating power up to 45 MW for long pulses (100 s) that will investigate innovative solutions for the divertor in various plasma configurations. It is presently being built at the Frascati ENEA laboratory(2).

4. – Materials that can withstand high neutron fluxes

The use of the DT reaction makes fusion conditions easier to achieve but it has the drawback that 80% of the fusion power is released in the form of 14 MeV neutrons. Part of the fusion-generated neutrons does not react with the lithium in the blanket to produce tritium but is absorbed by the structural materials of the reactor. This produces a degradation of structural properties and activation of wall materials. This degradation is mostly localized in the first few tens of centimetres (from the plasma exposed surface) of the plasma facing components. The vacuum vessel receives very low levels of neutron

(2) `https://www.dtt-project.it/`.

irradiation and it maintains its structural properties throughout the life of the reactor. On the contrary, the plasma facing components will need to be replaced every 3-4 years.

The damage suffered by structural materials can be quantified in terms of the displacements per atom (dpa). The nucleus in the structural material lattice that has absorbed the neutron energy (primary knock-on atom) releases its energy by displacing the surrounding atoms and producing point defects and dislocations. This effect is measured by the average number of displacements per atoms (dpa) in the lattice. Although deterioration of structural properties due to neutron irradiation is encountered also in fission reactors, the high energy of fusion neutrons (14 MeV compared with around 2 MeV for neutrons produced in fission reactors) induces a substantial amount of reactions forming H or He that accumulates in the lattice. These two effects produce embrittlement, swelling and irradiation creep and become important beyond the level of $\sim$ 10 dpa [19, 20]. As a result the temperature window for operation is limited between 350 °C and 550 °C. Typical numbers for radiation damage are 100–150 dpa in a fusion reactor, 30–70 dpa in DEMO and 2 dpa in ITER. Thus, there is no problem with material degradation in ITER. However, appropriate materials must be developed and qualified for fusion reactor application to avoid the frequent replacement of the plasma facing components and the blanket. For DEMO, it is possible to consider an initial exploitation with the presently qualified materials, but the second phase will require significant advances in material performance.

In view of the high-energy neutrons that characterize fusion, such a qualification will require a dedicated facility, the International Fusion Material Irradiation Facility (IFMIF). IFMIF will produce a neutron spectrum similar to that of a fusion reactor through various stripping reactions between two 40 MeV/125 mA deuteron beams and a liquid-lithium target. The engineering validation and engineering design activity are presently being finalized within the framework of a collaboration between the EU and Japan [21].

Activation is the second issue of neutron irradiation. Since the fusion reaction does not produce radioactive materials (the primary source of waste for fission), the issue of activation is limited to the structural materials of the plasma facing components. Among the structural materials so far investigated, reduced activation ferritic-martensitic steels (RAFM, such as EUROFER [22]), appears to be the near-term solution. RAFM steels differ from austenitic steels as molybdenum, nickel and niobium are replaced by tungsten, tantalum, vanadium and/or titanium that have a better behaviour under neutron irradiation. These alloys achieve a sufficiently low level of radioactivity in a sufficiently short time (say 100 years after the end of the reactor operation) in such a way that all the reactor materials can be easily recycled in a new reactor. Simple recycling techniques are expected to apply below contact dose rates $\leq$ 2 mSv/h, whereas for a contact dose rate $\leq$ 20 mSv/h recycling is still considered possible although through more complex remote handling systems. For comparison, the hands-on limit (*i.e.*, the contact dose rate that allows a maximum dose of 20 mSv/y for a radiation exposed worker) corresponds to 10 μSv/h.

In its ideal composition, EUROFER (Cr 9% W 1.1% Mn 0.4% V 0.2% C 0.11% Ta % Si 0.05% N 0.03% Ti 0.01% and Fe for the rest) would achieve a level of radiation below

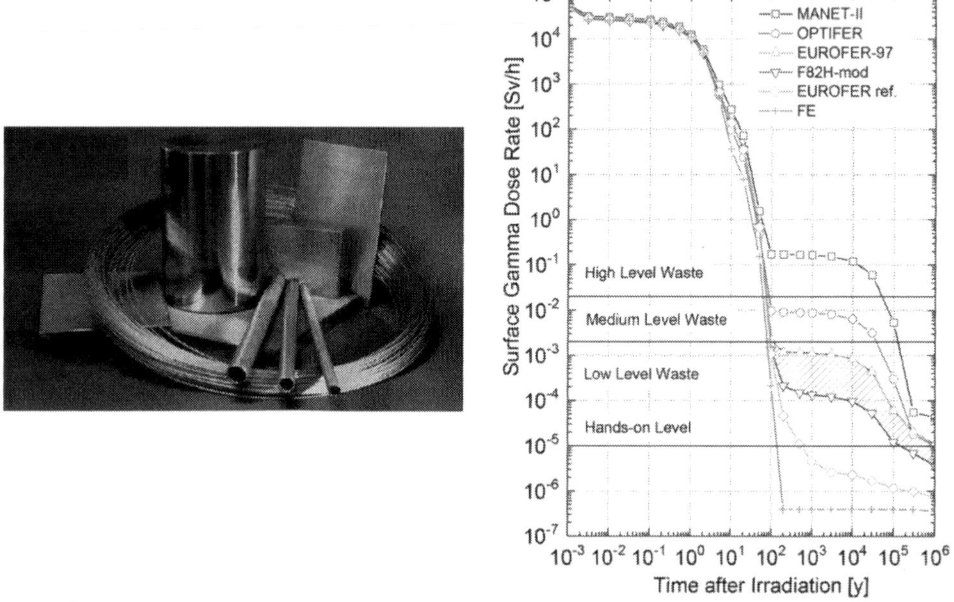

Fig. 6. – (a) Picture of EUROFER components (courtesy EUROfusion) and (b) contact dose rate *vs.* time for different steels from ref. [23].

2 mSv/h in less than 100 years and the hands-on limit in about 400 years (see fig. 6). The EUROFER produced today (the so-called EUROFER97) still contains some impurities but can already reach the simple recycling limit in $\sim$ 100 years after shut down [23]. Thus, for fusion power no geological repository will be necessary.

The development of advanced steels that would allow a wider operating temperature window is an active area of research. Oxide dispersion strengthened steel is expected to allow operation up to 650 °C. A review of the present status of materials for reactor applications can be found in ref. [24].

5. – Demonstration of tritium self-sufficiency

Tritium does not exist in nature. It must be produced inside the reactor. A 1.5 GWe reactor will consume about 0.5 kg of tritium per day and will need to produce the same amount every day. A tritium breeding ratio around 1.1–1.15, is the target for the self-sustainability of a reactor.

Tritium is produced through the reaction between neutron and lithium:

$$\mathrm{n} + {}^{6}\mathrm{Li} \rightarrow {}^{4}\mathrm{He}\ (2.1\,\mathrm{MeV}) + \mathrm{T}\ (2.7\,\mathrm{MeV}),$$
$$\mathrm{n} + {}^{7}\mathrm{Li} \rightarrow {}^{4}\mathrm{He} + \mathrm{T} + \mathrm{n} - 2.5\,\mathrm{MeV}.$$

The reaction involving ^{6}Li is exoergic and provides an additional contribution to the thermal power of the reactor. Its cross section at low energy is large (> 100 b at 1 eV).

The reaction involving ^{7}Li is endoergic. It has a threshold at 2.5 MeV so only high-energy neutrons can be used. Its maximum cross section is below 1 b. Therefore the lithium of a reactor must be enriched in ^{6}Li. The tritium production using these two reactions has been tested at low 14 MeV neutron fluence [25]. These show good agreement between the calculated and experimental production of tritium.

Some of the neutrons are lost due to various processes (absorption, streaming through the ports, etc.). Thus, neutron multipliers must be used. The main candidates are beryllium and lead. The $(n, 2n)$ reaction of Be has a lower cross section (~ 0.5 b) but also a lower neutron energy threshold (1.7 MeV). In the case of Pb the cross section is larger (~ 2 b) but the neutron energy threshold is at 7.4 MeV. The use of Be as a multiplier is made in connection to the solid breeder concept in which ceramic compounds (Li_4SiO_4, Li_2TiO_3 or others) are produced in the form of pebbles. The breeder is separated from the Be multiplier and the two are cooled by helium (water is avoided due to the reactivity with Be). The use of Pb is made in connection with the liquid breeder concept in which the breeder and the multiplier form a liquid eutectic compound ($Li_{17}Pb_{83}$) that flows at low velocity and is cooled by either He or water. There are advantages and disadvantages in both concepts and one of the ITER goals is to test blanket prototypes to verify the modelling assumptions [26].

6. – The intrinsic safety features of fusion

Fusion has a feature that makes it attractive within the nuclear technologies: its intrinsic safety. No chain reactions take place in a fusion power plant. The main difference between fission and fusion power plants is that a fission power plant is like a pile (the Fermi pile!): all the energy that will be released is stored in the fuel initially placed in the reactor. A recovery action following a malfunctioning during operations may need to cope with the release of a large amount of power. A fusion plant is like a normal gas boiler, in order to run it has to be continuously fuelled. If something goes wrong it is sufficient to stop fuel from entering the reaction chamber and the reactor stops. In both cases there is always a residual decay heat that has to be exhausted to avoid melting of the confinement structures (this was indeed the problem with the Fukushima accident). However, in the case of fusion the decay heat can be exhausted using only passive conduction. Indeed the incidental studies done so far show that the temperature of the in-vessel components in a fusion power plant always remains below the melting temperature [27].

7. – The integration of advanced technologies

Fusion power plants require the integration of different technologies. A large number of these technologies have been developed in connection with the ITER R&D activities such as those related with large superconducting magnets and heating and current drive systems. DEMO will have to integrate the technologies for tritium breeding (tested at the level of proof-of-principle in ITER) and for the production of electricity through a balance-of-plant (BoP) system, *i.e.* a system of heat exchangers and conventional turbines to convert into electricity the heat generated by the fusion reactions. The integration of different technologies motivates a focused effort on the solutions that can be really implemented in a reactor.

The technology of low-temperature superconducting magnets is today well established. The challenges of the ITER magnets [28] have been addressed within a dedicated R&D programme during the Engineering Design Activity [29]. The design and construction of the magnet system for DEMO requires limited extrapolations of the ITER design and ensures that a technical solution already exists while fusion-relevant high-temperature superconductors are developed in parallel. R&D activity is ongoing on more advanced low-temperature superconducting cables in order to reduce degradation under cyclic operation and cost. The integration of the magnet systems into DEMO is expected to largely build on the ITER experience. In parallel, the development of high temperature superconducting coils [13] will demonstrate the feasibility of more compact fusion reactors.

ITER will test three auxiliary heating and current drive systems: neutral beam injection, electron cyclotron heating and current drive and ion cyclotron heating. Their use in ITER and DEMO poses specific technical challenges such as the development of high-power, continuous source (gyrotrons) for millimetre wave radiation (170 GHz) for electron cyclotron heating and current drive and large negative ion sources and accelerator systems for negative ion injection ($1\,\mathrm{MeV}/200\,\mathrm{Am}^{-2}$). The choice of the system will have to be made mostly on the basis of the impact on tritium breeding (the opening in the vacuum vessel for the power injection are lost to the breeding blanket), on the degree of reliability and on the demonstration of sufficiently high efficiency to avoid large amounts of re-circulating power (see ref. [30] for a recent review and references therein).

A large impact on the design will come from the requirements of minimizing the plant down time through an effective remote maintenance system. The scheme presently under development foresees the use of large vertical sectors (vertical maintenance scheme) [31].

The integration of the DEMO components with the BoP has been investigated in the last few years. The choice of the BoP has a number of consequences on the choice of blanket coolant and materials [32].

8. – Electricity at low cost from fusion

The cost of fusion electricity has been the subject of several studies using the standard levelised cost of electricity approach by the IEA that includes the future expenditure (capital, operation and maintenance, replacements, fuel and decommissioning charges) all discounted to present day. With reasonable assumptions on the technology learning factor (based on the present experience with several energy technologies), on the plant availability (75%) and lifetime (40 years) the cost of the fusion kWh is estimated to be between 5 c€/kWh and 10 c€/kWh [27] in line with present market values. Most of the cost depends on the capital investment whereas fuel costs have a negligible impact.

DEMO is expected to be a relatively small extrapolation with respect to ITER. Its major radius should be at most 50% larger than ITER if the present physics and technological basis will be confirmed by future R&D. Therefore extrapolations to DEMO of the ITER experience in many areas can be made with confidence.

Nevertheless the experience with the ITER costs and with the costs of several nuclear power plants under construction shows that particular attention has to be devoted to this challenge. DEMO is not expected to produce electricity at a competitive price but should demonstrate that the capital costs can be contained to a level that makes fusion a competitive energy source in the long term.

9. – Concluding remarks

Fusion has the potential to become a major source of electricity. The research carried out in the last 50 years allows today to routinely produce plasmas at reactor relevant density and temperature. Substantial progresses have been also made in addressing the key technological challenges.

The demonstration of plasma operations with alpha particles being the dominant heating mechanism will be achieved on ITER. On the basis of the present theoretical and experimental evidence a fusion gain $Q = 10$ is expected. Methods to avoid, prevent or mitigate plasma disruptions and to control benign plasma instabilities have been developed and must now be tested in ITER.

The challenge of coping with the heat exhaust will be addressed in ITER through the use of high-heat flux components based on the tungsten monoblock technology and the use of partially detached divertor operation. In order to mitigate the risk that this solution cannot be extrapolated to DEMO, schemes based on the use of advanced divertor configuration and liquid metals as plasma facing materials are being investigated and a dedicated DTT facility is under construction.

The qualification of structural materials that can withstand the intense neutron flux of a fusion reactor and have benign activation properties needs to be further pursued. Possible materials (such as EUROFER) have already been produced. They could be used with limited extrapolations at the beginning of DEMO operation since the level of nuclear damage is expected to be below 20 dpa for the first phase of the DEMO operation. For the second phase of DEMO operation new materials need to be qualified. This requires a specific facility (IFMIF) with a neutron energy spectrum that simulates that of a fusion reactor.

Efficient tritium breeding is mandatory in a fusion power plant since about 0.5 kg of tritium are burned every day and an equal amount has to be produced in the blanket and re-circulated in the reactor. Blanket technologies based either on the liquid eutectic LiPb compound or the solid ceramic breeder with Be as neutron multiplier are under development and need to be integrated into a coherent design that accounts for the limited operating temperature window of the presently available materials, the need of minimizing tritium permeation outside the fuel cycle and the requirements of the balance of plant system to produce a net electrical power output.

ITER will demonstrate many of the critical technologies for a fusion power plant: magnets based on the use of low-temperature superconductivity, construction of a large-scale vacuum vessel with strict tolerances, actively cooled plasma facing components under high heat loads, high-efficiency auxiliary heating system, fuel cycle technologies capable of managing large tritium inventories. DEMO will largely benefit of this experience but will have to ensure reliable breeding blanket technologies and an efficient balance of plant systems at sufficiently low cost.

A successful R&D focused on the program priorities and the first operational experience on ITER may allow the construction of a DEMO reactor and the demonstration of fusion electricity around the middle of this century.

REFERENCES

[1] https://www.iea.org/reports/net-zero-by-2050.
[2] Vienna Standard Mean Ocean Water, Reference Sheet for International Measurement Standards, IAEA Vienna, Dec 2006; https://nucleus.iaea.org/rpst/documents/VSMOW_SLAP.pdf.
[3] Rosman K. J. R. and Taylor P. D. P. (Editors), *Pure Appl. Chem.*, **70** (1998) 217.
[4] Atzeni S. and Meyer-ter-Vehn J., *The Physics of Inertial Fusion* (Oxford University Press) 2004.
[5] Keilhacker M. on behalf of the JET Team, *Nucl. Fusion*, **39** (1999) 209.
[6] https://www.llnl.gov/news/national-ignition-facility-experiment-puts-researchers-threshold-fusion-ignition.
[7] Romanelli F. *et al.*, *Fusion electricity - A Roadmap to the realization of fusion energy*, ISBN 978-3-00-040720-8, https://www.euro-fusion.org/wpcms/wp-content/uploads/2013/01/JG12.356-web.pdf.
[8] https://www.euro-fusion.org/eurofusion/roadmap/.
[9] Wesson J., *Tokamaks*, third edition (Clarendon Press, Oxford) 2004.
[10] Ongena J., these proceedings, p. 143.
[11] "ITER Physics Expert Groups on Confinement and Transport and Confinement Modelling and Database, ITER Physics Basis Editors and ITER EDA", *Nucl. Fusion*, **39** (1999) 2175.
[12] Bruzzone P. *et al.*, *Nucl. Fusion*, **58** (2018) 103001.
[13] https://news.mit.edu/2021/MIT-CFS-major-advance-toward-fusion-energy-0908.
[14] Matsunami N. *et al.*, IPP-AM-32 report Nagoya (1983).
[15] Loarte A. *et al.*, *Nucl. Fusion*, **47** (2007) S203.
[16] Piras F. *et al.*, *Plasma Phys. Control. Fusion*, **51** (2009) 055009.
[17] Morris W. *et al.*, *IEEE Trans. Plasma Sci.*, **46** (2018) 1217.
[18] Albanese R. *et al.*, *Fus. Eng. Design*, **146** (2018) 194.
[19] Baluc N. *et al.*, *Nucl. Fusion*, **47** (2007) S696.
[20] Baluc N., *Phys. Scr. T*, **138** (2009) 014004.
[21] Knaster J. *et al.*, *Nucl. Fusion*, **57** (2017) 102016.
[22] van der Schaaf B. *et al.*, *J. Nucl. Mater.*, **283-287** (2000) 52.
[23] Lindau R. *et al.*, *Fus. Eng. Design*, **75-79** (2005) 95.
[24] Stork D. *et al.*, *J. Nucl. Mater.*, **455** (2014) 277.
[25] Batistoni P. *et al.*, *Fusion Eng. Des.*, **82** (2007) 2095.
[26] Federici G. *et al.*, *Fusion Eng. Des.*, **141** (2019) 30.
[27] Maisonnier D. *et al.*, *Nucl. Fusion*, **47** (2007) 1524.

[28] MITCHELL N., *Fusion Eng. Des.*, **66** (2003) 971.
[29] MITCHELL N. *et al.*, *Supercond. Sci. Technol.*, **20** (2007) 25.
[30] FRANKE T. *et al.*, *IEEE Trans. Plasma Sci.*, **46** (2018) 1633.
[31] KEEP J. *et al.*, *Fusion Eng. Des.*, **124** (2017) 420.
[32] BARUCCA L., *Fusion Eng. Des.*, **136** (2018) 1557.

Energy from biomass

Bioenergy resource base, provision technologies and integration concepts in energy systems with net zero climate gas emissions

DANIELA THRÄN

DBFZ Deutsches Biomasseforschungszentrum gemeinützige GmbH
Torgauer Straße 116, 04347 Leipzig, Germany

UFZ Helmholtz Center for Environmental Research GmbH, Department Bioenergie
Permoser Straße 15, 04318 Leipzig, Germany

Summary. — Bioenergy is the most relevant renewable energy source today. Many technologies and concepts are implemented to provide heat, power and transport fuels. While biomass is a limited resource, so bioenergy provision needs integration, not only in the energy system but also in sustainable supply systems. The use of residual and waste materials will increase and needs mobilisation strategies and utilisation technologies. Due to different biomass potentials, energy infrastructure and energy strategies in different countries, even under the transformation towards more and more renewable energy-based energy systems the areas of bioenergy application remain diverse. However, flexible and hybrid concepts are gaining importance in all energy sectors; further technology development and digitalisation are pioneers. Additionally, the interaction with hydrogen and negative emissions is relevant in the long term and needs to be researched now.

1. – Introduction

Bioenergy is defined as energy derived from biomass. Biomass is a raw material of biological origin excluding material embedded in geological formations or transformed to fossilized material. Biomass can be processed into solid, liquid or gaseous fuels or stored

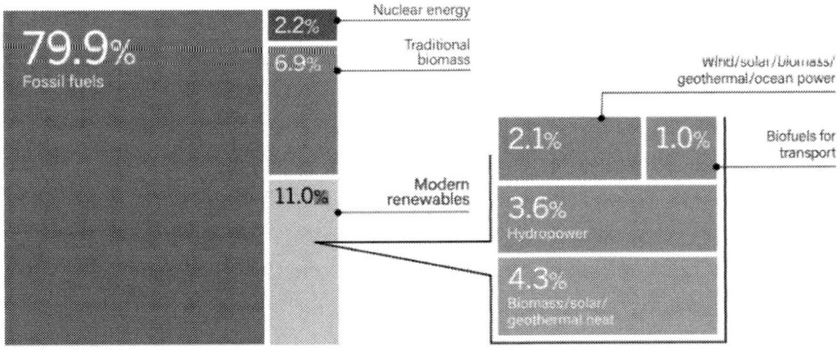

Fig. 1. – Estimated renewable share of total final energy consumption, 2018 [1].

energy in biomass can be directly converted into other forms of energy (*e.g.*, heat, light) (ISO 13065:2015).

Energy from biomass currently contributes on a worldwide basis 10% in 2018 to the gross primary energy demand (see fig. 1).

Due to geographical, economic, and climatic differences, the share of biomass energy in relation to total energy consumption differs widely between different countries, ranging from less than 1% in some industrialized countries like the United Kingdom to significantly more than 50% in some developing countries in Africa and Asia. Biomass is by far the most important renewable energy source, being significantly larger in energetic terms than the second largest, hydropower.

Cooking with biofuel is extremely rare today in rich countries although there is significant use of biofuel for space heating in regions with easy access to forests. However, because urban households tend to rely less on biofuels than rural households do. In a global context, the use of biomass in open fireplaces is still dominating the resource demand.

In parallel, the demand for electric energy is increasing in industrialized as well as in developing countries. Therefore, electricity production from biomass is often seen as an important future market for biomass worldwide. The same is also true for the production of transportation fuels from biomass.

2. – Biomass potentials

In principle, a wide range of resources can be used for energetic purposes, including biomass from agricultural and silvicultural primary production as well as by-products and residues from downstream industries and municipal wastes.

One important parameter for describing the biomass potential is the type of potential. Literature distinguishes between a multitude of different terms such as theoretical, technical, economic, sustainable, and realizable/available potential. The most frequently used types of potentials are:

"The theoretical potential is the theoretical energy supply in a given region that can be used physically within a specific period of time (*e.g.*, the energy

> stored in the whole plant mass). It is determined solely by the given physical utilisation boundaries and thus delineates the upper limit of the theoretically realisable energy supply" [2].
>
> "The technical potential describes the portion of the theoretical potential that can be used, taking into account the given technical restrictions (*e.g.*, recovery rates, conversion losses). Additionally, the given structural and legal environmental or other limitations are taken into consideration since they are regarded, similar to the technical restrictions, as "insurmountable" (*e.g.*, legally [nature] protection areas, legally/administratively: cross compliance regulation, societal: considering food production and material utilisation). As a result, it describes —primarily from a technical point of view— time and location dependent, possible contribution of the biomass to energy supply" [2].

Many studies deal with the following question: what amount of biomass is available for energetic use now and in the future?. However, those studies cover a wide range of results and also changed over the time. The main differences among these studies are the consideration of different biomass, and also the expectation on land availability. However, compared to former trends the expectation for biomass from arable land are decreasing [3] and focusing on degraded land.

Biogenic residues and wastes are seen as becoming more important for energy generation. The majority of those biomasses has a low energy density and is therefore not transported over long distances (there are exception, *e.g.*, used cooking oil or recovered wood). For Germany, the bioenergy potential of biogenic residues and wastes is estimated at 0.93–1.22 EJ (0.54 EJ already in use, 0.39–0.68 unused). This corresponds to at least 7% of the national primary energy consumption [4].

Beside a necessary resource quality, the question of the origin and the according life path arises for the supply of biogenic resources —for material as well as for energetic use— in the context of available resource potentials. For instance, there is a significant influence in this regard whether a certain resource comes as residual material or is specifically cultivated. The latter as well has influence on direct or indirect land use changes. In the case of biofuels, transparency in the sustainability of the supply has already started with the help of certification systems. This assessment and certification framework shall be extended beyond biofuels in the future and shall have influence on potential assumptions accordingly [5].

3. – Bioenergy provision

Biomass consists primarily of carbon, hydrogen and oxygen (as well as other elements) and is produced in the photosynthesis of plants. The overall principle of energy generation from biomass is —the same as from fossil fuels— the oxygenation of carbon hydrates, which is an exothermal reaction. Heat released from the oxygenation is the base for our energy provision. It can be used for heating or further processed to electricity or

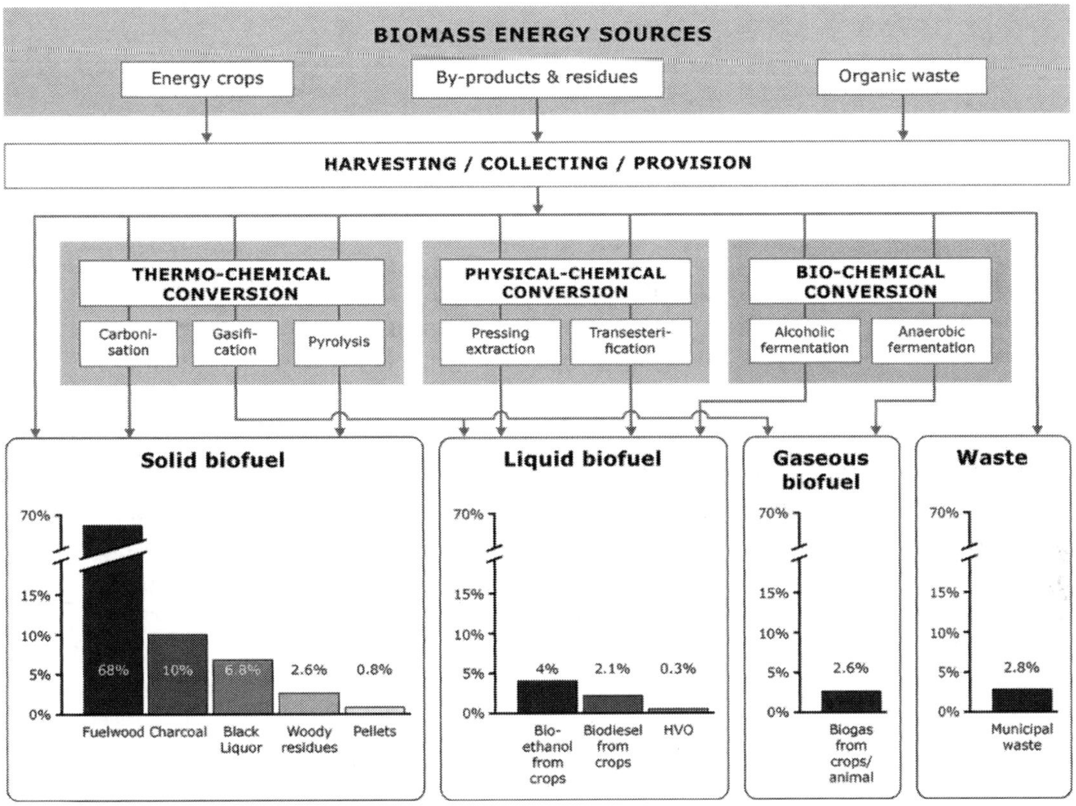

Fig. 2. – Bioenergy carriers [6].

mobility. The CO_2 from biomass or biofuel oxygenation does not have global warming effects if the biomass use is balanced with the biomass growing.

Most biomass used for energetic purposes is directly combusted to produce heat and/or power, but huge varieties of additional possibilities are available to provide environmentally sound heat and/or electricity as well as transportation fuels from organic material. The most important conversion routes available now or in the near future will be discussed according to the framework shown in fig. 2.

It is useful to consider two major forms of biomass fuel: unprocessed and processed. For unprocessed biofuel in which the material is used essentially in its natural form (as harvested) direct combustion usually supplies heat for cooking, space heating, or electricity production, although there are also small- and large-scale industrial applications for steam raising and other processes requiring low-to-medium temperature process heat.

Processed biofuels with clearly defined fuel characteristics allow for a technically easy and environmentally sound conversion into the desired useful energy. Such biofuels can then be traded easily and used with fewer problems to meet a supply task efficiently and comfortably. To ensure this, the following conversion routes are available (fig. 2).

- Thermo-chemical conversion summarizes all conversion processes of biomass into a solid, liquid or gaseous fuel based on heat. Therefore, gasification, pyrolysis and charcoal production count to these processes. From these possibilities, only charcoal production is state of the art and widely used so far. However, gasification in connection with electricity production is a promising option, which might become available in the near future.

- A physical-chemical conversion process provides liquid fuels based on physical (like pressing) and chemical (like esterification) processes. The most frequently used process so far is the vegetable oil production from oil seed and the transesterification of this vegetable oil to Fatty Acid Methyl Ester (FAME) as a substitute for petroleum-based diesel fuel.

- Bio-chemical conversion summarizes conversion processes based on biological processes. The most important possibilities are alcohol production from biomass containing sugar, starch and/or celluloses and biogas production from organic waste material. Both technologies are state of the art and widely used for energy provision.

Heat production is so far the most important use for biomass fuel worldwide. Direct combustion devices are widely distributed with thermal capacities ranging from a few kW in household stoves up to heating plants with several tens of MW. The conversion efficiencies vary from 8 to 18% for simple three stone stoves up to approximately 95% and above for modern heating units with high-end condensing boiler technology.

Within the biofuels, a distinction has been made between conventional and advanced biofuels. While conventional or first-generation biofuels are based on food and feed crops, the advanced biofuels use residues and waste biomass (2nd generation) or algae as feedstock (3rd generation). Apart from feedstock, conventional and advanced biofuels differ in technology maturity and GHG reduction. The technology of conventional biofuels such as biodiesel from oil crops, bioethanol from starch or sugar crops and biogas from anaerobic digestion is considered mature and is used commercially. The technology of advanced biofuels such as cellulosic ethanol, Btl, Bio-SG, methanol, bio butanol, DME or OME, hydrogen from gasification of woody biomass or biogas reforming is in research, demonstration or early commercial stage (see fig. 3).

Advanced biofuels additionally offer the prospect of requiring less land, lowering the competition between food/feed and fuel, improving the GHG balances due to the usage of residues and waste and thus reducing sustainability concerns for biofuels. Special interest for advance fuels is announced from the aviation sector. While road and trail transport can be electrified, directly by using net electricity or indirectly by using power-based fuels (Ptx fuels), the aviation sector will require liquid fuels in mid and long term. Thus, biofuels are one alternative to reduce GHG emissions in air transport.

A wide range of biojet fuels have been tested successfully. Technical standards/Certification for five biojet fuels have been established:

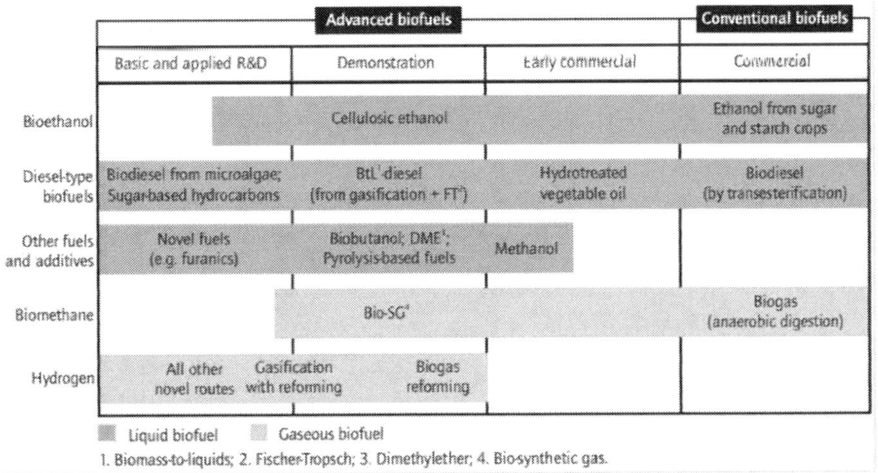

Fig. 3. – Maturity of biofuel technology [7].

- HEFA/HVO (Hydro processed Esters and Fatty Acids/Hydrogenated vegetable oil)
- FT-diesel (Fischer-Tropsch-diesel)
- SIP fuels (Renewable Synthesized Iso-Paraffinic fuel; renewable farnesane hydrocarbon)
- ATJ fuels (Alcohol to Jet Fuel)
- HPO (Hydrogenated Pyrolysis Oils; or: Hydrotreated Depolymerised Cellulosic Jet (HDCJ))

They can be applied as drop-in fuels without major changes in infrastructure or aircraft engines.

4. – Integration of bioenergy into the energy transition

The energy transition towards net zero energy supply systems is one of the biggest challenges for the next century. For bioenergy, the need for further development arises from the need of a sustainable biomasses sourcing and from increasing supply of fluctuating electricity from wind and PV [8].

The ongoing energy transition is mainly driven by reductions in the cost of wind and solar energy, and political efforts to reduce greenhouse gas emissions. Although substantial deployment of variable renewable energy (VRE) is an important part of the overall transformation, rapid changes in the energy mix may pose challenges to the resilience of the electricity grid, particularly in times of weather-related stress [9].

Flexible bioenergy can be arranged in different ways. However, biogas technologies already offer the possibility of short reaction times and buffering properties of gas storage that balance short- as well as long-term fluctuations. The system services comprise the provision of control energy and the offer of increased residual load coverage in times when fluctuating energies such as photovoltaic and wind power are not sufficiently available. On the energy markets, flexibly used bioenergy can thus achieve higher electricity market revenues. Additionally, innovative combined heat and power plants can combine an optimized heat supply with a stable power supply, even with a very high share of renewable energies.

In order to achieve these goals, flexibility requires the provision of corresponding capacities for power generation, remote control of the bioenergy plant via network-based communication and intelligent control concepts, also from the field of artificial intelligence. The efficiency of the flexibilization of block-type thermal power stations and biogas plants is described by the following dimensions, which can be determined for each unit: ramp (plant-specific load ramps), range (amplitude or control range between maximum and minimum plant output) and duration (maximum remaining in a certain load condition). In an "external" control loop, see fig. 4, external information from the higher-level network (*e.g.*, availability of other renewable energies, network utilization and energy market information) is digitally processed with internal plant information (*e.g.*, load situation, filling level and process parameters) in order to flexibly adapt the operating status of the plant along the dimensions ramp, range and duration to the current requirements via an "internal" control loop [10].

Bioenergy plants can primarily meet longer-term balancing requirements (fluctuations within 24 hours up to seasonal fluctuations), as they can use the fuel as a chemical energy store to supply electricity and heat as required. The associated information and communication technology have to meet a wide range of requirements, from network-based communication and standardized interfaces to decentralized and central information processing in order to optimally control the flexibility of the plant. Central data storage and digital information processing can take place via secure cloud services over the Internet.

Another approach is to build hybrid concepts of bioenergy with other renewable energies for certain supply systems. Especially in the heating sector many different combinations of bioenergy with heat pumps, with solar heating systems are possible [12].

Finally, also the interplay of hydrogen with bioenergy provides interesting options for the interplay of VRE and bioenergy. While the generation of hydrogen from biomass is only relevant for some niche applications, renewable hydrogen is an interesting option to increase biofuel yields or to build additional hybrid supply systems [13].

5. – BECCS

Finally, bioenergy can also contribute to accelerate for net zero climate gas emissions: Combining biomass for energy production purposes with CO_2 capture and storage (Biogenic Carbon, Capture and Storage) can be a cost-effective technology for reducing the atmospheric CO_2 concentration. For doing so, CO_2 capture can be realised 1) during biological processing (*e.g.*, fermentation), 2) in pre-combustion CO_2 separation when gasifying biomass to produce hydrogen-rich syngas ("producer gas"), 3) in post-combustion CO_2 capture as downstream unit in biomass combustion plants and oxyfuel-combustion

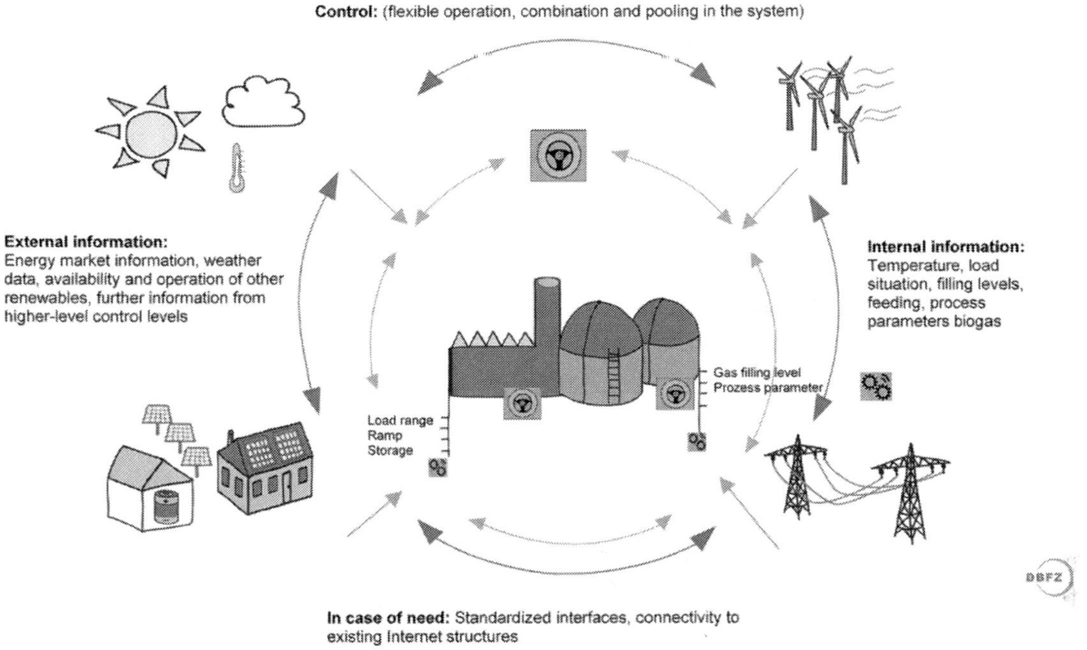

Fig. 4. – System elements and data flows for the flexibilization of bioenergy plants using modern information and communication technology [11].

CO_2 capture (*i.e.*, combustion of the energy carrier in pure oxygen rather than in air) when producing power and heat based [14].

This biogenic CO_2 is not burdened with climate-relevant emissions though the environmental impacts of BCCS technology pathways may vary greatly depending on design and characteristics of the processes. Once CO_2 is separated, it has to be compressed to obtain the necessary pressure for transporting the CO_2 to dedicated storage sites. Such suitable formations can be saline aquifers, depleted oil and natural gas fields, or unminable coal seams. Effective and safe long-term CO_2 storage is crucial with regard to storage site selection in order to prevent CO_2 leaking back into the atmosphere. Alternatively, instead of storing CO_2 underground, it can also serve as input to various applications that can commercially utilise CO_2 (*e.g.*, for greenhouses or algae cultivation; within enhanced commodity production (such as urea) or for the chemical industry). For BCCUS technologies, high-variability potential environmental impacts were found to be dependent on the choice of CCU technologies and CO_2 utilisation pathways [15].

The net-zero-energy road map of the IEA has included those different additional demands for bioenergy and presented a long-term vision in July 2021 [16].

6. – Conclusion

Bioenergy provision is embedded in national and regional energy systems, industrial infrastructure, land uses and value chains. Countries differ concerning biophysical conditions for bioenergy and other energy sources, geological CO_2 storage capacity, gas and electricity grids, public transport infrastructure, etc. A country-specific energy system transformation strategy for climate-friendly energy systems needs to frame the role of bioenergy. Due to different biomass potentials, energy infrastructure and energy strategies in different countries, the attractiveness of different bioenergy options differs among countries. We expect, that even under the transformation towards more and more renewable-energy–based energy systems the areas of bioenergy application remain diverse. However, flexible and hybrid concepts are gaining importance in all energy sectors. Further technology development (especially particulate matter) and digitalisation (predictive controls, interface standardisation) are pioneers. Additionally, the interaction with hydrogen and negative emissions is relevant in the long term and needs to be researched now.

REFERENCES

[1] REN21 (2020): Renewables 2020 Global Status Report, `https://www.ren21.net/wp-content/uploads/2019/05/gsr_2020_full_report_en.pdf` (retrieved 25.06.2021).

[2] Thrän D. and Pfeiffer D., "Methodenhandbuch – Stoffstromorientierte Bilanzierung der Klimagaseffekte. Methoden zur Bestimmung von Technologiekennwerten, Gestehungskosten und Klimagaseffekten von Vorhaben im Rahmen des BMWi-Forschungsnetzwerkes Bioenergie/BMWi-Förderbereich 'Energetische Biomassenutzung'" Schriftenreihe des BMU-Förderprogramms "Energetische Biomassenutzung", No. 4 (Deutsches Biomasseforschungszentrum gemeinnützige GmbH, Leipzig, Hrsg.) 2021, p. 233, `https://www.energetische-biomassenutzung.de/en/working-groups-methods/method-handbook`.

[3] Pfeiffer D. and Thrän D., "One century of bioenergy in Germany: wildcard and advanced technology", *Chem. Ing. Tech.*, **90** (2018) 1676.

[4] Brosowski A., Thrän D., Mantau U., Mahro B., Erdmann G., Adler P., Stinner W., Reinhold G., Hering T. and Blanke C., "A review of biomass potential and current utilisation – Status quo for 93 biogenic wastes and residues in Germany", *Biomass Bioenerg.*, **95** (2016) 257.

[5] Majer S., Wurster S., Moosmann D., Ladu L., Sumfleth B. and Thrän D., "Gaps and research demand for sustainability certification and standardisation in a sustainable bio-based economy in the EU", *Sustainability*, **10** (2018) 2455.

[6] Thrän D., Billig E., Brosowski A., Klemm M., Seitz S. B. and Witt J., "Bioenergy Carriers – From Smoothly Treated Biomass towards Solid and Gaseous Biofuels", *Chem. Ing. Tech.*, **90** (2018) 68.

[7] IEA: *Technology Roadmap – Biofuels for Transport, IEA Bioenergy ExCo Workshop, 2011*, `https://www.ieabioenergy.com/wp-content/uploads/2013/10/P02-Technology-Roadmap-Biofuels-for-Transport-Adam-Brown.pdf` (retrieved 08.04.2021).

[8] Thrän D., *Smart Bioenergy. Technologies and concepts for a more flexible bioenergy provision in future energy systems* (Springer, Heidelberg) 2015.

[9] IEA Bioenergy, *Task 44: Flexible Bioenergy and System Integration* (2021) `https://task44.ieabioenergy.com/` (retrieved 22.09.2021).

[10] Thrän D. *et al.*, script for the lecture *Bioenergy Technologies*, Summer term 2020 (Universität Leipzig) 2020.

[11] Thrän D., Lenz V., Liebetrau J., Krautkremer B., Kneiske T., Dreher A., Wille-Haußmann B., Dahmen M., Shu D. Y., Bau U., Kold T., Graf F., Köppel W. and Lehneis R., "Flexibler Einsatz von KWK, BHKW und Biogas-Anlagen durch Informations- und Kommunikationstechnik", *Die Energiewende - smart und digital: Jahrestagung 2018 des ForschungsVerbunds Erneuerbare Energien, 17-18 Oktober 2018, Umweltforum Berlin, FVEE-Themen 2018, ForschungsVerbund Erneuerbare Energien (FVEE), Berlin* (2019) pp. 35–40.

[12] Jordan M., Lenz V., Millinger M., Oehmichen K. and Thrän D., "Future competitive bioenergy technologies in the German heat sector: Findings from an economic optimization approach", *Energy*, **189** (2019) 116194.

[13] BMWi-Forschungsnetzwerk Bioenergie, *Stellungnahme. Biomasse und Bioenergie als Teil der Wasserstoffwirtschaft* (2021) `https://www.energetische-biomassenutzung.de/fileadmin/media/6_Publikationen/Stellungnahmen/Stellungnahme_FNBioE_H2-BM_final.pdf`.

[14] IEAGHG: "Potential for biomass and carbon dioxide capture and storage", unter Mitarbeit von Joris Koornneef, Pieter van Breevoort, Chris Hendriks, Monique Hoogwijk, Klaas Koops und Michéle Koper (2011) `http://hub.globalccsinstitute.com/sites/default/files/publications/102121/potential-biomass-carbon-dioxide-capture-storage.pdf` (retrieved 28.03.2018).

[15] Thrän D., Cowie A. L. and Berndes G., "Roles of bioenergy in energy system pathways towards a "well-below-2-degrees-Celsius (WB2)" world", workshop report and synthesis of presented studies, a Strategic Inter-Task Study carried out with cooperation between IEA Bioenergy Tasks 40, 43, 44 and 45, IEA Bioenergy (2020) p. 127.

[16] IEA, *Net Zero by 2050* (IEA, Paris) 2021, `https://www.iea.org/reports/net-zero-by-2050`.

EPJ Web of Conferences **268**, 00015 (2022) https://doi.org/10.1051/epjconf/202226800015
LNES 2021

Exploitation of energy geo-resources and their impacts on the environment

ELVIRA BATTIMELLI

Dipartimento di Fisica "E. R. Caianiello", Università degli Studi di Salerno
via Giovanni Paolo II, 132, 84084 Fisciano (SA), Italy

Summary. — The increase of the world population has led to a major demand for energy with a consequent huge exploitation of natural fossil resources. To build a sustainable future and reach the goals of the green economy, it is necessary to gradually use low-carbon energies taking into account the possible risks associated with each type of energy production. A much-debated incidental event is the anthropogenic or induced seismicity connected with the exploitation of energy geo-resources. It arises because of changes in the pore pressure of subsurface rocks that, in turn, create increased stress that can exceed the threshold value for fracture. This situation may verify from underground extraction or injection of fluids, geothermal energy production, storage of CO_2, natural gas, or hydrogen in underground cavities. The interest of the scientific community, politics and society towards this issue has grown in the last decade especially after the wide use of the fracking technique in the United States and the occurrence of the 2012 Emilia Romagna earthquake (Italy). Both conventional and unconventional geo-resources exploitation techniques can induce seismicity causing damage to the industrial plant itself and the surrounding buildings especially in highly urbanized areas and in extreme situations even generate loss of life. It is, therefore, necessary to carry out detailed studies of seismicity occurred in areas of geo-resources exploitation to expand knowledge of this phenomenon and develop safety protocols to prevent and mitigate the risks.

1. – Introduction

Today's greatest challenge facing science, technology, policymakers, and society is to make the world more sustainable. The increasing world population and the resulting growing demand for energy has led to a massive exploitation of fossil fuels (coal, oil, and gas). This trend has led the energy sector to be the major contributor to anthropogenic greenhouse gas emissions into the atmosphere [1]. Therefore, talking about sustainability inevitably means talking about energy transition and a change of approach regarding energy production and consumption to avoid the most serious impacts of climate change. The European Union's strategy for a sustainable future is via the Green Deal, which contemplates a 55% reduction in greenhouse gas emissions by 2030 up to net-zero emissions in 2050 [2]. To achieve this ambitious result, it is necessary to gradually abandon fossil sources in favor of low-carbon energy and plan the development of reliable technologies for carbon capture and storage. At the same time, in accordance with the objectives of the green economy it is necessary to consider the risks that each type of industrial exploitation of geo-resources is able to generate on the environment, society and economy. It is important to remember that the use of any energy source, whether fossil or renewable, inevitably involves risks. Reducing the impacts is possible through detailed risk analyses that take into account each phase of energy production: from raw material sourcing and plant construction to decommissioning. The accidents that produce the greatest consequences have a low frequency of occurrence and happen especially when the industrial and natural worlds collide. These incidental events can be classified into TechNa (Technological Hazard Triggering Natural Disaster) and NaTech (Natural Hazard Triggering Technological Disaster) events depending on whether the natural phenomenon is a consequence of anthropogenic industrial-type activity or acts as a hazard for an accidental industrial event. In the first case a technological activity can generate a natural disaster, such as the occurrence of an earthquake, the pollution of air, seas, or groundwater. Conversely, a NaTech event can be originated when a natural event (*e.g.*, earthquake, tsunami, extreme weather events, etc.) impacts on the industrial plant causing the release of toxic substances giving rise to a domino effect [3]. The study of NaTech and TechNa events is of particular importance to develop a deeper knowledge and elaborate strategies for risk prevention and mitigation in the perspective of sustainable development. In the following, an overview of a particular TechNa event, known as anthropogenic or induced seismicity is presented.

2. – Anthropogenic seismicity

Anthropogenic activities related to the exploitation of energy geo-resources can be risky for the surrounding environment. Although the best-known environmental consequence is air pollution with the emission of greenhouse gases, the attention of the scientific community but also of politics and society has in recent years also focused on induced seismicity due to the damage that this event can generate [4]. Technological activities such as extraction of fossil fuels (oil and gas), geothermal energy production, storage of CO_2, natural gas, or hydrogen in underground cavities are all potential causes of anthropogenic seismicity. Public interest in this topic has increased especially following specific seismic events, such as the 2011 Oklahoma earthquake that occurred after wastewater injection following a fracking operation and the much-debated 2012 Emilia-Romagna (Italy) seismic sequence [5]. Because of increasing numbers of induced seismicity cases, the Human-Induced Earthquake Database (HiQuake) was created in 2017 [6]. This freely accessible database was initially funded by Nederlandse Aardolie Maatschappij BV (NAM) to Durham University (UK) and Geoenergy Durham Ltd, containing to date the largest collection of anthropogenic projects worldwide that have reported induced seismicity (approximately 1231 cases) [7].

To understand why the technological activities mentioned above are potential causes of seismicity, it is necessary to remember that an earthquake occurs when an effort acting on a point of the Earth's lithosphere exceeds a threshold value, generating a fracture that propagates as long as the stress and friction conditions allow it. Through the extraction or injection of fluids in the underground these industrial activities are able to change the physical parameters of the rocks leading to breakage. The critical condition is quantified by the fracture criterion established by Coulomb in 1773. It is expressed by the following equation:

$$\tau_{\text{crit}} = \mu(\sigma_n - P) + \tau_0,$$

where τ_{crit} is the shear stress that generates the fracture, μ is the internal friction coefficient, $(\sigma_n - P)$ the difference between the normal stress and the pore pressure, called the effective normal stress. Finally, τ_0 is the cohesion or cohesive strength. A small variation of one of these parameters can perturb the system generating an earthquake.

The industrial activities that have attracted more attention than others on this topic are surely the extraction and injection of fluids in the subsoil, in particular to produce oil and gas. The extraction process can cause a contraction of the reservoir giving rise to changes in stress in the surrounding rocks and neighboring fault systems. To increase production, it is often necessary to inject fluids in the subsurface [8, 9]. The most common method is water flooding which involves a controlled injection of water to avoid exceeding the initial reservoir pore pressure [10]. Nevertheless, it may happen that the threshold value is exceeded and therefore the conditions of the stress field can vary. In addition, in the last decades, the well-known unconventional technique of fracking has come to the fore, raising new questions about induced seismicity in areas of low seismic risk [4]. This technique widely used in North America and Canada permits to extract oil and gas by hydro-fracturing rock formations with low permeability (*e.g.*, oil shales or black shales). Fluid pumped at high pressure fractures the rock creating pathways into which the hydrocarbon can pass. To prevent fracture closure, a solid substance (quartz sand or ceramic microspheres) called proppant is added to the injected water. The fracking process lasts from hours to days during which a network of horizontal fractures connected to the well bore are created [9]. In this way a microseismicity ($M_w < 1$) [4] that serves as a guide to understand the geometry of the fractured area is generated.

Upon completion of fracturing, the well is opened and the flowback phenomenon begins. Previously injected water or produced water extracted with hydrocarbons and other substances (*e.g.*, chemicals, debris) collected during the process return to the surface where they are cleaned and used for other purposes. In some cases, the wastewaters are re-injected into the subsurface at depths greater than those of the extraction wells and in nearby areas. The re-injection process takes years or decades in new wells without a balance between extracted and injected fluids, leading to an increase in the pore pressure and a subsequent variation of the stress field. Therefore, this technique has a high potential to generate seismicity.

Unfortunately, induced seismicity is not exclusive to industrial activities that exploit fossil resources but is also potentially caused by some of the technologies that will be widely used in a sustainable future. Among them is geothermal energy that exploits the internal heat of the Earth and is known to be a low-carbon green energy. The increasing adoption of unconventional methods such as Enhanced Geothermal Systems (EGS) challenges the scientific community and civil society about the possible consequences that such facilities may generate. The technique involves pumping cold fluids into hot, dry rocks to create artificial fractures or the opening of pre-existing faults to reduce reservoir permeability and form pathways in which water can circulate [5]. As fluids reach deep wells, they heat up in contact with hot rock and then rise to the surface. This process deliberately generates microseismicity that permits to understand the path of water in the subsurface, also indicating areas where an increase in permeability occurs. Seismic events induced with this technique are generally of low magnitude but especially in densely populated or already seismically active areas could be felt by the population and generate significant damages.

In addition to renewable forms of energy, to achieve net-zero emissions by 2050 it is necessary to develop advanced technologies to capture and store CO_2. The Carbon Capture and Storage (CCS) has the aim to capture the carbon dioxide produced by industries before it reaches the atmosphere and then storage it in underground geological formations suitable for this purpose [9]. This process is not immune to the risk of inducing seismicity since the injection of high rates of CO_2 at high pressure for long periods of time could lead to increases in pore pressure and changes in the stress field [11,12]. Since this technology will be increasingly important and more used in the future, it is necessary to carry out careful studies of induced seismicity related to this type of activity and risk analysis to mitigate the impacts and design facilities that can maximize the benefits.

3. – Main cases of induced seismicity in the world

There are numerous episodes of induced seismicity confirmed through detailed studies. Among the best documented cases of induced seismicity by gas extraction there is the Gazli gas field in Uzbekistan [9]. The gas production began in 1962. In the following years large quantities of water were injected into the reservoir to facilitate extraction, but, despite this, there was a consistent decrease in pressure (from 70 atm in the 1960s to about 30–35 atm since 1976 and to 15 atm since 1985). This situation caused three seismic events of magnitude 7 that occurred in 1976 and 1984 in the northern area of the production field that was previously seismically silent. Even the northern area of the Netherlands, naturally at low seismic risk, has had to deal with the increase in seismicity due to the extraction of gas using the conventional method. In the Groningen field, in operation since 1963, the first earthquake felt by the population occurred in 1991 with a local magnitude 2.4. In 2012, an earthquake with M_L 3.6 caused significant damage to homes in the neighborhood areas [13]. In conjunction with oil and gas production, wastewater is produced, often re-injected into the underground. The first documented and known earthquake caused by wastewater injection occurred in Denver, Colorado, in the 1960s in an area considered to be of low seismicity. The injection activity began in 1962 and lasted until 1966 despite the hundreds of tremors recorded in that period by the Bergen Park seismic station. The three strongest events of magnitude 5, however, occurred in 1967, more than a year after the end of the fluid pumping activities, causing extensive structural damage [9]. In 2011, seismic events induced by wastewater injection occurred in Trinidad, Colorado, and Prague, Oklahoma, with magnitudes 5.3 and 5.6, respectively [14]. It is precisely this re-injection following fracking operations that has led Oklahoma to an increase in seismicity in the last decade. Considering the geothermal energy production, a major unconventional project suspended due to the occurrence of seismicity was launched in Basel, Switzerland, in 2006. On December 2 of the same year high-pressure water was injected into the crystalline basement rock at a depth of about 5 km with the aim to create a geothermal reservoir and increase its permeability. Over the following days, increased injection and pressure rates of injected fluids resulted in a concomitant increase in seismicity that generated a seismic event of local magnitude 2.6, exceeding the expected safety threshold during well stimulation [15]. On December 8, the project was stopped but a few hours later another seismic event (M_L 3.4) was felt by the population.

4. – Conclusions

In order to achieve the green economy goals, it is necessary to reduce the impacts that human activities related to the exploitation of geo-resources produce on the environment. The development of innovative technologies and a careful risk assessment at each phase of the industrial process can help prevent and mitigate the impacts (environmental, economic, and social) associated with these industrial operations. This kind of studies are not simple because they require multidisciplinary knowledge. More specifically, to elaborate detailed studies of induced seismicity being able to discriminate between a natural or anthropogenic earthquake, seismological and industrial data (*e.g.*, rate and pressure of fluids extracted or injected) are needed. The last are often inaccessible to researchers because they are only available for energy companies. For this purpose, it is essential to overcome this barrier promoting the cooperation between scientists and energy leaders to carry out site-specific studies to analyze and understand in detail the mechanisms related to this type of seismicity especially in tectonically active areas. In addition, based on these analyses, it is very important to develop safety protocols and emergency plans to deal with the occurrence of these events and finally educating and informing in an appropriate way the population living in the surroundings of industrial plants.

REFERENCES

[1] IEA, *Net Zero by 2050* (IEA, Paris) 2021, `https://www.iea.org/reports/net-zero-by-2050`.
[2] REGULATION (EU) 2021/1119 of the European Parliament and of the Council of 30 June 2021 establishing the framework for achieving climate neutrality and amending Regulations (EC) No 401/2009 and (EU) 2018/1999 ("European Climate Law").
[3] KRAUSMANN E. and CRUZ A. M., "Past Natech Events", in *Natech Risk Assessment and Management – Reducing the Risk of Natural-Hazard Impact on Hazardous Installations*, edited by KRAUSMANN E., CRUZ A. M. and SALZANO E. (Elsevier) 2017, chapt. 2, pp. 3–31.
[4] ELLSWORTH W. L., "Injection-Induced Earthquake", *Science*, **341** (2013) 1225942.
[5] GRIGOLI F. *et al.*, "Current challenges in monitoring, discrimination, and management of induced seismicity related to underground industrial activities: A European perspective", *Rev. Geophys.*, **55** (2017) 310.
[6] WILSON M. P. *et al.*, "HiQuake: The Human-Induced Earthquake Database", *Seismol. Res. Lett.*, **88** (2017) 1560.
[7] The Human-Induced Earthquake Database (HiQuake), `https://inducedearthquakes.org/`.
[8] SHEPHERD M., "Factors influencing recovery from oil and gas fields, in Oil field production geology", *AAPG Memoir*, **91** (2009) 37.
[9] National Research Council, *Induced Seismicity Potential in Energy Technologies* (The National Academies Press, Washington, DC) 2013, `https://doi.org/10.17226/13355`.
[10] AHMED T.,, in *Reservoir Engineering Handbook - Fourth Edition* (Gulf Professional Publishing) 2010, pp. 909–1095, `https://doi.org/10.1016/B978-1-85617-803-7.50022-5`.
[11] EIKEN O. *et al.*, "Lessons learned from 14 years of CCS operations: Sleipner, in Salah and Snøhvit", *Energy Proc.*, **4** (2011) 5541.
[12] NICOL A. *et al.*, "Induced seismicity; observations, risks and mitigation measures at CO_2 storage sites", *Energy Proc.*, **37** (2013) 4749.
[13] VAN THIENEN-VISSER K. and BREUNESE J., "Induced seismicity of the Groningen gas field: History and recent developments", *Lead. Edge*, **34** (2015) 664.
[14] RUBINSTEIN J. and MAHANI A. B., "Myths and Facts on Wastewater Injection, Hydraulic Fracturing, Enhanced Oil Recovery, and Induced Seismicity", *Seismol. Res. Lett.*, **86** (2015) 1060.
[15] DEICHMANN N. and GIARDINI D., "Earthquakes Induced by the Stimulation of an Enhanced Geothermal System below Basel (Switzerland)", *Seismol. Res. Lett.*, **80** (2009) 784.

Robust, distributed and optimal control of smart grids

Juan E. Machado, Saeed Ahmed and Jacquelien M. A. Scherpen

Jan C. Willems Center for Systems and Control,
Engineering and Technology Institute Groningen,
Fac. Science and Engineering, University of Groningen
Nijenborgh 4, 9747 AG Groningen, The Netherlands

Michele Cucuzzella

Department of Electrical, Computer and Biomedical Engineering,
University of Pavia - via Ferrata 5, 27100 Pavia, Italy

Jan C. Willems Center for Systems and Control,
Engineering and Technology Institute Groningen,
Fac. Science and Engineering, University of Groningen
Nijenborgh 4, 9747 AG Groningen, The Netherlands

Summary. — These lecture notes provide an overview of recent research on the modeling and control of smart grids using distributed algorithms. In particular, energy-based modeling of general AC power networks using the framework of port-Hamiltonian systems theory is presented, and the relevance of such a formulation for stability analysis and control design is discussed. Low-level control design aspects (at a physical layer) for DC microgrids are also considered, achieving objectives such as fair load sharing among distributed generation units and (average) voltage regulation using limited data and measurements from the system. Finally, general frameworks for the optimal control of smart grids are introduced to consider both physical and economic constraints and exploit the flexibility brought up by storage devices and demand response from the grid's prosumers.

1. – Introduction

Energy system operators and stakeholders worldwide are currently striving to make the energy sector more sustainable so that future generations have access to secure and affordable energy [1]. A promising approach for achieving sustainability consists in substantially increasing the presence of renewable energy sources, such as solar and wind. However, renewable sources' uncertain and widely variable nature drastically contrasts with those conventionally used for energy production —usually large-scale thermal plants based on coal, gas, and oil— imposing great design and operational challenges for the efficient and reliable supply of energy to end-users. Critically, existing tools to monitor, control, and optimize large-scale energy systems are becoming progressively inadequate to guarantee optimized and robust system operation [2].

We provide an overview of recent contributions on robust, distributed, and optimal control of *smart grids*. Smart grids exploit information technologies as well as scalable and re-configurable control infrastructures to guarantee secure, resilient, and cost-optimized grid operations [3] (see also [4]). In particular, we report on the following contributions:

C1: The first contribution, originally reported in [5], refers to the development of a systematic framework for modeling AC power networks. A dynamic model of a general power system composed of synchronous machines, dynamic power lines, and nonlinear loads is obtained using graph theory and the framework of port-Hamiltonian systems. In addition, a discussion is presented on how such a model can be simplified so that open-loop stability analysis can be conducted.

C2: The second contribution, initially reported in [6], pertains to the design of a distributed control scheme that achieves (average) voltage regulation as well as proportional current sharing among DGUs of a DC microgrid containing linear loads. Some of the main contributions of [6] are: i) the consideration of an arbitrary meshed microgrid topology with resistive-inductive (and hence dynamic) power lines; ii) the satisfaction of a notion of voltage regulation using only current measurements.

C3: The third contribution, firstly reported in [7], concerns a price-based approach for achieving congestion management in AC distribution power systems with distributed, renewable generation units. The main goal is to maximize a social welfare function, which includes net revenues of end-users and the cost of power losses, while guaranteeing that voltage magnitudes are within prescribed bounds and that the grid operates within its capacity. Some simulation results on a modified IEEE-37 bus benchmark, which uses realistic data on photovoltaic generation units, are briefly discussed.

C4: The final contribution is concerned with the study of a distributed, optimal control framework to minimize the imbalance between supply and demand in smart grids with high penetration of renewable sources. One of the main features of this contribution is that it relies on *demand response* thereby shifting the production or consumption of energy assets in time, allowing to reduce or avoid load peaks. This contribution is reported in [8].

The rest of this document is structured as follows. Contributions C1–C4 are discussed in sects. **2–5**, respectively. Some concluding remarks appear in sect. **6**.

Notation: $\mathbb{R}$ denotes the set of real numbers; $\mathcal{C}^1$ is the set of continuously differentiable functions; ∇f is the transpose of the gradient of a function $f : \mathbb{R}^n \to \mathbb{R}$; $A^\top$ is the transpose of a matrix $A \in \mathbb{R}^{n\times n}$; graphs are denoted by $G = (N, E)$, where N represents its nodes, E its edges, $\mathcal{B}$ its incidence matrix, $\mathcal{L}$ its Laplacian matrix, and $\mathcal{E}_i$ the edges incident to any $i \in N$. We let $\mathrm{diag}(x_i)_{i\in\mathcal{I}}$ denote a diagonal matrix with diagonal elements equal to x_i ordered according to the index set $\mathcal{I}$, and $\mathrm{col}(x_i)_{i=1}^n = [x_1, x_2, \ldots, x_n]^\top \in \mathbb{R}^n$.

2. – Port-Hamiltonian modeling of AC power grids

The vast energy demand of modern societies has urged the development of very complex power systems for the production, transmission, and distribution of electricity [9]. Even though production has been mostly based on fossil fuels, there is currently a transition towards a more sustainable energy sector characterized by the increasing presence of renewable sources, such as photovoltaic arrays and wind farms [10, 11]; see fig. 1.

One of the main goals of a power system is to guarantee the supply of energy to end-users reliably, with minimum costs and minimum environmental impact [9, 12]. In particular, reliability is strongly related to the notion of system stability around desired operation regimes, which is tightly coupled with control systems design and implementation. A fundamental step towards the conceptualization of optimization and control algorithms is the development of a model that qualitatively and quantitatively describes the behavior of the devices and physical phenomena occurring during system operations [2, 13]. In this regard, dynamic models are useful to investigate the response

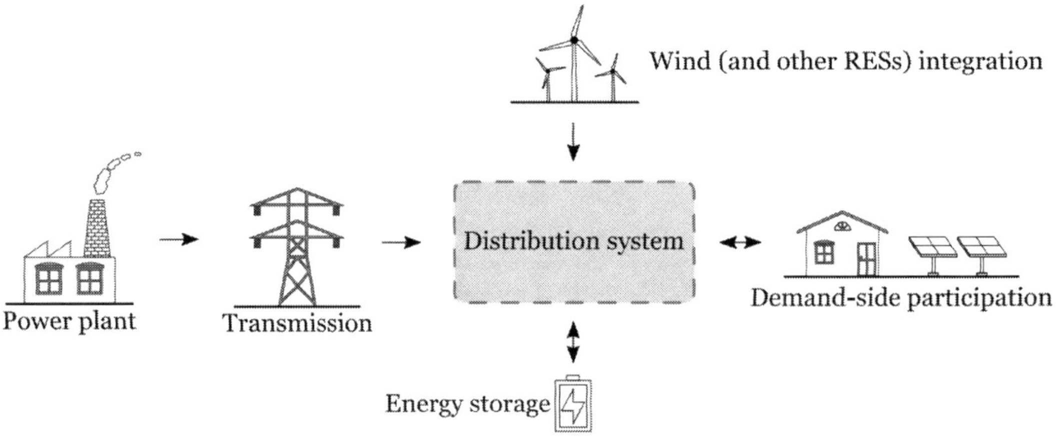

Fig. 1. – Schematic diagram based on modern electric power distribution systems, [14].

against disturbances, *e.g.*, power outages, whereas static (steady-state) models are more often employed in optimization analysis (see, *e.g.*, [13, 15]).

A well-established theory for dynamic modeling of complex physical systems is referred to as *port-Hamiltonian systems theory*. This theory provides a framework for modeling, analysis, and control of multi-physics systems in general [16] and allows to view complex systems as the interconnection of a small set of elements that store, dissipate, and transport/convert energy [17]. Notably, *energy* becomes a common language to interrelate systems from different domains, such as mechanical, electrical, hydraulic, and thermal ones [18]. Furthermore, physical systems with different time and spatial scales can be naturally integrated as the port-Hamiltonian representation is preserved during the interconnection of individual subsystems (see [19, 18]). For these reasons, port-Hamiltonian systems theory is suitable to model large-scale, networked multi-energy systems.

In this section, following the exposition in [5], port-Hamiltonian modeling of AC power systems is discussed. For the sake of brevity, a general class of port-Hamiltonian systems is defined next which is suitable to represent the class of energy systems treated in these lecture notes. Following [20], let $\mathcal{T}$ be a real interval, let $x : \mathcal{T} \to \mathbb{R}^n$ be the state and let $u, y : \mathcal{T} \to \mathbb{R}^m$ be the input and output, respectively. Then, the following dynamic equation describes a port-Hamiltonian system (see also [18]):

$$\begin{aligned} \dot{x} &= (J(x) - R(x))\, \nabla\mathcal{H}(x) + g(x)u, \\ y &= g^\top(x)\nabla\mathcal{H}(x). \end{aligned} \tag{1}$$

Each component of the state x represents a variable that determines the energy stored on a given physical device present in the system, *e.g.*, the electric charge of a capacitor. Also, $\mathcal{H} : \mathbb{R}^n \to \mathbb{R}^+$ continuously differentiable is the *Hamiltonian*, which represents the total energy stored in the devices conforming the overall physical system. Moreover, the matrices $J(x) = -J^\top(x) \in \mathbb{R}^{n\times n}$ and $R(x) = R^\top(x) \in \mathbb{R}^{n\times n}$, which have entries depending smoothly on x, are respectively referred to as *interconnection* (or *structure*) and *dissipation* matrices. Also, $g(x) \in \mathbb{R}^{n\times m}$ can be referred to as the *port* or *input* vectorfield and the specific choice of y is referred to as the *natural passive output*.

Next some instrumental steps that lead to representing AC power system models as in (1) are explained. Before that, we would like to remark that any solution of (1) satisfies the following power balance equation:

$$\frac{\mathrm{d}}{\mathrm{d}t}\mathcal{H}(x(t)) = -\nabla\mathcal{H}(x)^\top R(x)\nabla\mathcal{H}(x) + y^\top u. \tag{2}$$

Note that the left-hand side of (2) represents of the rate of change of the stored energy in the system, whereas the right-hand side is the total power provided by the external input u minus the total energy dissipated, *e.g.*, by electric resistors. Under some convexity assumptions on $\mathcal{H}$, and with $R(x) \geq 0$, eq. (2) implies that port-Hamiltonian systems represented as in eq. (1) are passive [18] (see also [16]) with storage function being the Hamiltonian, and with respect to the input output pair (u, y), *i.e.*, $\dot{\mathcal{H}} \leq u^\top y$. Even

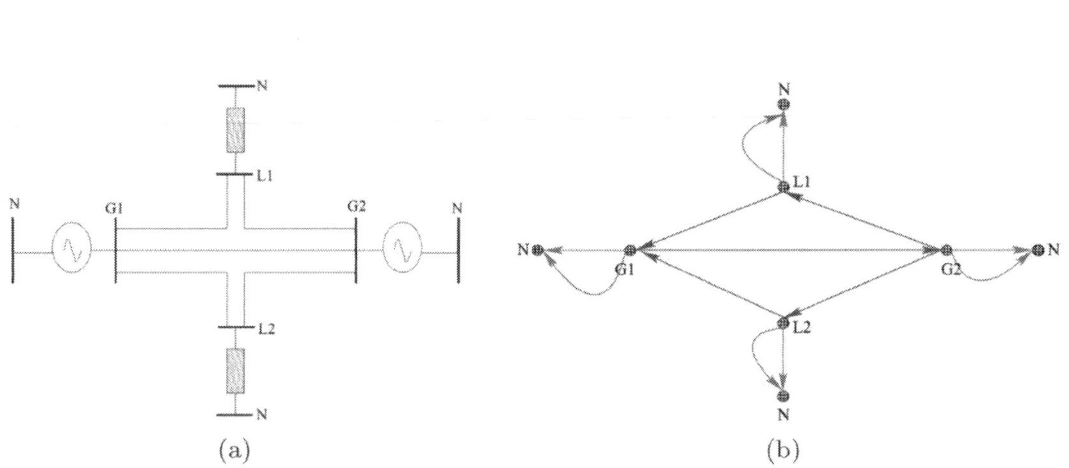

Fig. 2. – (a) Scheme of an exemplary power system with two generation buses, two load buses and five power lines. (b) Graph-based representation of the same system. Figure courtesy of [5].

though this is sufficient to guarantee open-loop stability when $u = 0$, such conclusions are not enough to understand stability properties of power systems in general, in which the *forced* equilibrium of interest is not the origin (see sect. 4.1 in [5]).

2·1. *AC grids as graphs.* – Consider a power network represented as a directed (and connected) graph $G = (N, E)$. The set of nodes N are the network's buses and the edges E are the generators, loads and power lines. It is assumed there is a single reference bus which is at ground potential, and that one node of each generator or load edge is necessarily the reference bus. For simplicity, the presence of interior nodes is neglected, *i.e.*, each node in N is either an end node of a generator or a load. Without loss of generality, the sets of nodes can be split as $N = N_{\mathrm{g}} \cup N_{\ell} \cup N_{\mathrm{r}}$, where N_{g} and N_{ℓ} would be the generation and load buses, and N_{r} the reference bus.

For the transmission lines a Π-model is adopted (see [9]). Then, each line consists of a series connection of an inductor and a resistor and, at each end node of the line, it is considered that there is a capacitor. One end node of such a capacitor is assumed to be the grid's reference node. In the case that more than one transmission line is incident to a given load or generation bus, then all capacitive effects are lumped into a single capacitor per generation or load bus. In view of the Π-model, the set of edges are conveniently split as $E = E_{\mathrm{g}} \cup E_{\ell} \cup E_{\mathrm{c}} \cup E_{\mathrm{T}}$, where E_{g} and E_{ℓ} are generation and load edges, E_{c} are the capacitor edges, and E_{T} are the $R - L$ components of each power line. The diagram of an exemplary power system and its graph representation is shown in fig. 2.

Note that a 3-phase power system is considered, then each node or edge associated to any generator, load or transmission line corresponds in fact to three different nodes and edges representing three different phases. Next, the dynamics of each edge of the system is written in port-Hamiltonian form (see eq. (1)). After that, algebraic interconnection equations are established (based on Kirchhoff's laws) from which the overall power system model can be obtained.

2·2. *Synchronous generator*. – Let $i \in E_{\mathrm{g}}$ denote an arbitrary synchronous generator in the network and let $\Psi_{\mathrm{s},i} = [\psi_{\mathrm{a},i}, \psi_{\mathrm{b},i}, \psi_{\mathrm{c},i}]^\top \in \mathbb{R}^3$ and $\Psi_{\mathrm{r},i} = [\psi_{\mathrm{f},i}, \psi_{\mathrm{kd},i}, \psi_{\mathrm{kq},i}]^\top \in \mathbb{R}^3$ denote its stator and rotor flux linkages. Also, let $\theta_i \in \mathbb{R}$ and $p_i \in \mathbb{R}$ denote the rotation angle and momentum of the rotor, respectively. Then, the total energy of the synchronous machine, which is the sum of the electric and mechanical energy, can be written as follows:

$$\mathcal{H}_i = \frac{1}{2}\begin{bmatrix}\Psi_{\mathrm{s},i}^\top & \Psi_{\mathrm{r},i}\end{bmatrix} L_i^{-1}(\theta_i)\begin{bmatrix}\Psi_{\mathrm{s},i}^\top & \Psi_{\mathrm{r},i}\end{bmatrix}^\top + \frac{1}{2J_i}p_i^2, \quad i \in E_{\mathrm{g}}, \tag{3}$$

where $L_i(\theta_i) \in \mathbb{R}^{6\times 6}$ is the generator's inductance matrix, the components of which codify self and mutual inductances between the stator and rotor windings (see [9]). Also, $J_i > 0$ is the rotational inertia of the rotor. Using basic physical laws, such as Maxwell's and Newton's laws, in combination with port-based modeling [17] it is possible to arrive to the following port-Hamiltonian model for the generator (further modeling details appear in [9]):

$$\dot{x}_i = (\mathcal{J}_i - \mathcal{R}_i)\nabla\mathcal{H}_i(x_i) + g_i(x_i)u_i, \quad i \in E_{\mathrm{g}}, \tag{4a}$$

$$y_i = g_i^\top(x)\nabla\mathcal{H}_i(x_i), \tag{4b}$$

where $x_i = [\Psi_{\mathrm{s},i}, \Psi_{\mathrm{r},i}, p_i, \theta_i]^\top$, $u_i = [V_i, E_{\mathrm{f},i}, T_{\mathrm{m},i}]$ and $y_i = [I_i, -I_{\mathrm{f},i}, -\omega_i]^\top$ are the state, input and output vectors of each generator, respectively. The input vector u_i is comprised by the stator voltages $V_{\mathrm{s},i}$, the voltage applied across the rotor windings $E_{\mathrm{r},i}$, and the mechanical torque $T_{\mathrm{m},i}$ acting on the rotor. The output vector y_i is conformed by the stator currents I_i, the current through the rotor windings $I_{\mathrm{f},i}$ and rotor's angular velocity ω_i. Note that u_i and y_i are conjugated in the sense that their product $u_i^\top y_i$ represents power. The interconnection and damping matrices in (4) are, respectively, given by

$$\mathcal{J}_i = \begin{bmatrix}0_{3\times 3} & -e_3\\ e_3^\top & 0\end{bmatrix}, \quad \mathcal{R}_i = \mathrm{diag}\left\{R_{\mathrm{s},i},\ R_{\mathrm{r},i}, d_i, 0\right\}, \quad i \in E_{\mathrm{g}}, \tag{5}$$

where $e_3 \in \mathbb{R}^3$ denotes the 3rd element of the canonical basis of $\mathbb{R}^3$, $R_{\mathrm{s},i}, R_{\mathrm{r},i} \in \mathbb{R}^{3\times 3}$ are positive definite diagonal matrices containing in their main diagonal the resistances associated with the stator and rotor windings, respectively. The term d_i is the total mechanical damping (due to viscous friction). Due to space constraints $g_i(x_i)$ is not shown here, however its explicit form appears in [5], eq. (7).

2·3. *Load model*. – Let $i \in E_\ell$ be an arbitrary load edge. For the sake of simplicity, loads are represented by static (algebraic) relations as follows:

$$\gamma_i(V_i, I_i) = 0, \quad i \in E_\ell, \tag{6a}$$

$$V_i I_i \geq 0, \tag{6b}$$

where V_i and I_i are the voltage across, and the current through, the load, respectively. Note that the second inequality imposes that each load corresponds to a dissipative device, *i.e.*, it drains power from the network, and it cannot store energy.

2·4. *Power line dynamics.* – Let $i \in E_{\mathrm{T}}$ be an arbitrary power line, which we recall is conformed by a series connection of a resistor and an inductor (one per phase). Let $I_i, \Psi_i \in \mathbb{R}^3$ denote the current through, and the flux linkages of, the inductors. Also, let R_i and L_i denote 3×3 diagonal, positive definite matrices with the line resistances and inductances in their main diagonals. Then, the total energy stored in the power line is given by $\mathcal{H}_i = \frac{1}{2}\Psi_i^\top L_i^{-1}\Psi_i$, which represents the line's total magnetic energy. Through Kirchhoff's laws it is possible to obtain the following line dynamics:

$$\dot{x}_i = -\mathcal{R}_i \nabla \mathcal{H}_i(x_i) + u_i, \quad i \in E_{\mathrm{T}}, \tag{7a}$$

$$y_i = \nabla \mathcal{H}_i(x_i), \tag{7b}$$

where $x_i = \Psi_i$ is the state and $y_i = I_i$ is the output. The input $u_i = V_i$ represents the voltage across the $R - L$ component of the line. Also, $\mathcal{R}_i = R_i$. Note that mutual inductance among the lines of the different phases is neglected.

Let $i \in E_{\mathrm{c}}$ be an arbitrary capacitor at some end node of a given transmission line. Let $Q_i, V_i, I_i \in \mathbb{R}^3$ denote its electrical charge, voltage and current (per phase). If C_i is a 3×3 diagonal matrix with the capacitances, then the total energy stored at the capacitor is given by $\mathcal{H}_i = \frac{1}{2}Q_i^\top C_i^{-1} Q_i$. It follows from the constitutive relations for the capacitor that the following dynamic equation holds:

$$\dot{x}_i = u_i, \quad i \in E_{\mathrm{c}}, \tag{8a}$$

$$y_i = \nabla \mathcal{H}_i(x_i), \tag{8b}$$

where $x_i = Q_i$ is the state, $u_i = I_i$ is the input and $y_i = V_i$ is the output.

2·5. *Overall AC grid model.* – Until this point dynamic equations for any edge of the graph have been defined. Next diverse algebraic relationships are introduced, based on Kirchhoff's laws (see [21] for more details), to couple the edges' dynamics according to the physical interconnection of the devices in the network. Note that since the reference node is at ground potential, then $V_i = [0, 0, 0]^\top$ for $i \in N_{\mathrm{r}}$. In addition, since at each load or generation bus there is a capacitor, whose other terminal is connected to ground potential, it follows that for each $i \in E_{\mathrm{g}} \cup E_\ell$ there exists a unique $j \in E_{\mathrm{c}}$ such that $V_i = y_j = \nabla \mathcal{H}_j(x_j)$. Moreover, for each $i \in E_{\mathrm{T}}$ there exists a unique pair $(j, k) \in E_{\mathrm{c}} \times E_{\mathrm{c}}$, with $j \neq k$, such that $V_i = V_j - V_k = \nabla \mathcal{H}_j(x_j) - \nabla \mathcal{H}_k(x_k)$. On the other hand, the law of conservation of electrical charge at each load or bus node dictates that for each $i \in E_{\mathrm{c}}$ there exists a unique $j \in E_{\mathrm{g}} \cup E_\ell$ and a subset $\mathcal{E}_{\mathrm{T},i} \subset E_{\mathrm{T}}$ of transmission lines incident to the i-th generation or load bus such that $I_i + I_j + \sum_{m \in \mathcal{E}_{\mathrm{T},i}} I_m = 0$, or equivalently, that $I_i = -\alpha_i \nabla_{\Psi_{\mathrm{g},j}} \mathcal{H}_j(x_j) - (1 - \alpha_i) I_j - \sum_{m \in \mathcal{E}_{\mathrm{T},i}} I_m$, where $\alpha_i = 1$ if i is associated to a generation bus, and $\alpha_i = 0$ if i is associated to a load bus.

By substituting the above established relations for V_i, $i \in E_{\rm g} \cup E_\ell$, into (4), for V_i, $i \in E_{\rm T}$ into (7), and for I_i, $i \in E_{\rm c}$, into (8), the overall, ODE-based model of the power network is obtained. The explicit form of the overall power network's dynamics, which can be found in [5], eq. (21), can be shown to be written as in eq. (1) through a suitable definition of a state vector x, a total Hamiltonian $\mathcal{H} = \sum_{i \in E_{\rm g} \cup E_{\rm T} \cup E_{\rm c}} \mathcal{H}_i$ and matrices $\mathcal{R}$, $\mathcal{J}$ and $g(x)$. Additional contributions reported in [5] refer to the specialization of the load model (6) to consider linear, resistive loads. The resulting model is further simplified by virtue of the $Dq0$ transformation (see, *e.g.*, [22]) through which the dependence of the inductance matrix in (3) with respect to the rotor angle θ_i can be eliminated. This allows a construction of an equivalent *quotient system* whose steady states are characterized by equilibrium points and not by periodic solutions as for the primal model. An open-loop stability analysis is carried out for the quotient system, which relies on a key result from [23] for the stability analysis of a general port-Hamiltonian around non-zero equilibria. The conclusion drawn from this analysis, which imply explicitly finding a Lyapunov function (see [5], eq. (68)), is that the a single synchronous machine connected to a linear load is stable. However, the results are valid under the assumption that there is no energy dissipation in the stator windings, which should be relaxed for representing synchronous machines more accurately.

This section is concluded by noting that even though control design has not been covered, the description of a physical system using the port-Hamiltonian formalism can be considered as a starting point for the design of nonlinear, robust control algorithms [16]. Indeed, it is now widely recognized that physical relationships that are intrinsic to the system under consideration, *e.g.*, balance and conservation laws, can be exploited in the design of control schemes [18]; see also [24] and [25]. The latter aspect can be instrumental in providing closed-loop stability guarantees, which would be fundamental for the reliable operation of power systems. The interested reader is referred to [26-28] for the design and analysis of different control algorithms based on Lyapunov, passivity, sliding mode and output regulation theory to regulate the frequency in power networks, where each area is modelled by an equivalent generator including turbine-governor dynamics, and where the areas are nonlinearly coupled.

3. – DC microgrids

The concept of *microgrids*, firstly introduced in [10], refers to relatively small-scale (AC or DC) power networks that are mainly composed of distributed generation units (DGUs), storage devices, and loads. These networks can operate autonomously with respect to the main grid, and in many contexts, DC microgrids may be more advantageous than their AC counterpart (see [29]). For example, a large proportion of DC loads in data centers can be directly energized by any readily available DC-based DGU (such as photovoltaic panels), provided that a small-scale distribution network is available. Is it noteworthy that so far DC microgrids have found wide implementation in trains and aircraft and further placement can be expected due to the increasing presence of DGUs based on DC power (more details can be found in [29-38] and references therein).

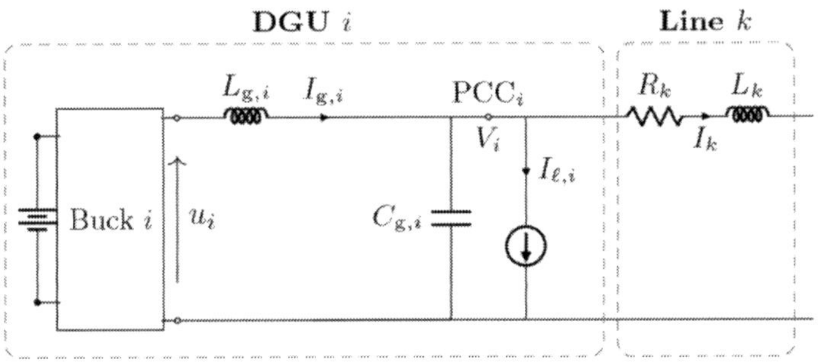

Fig. 3. – Schematic diagram based on [1] of modern electric power distribution systems.

3·1. *Modeling and problem formulation.* – Following [6], consider a DC microgrid composed of n DGUs and m resistive-inductive power lines, which are respectively viewed as the nodes N and the edges E of a connected graph $G = (N, E)$, as done in sect. **2** for AC power networks. As depicted in fig. 3, each DGU's voltage source u_i represents the main source of energy. Part of this energy is supplied through a Buck converter to a local load whose current is denoted by $I_{\ell,i}$ and whose measurement is not available for control design purposes. For any $i \in N$, the dynamic behavior of the current through its inductor $L_{g,i}$ and the voltage across the capacitor $C_{g,i}$ is described by the following equations:

(9a) $$L_{g,i}\dot{I}_{g,i} = -V_i + u_i,$$

(9b) $$C_{g,i}\dot{V}_i = I_{g,i} - I_{\ell,i} - \sum_{k\in\mathcal{E}_i} I_k,$$

where $\mathcal{E}_i \subset E$ denotes the power lines that are incident to $i \in N$ and through which the i-th DGU can exchange power with neighboring DGUs. Also, the current through any line $k \in E$ is represented by I_k. In addition, let L_k and R_k be the inductance and resistance of the power line $k \in E$. Then, the dynamic behavior of I_k is described by

(10) $$L_k\dot{I}_k = (V_i - V_j) - R_k I_k,$$

where $i, j \in N$ are the end nodes of k.

Let $\mathcal{B}$ represent the incidence matrix of G. Then, eqs. (9) and (10) can be gathered and written in vector form as follows:

(11a) $$L_g\dot{I}_g = -V + u,$$

(11b) $$C_g\dot{V} = I_g + \mathcal{B}I - I_\ell,$$

(11c) $$L\dot{I} = -\mathcal{B}^T V - RI,$$

where $L_g = \mathrm{diag}(L_{g,i})_{i\in N}$, $C_g = \mathrm{diag}(L_{g,i})_{i\in N}$ and $L = \mathrm{diag}(L_k)_{k\in E}$ have appropriate size.

Before stating the control objectives of interest, note that since the DC microgrid's graph G is connected, it follows that $\mathbf{1}^\top \mathcal{B} = \mathbf{0}$. One implication of this is that, at the equilibrium, the equality $\mathbf{1}^\top \bar{I}_{\mathrm{g}} = \mathbf{1}^\top I_\ell$ must hold, which indicates that the overall current generated in the microgrid by the DGUs must match the overall current demand from the loads. However, note that there are degrees of freedom on the actual current each DGU can inject. Motivated by this observation, the first control objective is stated next:

Objective 1:

$$\lim_{t\to\infty} I_{\mathrm{g}} = \bar{I}_{\mathrm{g}}, \quad \text{where} \tag{12a}$$

$$w_i \bar{I}_{\mathrm{g},i} = w_j \bar{I}_{\mathrm{g},j}, \quad \forall i, j \in N, \tag{12b}$$

where $w_i > 0$ are weights specified by each DGU. The above objective implies the regulation of the DGU's currents towards a proportional *current sharing* regime defined by the weighting factors w_i. Note that if the value of w_i, $i \in N$, is chosen to be inversely proportional to its respective DGU's capacity, then the DGU's with higher capacities will have higher current values in steady state.

It can be verified from the equilibrium equations associated with (11) that achieving Objective 1 imposes a restriction on the DGUs' steady-state voltage values that are admissible. Nonetheless, it can be shown that, at any equilibrium, the vector of voltages V satisfies $\mathcal{B}^\top \bar{V} = \mathcal{B}^\top(\bar{V} + a\mathbf{1})$, where $a \in \mathbb{R}$ is any scalar. This motivates a second control objective, which pertains the attainment of *average voltage regulation*. More precisely, let $V^\star \in \mathbb{R}^n$ represent a vector of desired DGU voltages. Then, it is of interest to satisfy the following:

Objective 2:

$$\lim_{t\to\infty} \mathbf{1}^\top W^{-1} V = \mathbf{1}^\top W^{-1} V^\star, \tag{13}$$

where $W = \mathrm{diag}(w_i)_{i\in N}$ contains in the main diagonal the same weights used in (12).

3·2. *Control design and simulation results.* – Following [6], in this subsection a distributed, dynamic control law is presented for the input u such that when the system (11) is in closed loop with it Objectives 1 and 2 specified above are met. It is assumed that the measurement of the current $I_{\mathrm{g},i}$ of each DGU $i \in N$ is available for control purposes. Also, it is assumed that there is a connected, undirected —and weighted— communication graph $G^{\mathrm{com}} = (N^{\mathrm{com}}, E^{\mathrm{com}}, \Gamma^{\mathrm{com}})$ involving all DGUs, *i.e.*, $N^{\mathrm{com}} = N$ but it does not have (necessarily) the same topology of the microgrid's graph G. Then, the following controller makes the considered DC microgrid attain Objectives 1 and 2:

$$T_\theta \dot{\theta} = -\mathcal{L}^{\mathrm{com}} W I_{\mathrm{g}}, \tag{14a}$$

$$T_\phi \dot{\phi} = -\phi + I_{\mathrm{g}}, \tag{14b}$$

$$u = -K(I_{\mathrm{g}} - \phi) + W\mathcal{L}^{\mathrm{com}}\theta + V^\star, \tag{14c}$$

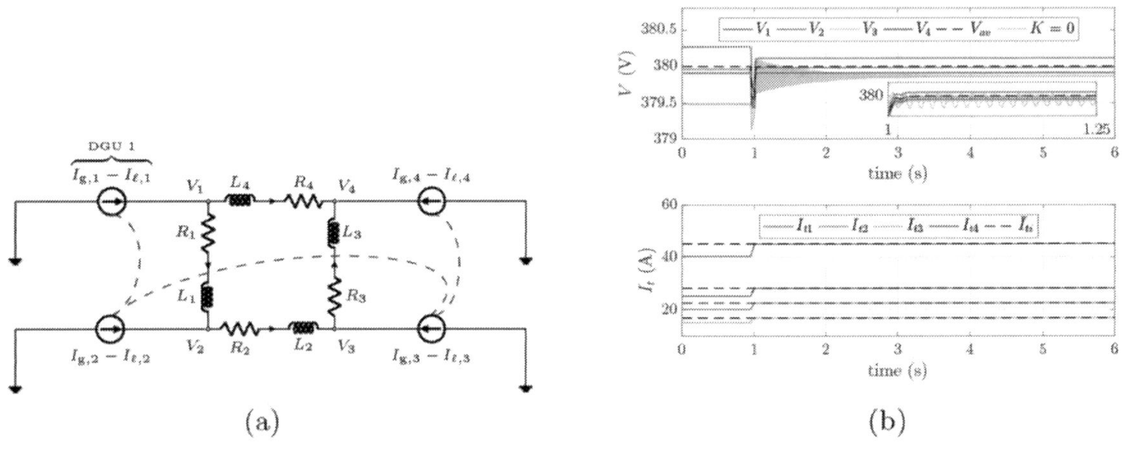

Fig. 4. – (a) Scheme of the considered DC microgrid with 4 power converters. The dashed lines represent the communication network; (b) evolution of the DGU voltages (top) and currents (bottom) under the proposed control scheme. For clarity, the satisfaction of Objectives 1 and 2 is enabled after $t = 1$ s. The gray line represents operation with the gain $K = 0$, which exhibits an oscillatory behavior. Figure courtesy of [6].

where $T_\theta, T_\phi, K \in \mathbb{R}^{n\times n}$ are positive-definite diagonal matrices whose components are control parameters to adjust, *e.g.*, the closed-loop system's transient response. Also, $\mathcal{L}^{\text{com}}$ is the (weighted) Laplacian matrix associated to G^{com}.

Observe that the first subequation of (14) is a consensus dynamics. It is often used in power systems control design to achieve either *fair* power or current sharing among DGUs: note that the equilibrium current $\bar{I}_{\text{g}}$ necessarily satisfies $W\bar{I}_{\text{g}} \propto \mathbf{1}$, which is equivalent to (12b). The introduction of the variable ϕ, which acts as a filter on I_{g} without altering its desired equilibrium value, showed in simulations to be useful to attenuate oscillations. In addition, the proportional term $-K(I_{\text{g}} - \phi)$ in u is fundamental to guarantee convergence of solutions of the closed-loop system towards a *constant* steady state. Finally, the term $W\mathcal{L}^{\text{com}}\theta + V^\star$ stems from the desire to establish a *passive* interconnection with the state θ and simultaneously address Objective 2: in closed loop, the steady-state relation $-\bar{V} + W\mathcal{L}^{\text{com}}\bar{\theta} + V^\star = 0$ holds, which after (pre) multiplication by $\mathbf{1}^\top W^{-1}$ leads to (13). In [6], the steady-state properties discussed above are complemented with a formal stability analysis —based on LaSalle's stability theorem— from which the solutions of the overall closed-loop system are shown to exponential convergence towards a steady-state satisfying (12b) and (13).

The transient performance of the controller (14) was assessed in [6] through simulations about a microgrid with four DGUs and four power lines as depicted in fig. 4(a). The numerical values of the physical parameters can be found in tables II, III of [39] and all control parameters, such as T_θ and T_ϕ, in sect. VI of [6]. For the simulation results displayed in fig. 4(b) the system is initialized at a steady state associated to the loads' currents $I_\ell = [30, 15, 30, 26]^\top$ A and at $t = 1$ s a (step) variation of $\Delta I_\ell = [10, 7, -10, 5]^\top$ A is introduced. It can be appreciated that the system responds to the disturbance by establishing a new steady-state and asymptotically converging to it. Note that, except for the scenario in which $K = 0$ (gray plots), there are no significant oscillations nor overshoots and Objectives 1 and 2 are clearly met.

The following are some extensions to the aforementioned work on DC microgrids. In [40], a slightly modified version of the controller (14), which makes it suitable to handle constant impedance and constant power loads (and not only constant current loads), was satisfactorily validated through experiments on a real-world microgrid with four DGUs connected through a common bus to a single load: both Objectives 1 and 2 were met with adequate transient performances. In [41] the problem of regulating DGU voltages in DC microgrids containing a widely used class of nonlinear loads and subject to time-varying disturbances is considered. The problem was addressed by designing two decentralized, passivity-based control schemes whose conception strongly relied on representing the microgrid's dynamics in port-Hamiltonian form (see sect. **2**) and which probably makes the overall system admit an exponentially stable equilibrium point (with specified voltage values). Finally, in [42], the control strategy developed in [6] is extended to AC microgrids comprising electrical loads with both constant impedance and constant current components. Unlike work on DC networks, the additional objective of decoupling the DGUs' active and reactive powers is achieved, whose satisfaction permits a more straightforward manipulation of the active power injected by each DGU.

4. – Utility optimization

Following [7], consider an AC power distribution network that is composed of n buses and m power lines. As before, the network is represented as a graph $G = (N, E)$, where the nodes N are the buses and the edges E the power lines. Then, the active and reactive power flows in the network are described by the following equations [9, 12]:

$$P_i = \sum_{j\in\mathcal{E}_i} G_{i,j}V_i^2 - G_{i,j}V_iV_j\cos(\theta_{i,j}) - B_{i,j}V_iV_j\sin(\theta_{i,j}), \tag{15a}$$

$$Q_i = \sum_{j\in\mathcal{E}_i} -B_{i,j}V_i^2 + B_{i,j}V_iV_j\cos(\theta_{i,j}) - G_{i,j}V_iV_j\sin(\theta_{i,j}), \tag{15b}$$

where P_i, Q_i, V_i and θ_i represent active power, reactive power, voltage magnitude and phase angle of $i \in N$. Also, $\mathcal{E}_i \subset E$ is the subset of power lines which are incident to $i \in N$. $\theta_{i,j} := \theta_i - \theta_j$ denotes the relative angle between any two $i, j \in N$, and $G_{i,j}$, $B_{i,j}$ are the conductance and susceptance of line $(i, j) \in E$, respectively.

At each $i \in N$ it is assumed there is an end user with the ability to generate, consume and store active power at rates $P_{g,i}$, $P_{\ell,i}$ and $P_{s,i}$, respectively. Such an end user can also generate and consume reactive power at rates $Q_{g,i}$ and $Q_{\ell,i}$, respectively. It follows that P_i and Q_i in (15) can be written as follows:

$$P_i = P_{g,i} - P_{\ell,i} + P_{s,i}, \quad Q_i = Q_{g,i} - Q_{\ell,i}. \tag{16}$$

It is assumed that loads are constant all the time, whereas power generation (active and reactive) will be viewed as decision variables of an optimization problem to be defined. Moreover, it is assumed that the energy stored by the end user evolves in time according to the following discrete model:

$$E_{s,i}(k+1) = E_{s,i}(k) - \eta_{s,i} P_{s,i}, \quad i \in N, \tag{17}$$

where $E_{s,i}(k)$ is the amount of stored energy at instant k and $\eta_{s,i}$ is the efficiency of the storage system: storage units are considered to be charging if $P_{s,i} \geq 0$, and discharging otherwise.

Next, the power losses incurred in during normal operation of the power system are computed. One source of losses is due to dissipation in the power lines, which implies costs to the network operator and whose total value can be computed from (16):

$$P^{r}_{loss} = \sum_{i \in N} P_{g,i} - P_{\ell,i} + P_{s,i}. \tag{18}$$

In addition, the losses associated to the production of electric power at each end user are also considered. Such losses are attributed to the operation of power electronic components used in, for example, photovoltaic inverters in the case of end users with solar panels. The following model is adopted to represent the generation losses:

$$P_{inv,i} = \rho_{inv,i}\left(P^2_{g,i} + Q^2_{g,i}\right), \tag{19}$$

where $\rho_{inv,i}$ is a coefficient which will be assumed to be dependent on the price of the energy (more details ahead). Note that the costs associated with energy dissipation in each storage unit are neglected in this work.

4·1. *Social welfare maximization.* – Having established the model for the power network, the optimization problem of interest is now defined. Let $x = [x_1^\top, x_2^\top, \ldots, x_n^\top]^\top \in \mathbb{R}^{3n}$, with $x_i = [P_{g,i}, P_{s,i}, Q_{g,i}]^\top \in \mathbb{R}^3$ for each $i \in N$. Let $f_0 : \mathbb{R}^{3n} \to \mathbb{R}$ denote the cost due to power line losses, which are covered by the network operator, and let $f_i : \mathbb{R}^n \to \mathbb{R}$ be the profit of end user i for selling electric energy. Consider then the following problem:

$$\max_x \quad -f_0(x) + \sum_{i \in N} f_i(x_i), \tag{20a}$$

$$\text{s.t.} \quad x_i \in X_i. \tag{20b}$$

In particular, f_0 is characterized as $f_0(x) = \gamma^{p} P^{r}_{loss} / |\sum_{i \in N} P_{\ell,i}|$, where γ^{p} is the exogenous price for active power and P^{r}_{loss} is given in (18). On the other hand, $f_i(x_i) = \gamma^{p} P_{g,i} + \gamma^{q} Q_{g,i} + \gamma^{s} P_{s,i} - C_i(P_{g,i}, P_{s,i}, Q_{g,i})$, where γ^{q} and γ^{s} are the exogenous price signals for reactive and stored (active) power. Also, C_i is each end user's cost for producing and storing energy: energy production is directly taken as in (19) and a strictly convex function is considered to represent storage costs. Concerning the constraints X_i in (20), they are defined for each $i \in N$ as

$$X_i = \left\{ x_i : \begin{array}{c} \underline{\mathrm{P}}_{\mathrm{g},i} \le P_{\mathrm{g},i} \le \bar{P}_{\mathrm{g},i},\ \underline{\mathrm{Q}}_{\mathrm{g},i} \le Q_{\mathrm{g},i} \le \bar{Q}_{\mathrm{g},i},\ \underline{\mathrm{P}}_{\mathrm{s},i} \le P_{\mathrm{s},i} \le \bar{P}_{\mathrm{s},i}, \\ \underline{\mathrm{V}}_i \le V_i \le \bar{V}_i,\ \underline{\phi} \le \theta_{i,j} \le \bar{\phi},\ \underline{\mathrm{c}} \le \sum_{i \in N} P_{\mathrm{g},i} + P_{\mathrm{s},i} - P_{\ell,i} \le \bar{c} \end{array} \right\} \tag{21}$$

and restricts the end users' generated active, reactive and stored power to be within predefined intervals. The same applies to the voltage magnitudes and to the relative angle between any two nodes $i, j \in N$. With abuse of notation, the last constraint shown in (21) considers the capacity limits of the grid, which is actually a *coupling* constraint.

Part of the contributions reported in [7] refers to finding a solution to the problem (20). One key step for achieving this involves the linearization of (15) to attain an approximate yet convex problem. Then, an associated Lagrangian dual problem can be defined, together with explicit necessary and sufficient optimality (K.K.T.) conditions [43]. In particular, the dual problem allows for decomposing the (approximate) optimization problem into several sub-problems with a hierarchical feature, leading to a scheme to solve the overall optimization problem through a distributed algorithm (which is highly desirable in large-scale networks). The decomposition is also beneficial to ensure a certain level of privacy since it let end users not share all their data with the network operator. Although all details about the proposed solution algorithm can be found in [7], Prop. 1, it is emphasized that the theoretical developments described above were tested in [7] on a numerical case study based on the IEEE 37 bus distribution system containing photovoltaic power generation, active and reactive power loads and storage systems, as modeled in [44] and references therein. The results depict (see a sample in fig. 5) that giving incentives to end users to adapt their network usage, *e.g.*, by actively using storage units, is effective in maintaining the desired grid operation.

More recently in [37], a novel optimization problem has been formulated which also seeks to maximize a social welfare function (but in DC grids) and integrates prosumers' motives in the utility function; the design of a distributed control scheme with convergence guarantees is also reported therein. Moreover, in [45] second-order behavior and personal norm-based models have been developed, which are consistent with some studies on opinion dynamics for describing and predicting human activities related to the final use of energy in AC grids, where psychological variables, financial incentives, and social interactions are considered.

5. – Distributed optimal control of power systems, market layer

Weather conditions strongly impact the power output of renewable sources, which can lead to large fluctuations in supply causing a mismatch between the forecast (day ahead) and the actual supply and demand of energy. Thus, minimizing the imbalance in supply and demand is of utmost importance in smart grids to avoid network overload and commercial losses. Smart grids utilize flexible appliances that can shift their load to overcome this challenge, thus changing the electricity production or consumption in time.

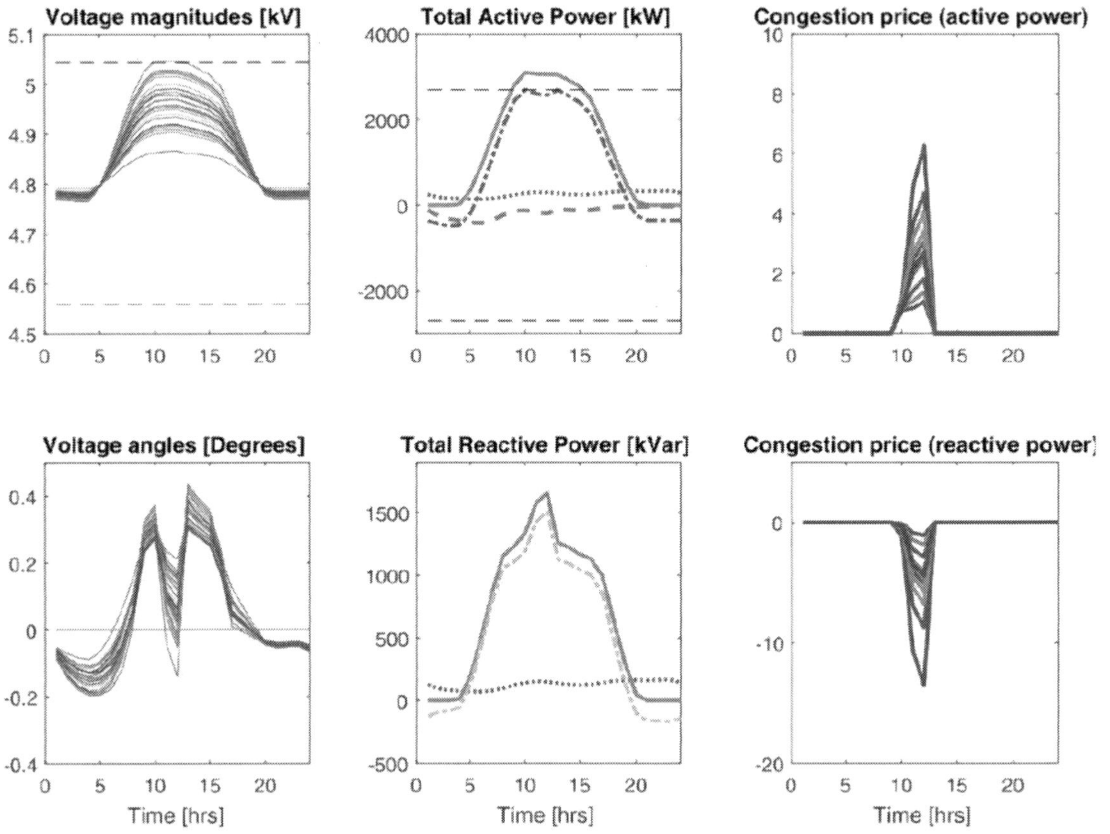

Fig. 5. – Simulation results of the distributed algorithm in [7], Prop. 1, to solve the problem (20) on a tailored IEEE 37 bus benchmark. Note that the net power and voltage magnitudes are within prescribed bounds (in dashed gray). Figure courtesy of [7].

The contribution to the balancing problem from the end-user side is generally referred to as *demand response*.

In this section, following the exposition in [8], we discuss about supply-demand matching in smart grids. Then, an optimal control problem is formulated which seeks to minimize imbalance through demand response regulation and employing distributed model predictive control (MPC) algorithms for its solution. We note that, as discussed in [46], the distributed structure of MPC conforms the decentralised nature of renewable power generation, moreover, it also improves the robustness of the system. The embedding of demand response regulation in the market structure of the Universal Smart Energy Framework (USEF) is also covered, following the ideas of [47].

5·1. *Demand response regulation.* – One way to achieve demand response is to control the flexible devices of the households, both on the production and the consumption side. Here the problem of minimising the overall imbalance (between supply and demand) in a network of households is considered, exploiting flexibility in a cooperative manner.

Let I be the set of prosumers, and let $f_i[k]$ and $g_i[k]$ be the flexible and the fixed load of prosumer $i \in I$, respectively, at discrete time step k. Load is the sum of supply (production) and demand (consumption), with the convention of using positive sign for the supply and negative for the demand. The imbalance of a prosumer, denoted by $\tilde{x}_i[k]$, is the sum of its net flexible and net fixed load,

$$\tilde{x}_i[k] = f_i[k] + g_i[k]. \tag{22}$$

As a consequence, the evolution of imbalance can be described by $\tilde{x}_i[k+1] = \tilde{x}_i[k] + u_i[k] + w_i[k]$, where $u_i[k] = f_i[k+1] - f_i[k]$ and $w_i[k] = g_i[k+1] - g_i[k]$ is the change in the flexible and the fixed load, respectively.

The prosumers are dynamically linked in order to exchange information which can be represented by a weighted, directed graph, in which an existing edge (i, j) means that information is sent from prosumer i to prosumer j. The weight on the edge characterises the importance of the shared information. The system model can be incorporated in the extended dynamics as

$$x_i[k+1] = A_{ii}x_i[k] + \sum_{j \neq i} A_{ij}x_j[k] + u_i[k] + w_i[k], \tag{23}$$

where the tilde notation is omitted as $x_i[k]$ is no longer the physical imbalance but rather information about imbalance. The difference is that the latter is a weighted accumulation of the self-imbalance and the imbalances of the connected neighbours. The total imbalance information should be equal to the total physical imbalance, *i.e.*, $\sum_{i \in I} x_i[k] = \sum_{i \in I} \tilde{x}_i[k]$. This connection structure was proposed in [48-50]. The condition is satisfied if coefficients A_{ij} are elements of a stochastic matrix $\boldsymbol{A}$ with properties: i) $A_{ij} \geq 0$, ii) $A_{ij} = 0$, if no information is sent from agent i to j, iii) $\sum_i A_{ij} = 1$. This matrix corresponds to the graph of the network, which is required to be strongly connected, *i.e.*, there exists a directed path between every pair i and j in both directions.

A quadratic cost function to minimize overall imbalance is defined as $L(x[k], u[k]) = \sum_{i \in I} x_i^2[k]$. Then the optimal control problem is to minimise $\sum_{k \in \mathcal{K}} \sum_{i \in I} x_i^2[k]$, subject to (22), (23), and boundary conditions $u_i \in U_i$, $w_i \in W_i$. This problem is solved using the distributed MPC approach as discussed in [46].

5·2. *Embedding of demand response regulation in the market structure.* – In this subsection, the initiative USEF is described and an explanation about how the demand response regulation can be embedded into this market structure is provided.

5·2.1. USEF [47]. USEF is an initiative by a collective of top sector companies to standardize smart grid solutions for the European energy market. Their aim is to create a platform to drive the fastest, most cost-effective route to an integrated smart energy future. USEF delivers a common standard on which to build all smart energy products and services. It unlocks the value of flexibility by making it a tradable commodity, and delivering a market structure, associated rules, and tools to make it work effectively. Flexibility can be invoked for grid capacity management to avoid or reduce peak loads, and allows for active balancing through optimization between supply and demand. The framework is designed to offer fair market access and benefits to the stakeholders, and is accessible to anyone internationally.

The following USEF stakeholders are considered: the Balance Responsible Parties (BRP), aggregators, prosumers, and the Distribution System Operator (DSO). Electricity is traded between the suppliers and the BRPs over the wholesale energy market (day ahead) and imbalance market (operation time). The BRPs dispatch the electricity to the aggregators, which in turn deliver to the prosumers. The aggregator is a new stakeholder in energy grids that groups the prosumers into clusters. Its responsibility is to accumulate and offer flexibility on behalf of the connected prosumers, with the aim of maximizing the value of flexibility. The DSO is responsible for the distribution of power and to resolve any disturbances that might interfere with that task. In this context, the main task of the DSO is to detect and resolve any congestion that might occur in the distribution lines. USEF employs a market-based control mechanism, which consists of five phases: contract, plan, validate, operate, and settlement. Contractual agreements between the stakeholders are established in the first phase. In the plan phase, a day-ahead forecast of the energy consumption is made, which is then validated by the DSO in the validate phase. The two phases are iterated until an agreement is reached on the forecast. In the operate phase, the system aims to follow the plan that has been created in the first two phases, and balances between the forecast and actual electricity load by procuring flexibility. Financial reconciliation is completed in the settlement phase. An overview of the USEF structure and market-based control mechanism is shown in fig. 6.

5˙2.2. Demand response in USEF. The distributed MPC method is embedded in the operation phase of USEF, assuming that the energy portfolio has already been forecasted and agreed on by the aggregators, BRP, and DSO. Each prosumer is equipped with one appliance, either a heat pump (representing consumption) or a μCHP (representing production). The algorithm is implemented on both the prosumer and aggregator levels, with the aggregators accumulating the imbalance and the flexible and fixed loads of their prosumers. The objective of minimising the total imbalance is now extended to the aggregators, hence the optimal control problem is formulated as minimization of $\sum_{k\in\mathcal{K}}\sum_{l\in\mathcal{L}}\sum_{i\in I} x_i^2[k]$ subject to

$$x_i[k+1] = A_{ii}x_i[k] + \sum_{j\neq i} A_{ij}x_j[k] + u_i[k] + w_i[k] - \Delta\text{goal}_i[k],$$

and other constraints from sect. **5**˙1, where $\mathcal{L}$ is the set of aggregators. The goal function is introduced to couple the two levels and intuitively reflects the flexibility invoked by the aggregators from their prosumers (*c.f.* [47]). This problem is solved using the distributed MPC approach as discussed in [46].

Distributed MPC and dualization for embedding hydrogen in the production and storage side of the power and gas grid resulting in two level optimization problems can be found in [51, 52].

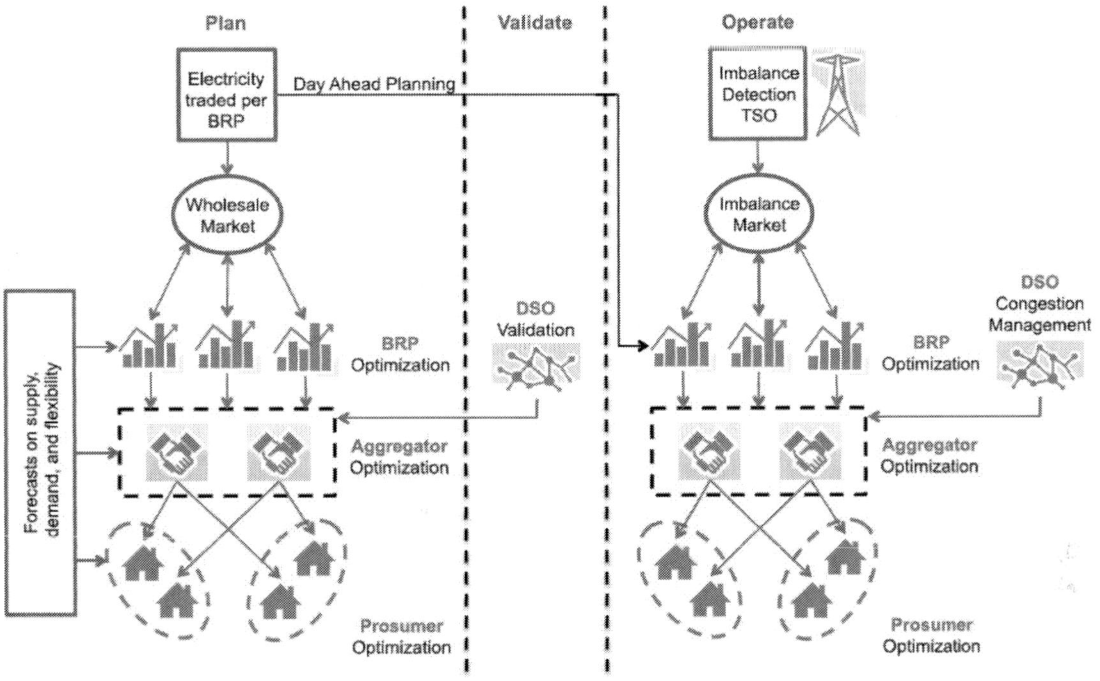

Fig. 6. – Hierarchical organization of USEF, and main interactions between its stakeholders. Figure courtesy of [47].

6. – Concluding remarks

Energy systems worldwide are gradually becoming more sustainable through innovations such as implementing demand response strategies, developments in materials aimed at improving energy efficiency, and a significant increase of renewable sources in the overall energy mix. Adversely, the unpredictable and highly varying nature of renewable sources brings about several challenges for the robust and optimal operation of energy systems with a high penetration of these sources, making, in many cases, the control schemes intended for conventional power systems inadequate.

In order to address these challenges, new modeling and control techniques have been developed in recent years. A notable example is the framework of port-Hamiltonian systems, which allows for seamlessly integrating models of diverse physical phenomena, of multiple domains, (*e.g.*, mechanical, electrical), and with various time and space scales into a common energy-based mathematical representation. Such traits have been shown in the literature to be particularly relevant in designing robust control algorithms that exploit intrinsic physical laws to achieve stable closed-loop system operations. In these lecture notes, we described several seminal contributions in these regards, covering modeling aspects of AC power systems using a port-Hamiltonian formulation, the design of distributed and robust controllers for modern DC microgrids, as well as the design of control frameworks that take into consideration economic aspects for optimally managing energy assets in smart grids, such as storage devices and controllable loads, while maintaining physically sensible values for signals of interest such as electric frequencies and voltage magnitudes.

Although substantial advances have been made towards the realization of smart grids, many open problems remain, among which we can enlist two we believe to be of primal importance. One of them pertains to the co-optimization and control of hybrid AC and DC networks that appear in, *e.g.*, smart parking lots that allocate electric vehicles, which can have significant benefits in terms of providing flexibility to the main grid (see, *e.g.*, [53]). A second open problem corresponds to the joint design and operation of the three main energy carriers —namely, electricity, gas, and heat— to exploit the influence these exert on each other to obtain socioeconomic, technical, and environmental benefits stemming from the potential flexibility to incorporate renewable sources further (see [54]) and storage devices in each of these domains; the reader is referred to [55-59, 19] for comprehensive coverage on modeling and control of heating networks and to [20] and references therein for gas networks.

REFERENCES

[1] MANCARELLA P., "Mes (multi-energy systems): An overview of concepts and evaluation models", *Energy*, **65** (2014) 1.

[2] KROPOSKI B. D., DALL-ANESE E., BERNSTEIN A., ZHANG Y. and HODGE B. S., *Autonomous energy grids*, technical report (National Renewable Energy Lab. (NREL), Golden, CO (USA)) 2017.

[3] FANG X., MISRA S., XUE G. and YANG D., "Smart grid—the new and improved power grid: A survey", *IEEE Commun. Survey Tutor.*, **14** (2011) 944.

[4] KROPOSKI B. D., BERNSTEIN A., KING J., VAIDHYNATHAN D., ZHOU X., CHANG C.-Y. and DALL'ANESE E., "Autonomous energy grids: Controlling the future grid with large amounts of distributed energy resources", *IEEE Power Energy Mag.*, **18** (2020) 37.

[5] FIAZ S., ZONETTI D., ORTEGA R., SCHERPEN J. M. A. and VAN DER SCHAFT A. J., "A port-Hamiltonian approach to power network modeling and analysis", *Eur. J. Control*, **19** (2013) 477.

[6] TRIP S., CUCUZZELLA M., CHENG X. and SCHERPEN J. M. A., "Distributed averaging control for voltage regulation and current sharing in DC microgrids", *IEEE Control Syst. Lett.*, **3** (2018) 174.

[7] JAFARIAN M., SCHERPEN J. M. A., LOEFF K., MULDER M. and AIELLO M., "A combined nodal and uniform pricing mechanism for congestion management in distribution power networks", *Electr. Power Syst. Res.*, **180** (2020) 106088.

[8] NGUYEN D. B., ALKANO D. and SCHERPEN J. M. A., "The optimal control problem in smart energy grids", in *Smart Grids from a Global Perspective* (Springer, Cham, Switzerland) 2016, pp. 95–111.

[9] KUNDUR P., *Power System Stability and Control*, Vol. **10** (McGraw Hill, New York, USA) 1994.

[10] LASSETER R. H., "Microgrids", *IEEE Power Engineering Society Winter Meeting, Conf. Proc.*, **1** (2002) 305.

[11] SCHIFFER J., *Stability and Power Sharing in Microgrids*, PhD Thesis, Technical University of Berlin (2015).

[12] MACHOWSKI J., BIALEK J. W. and BUMBY J. R., *Power System Dynamics: Stability and Control*, 2nd edition (Wiley, UK) 2008.

[13] MANCARELLA P., ANDERSSON G., PEÇAS-LOPES J. A. and BELL K. R. W., "Modelling of integrated multi-energy systems: Drivers, requirements, and opportunities", in *IEEE Power Systems Computation Conference (PSCC)* (IEEE) 2016, pp. 1–22.

[14] European Distribution System Operators (EDSO), `https://www.edsoforsmartgrids.eu/home/why-smart-grids/`, accessed: 2022-04-25.

[15] GENG S., VRAKOPOULOU M. and HISKENS I. A., "Optimal capacity design and operation of energy hub systems", *Proc. IEEE*, **108** (2020) 1475.

[16] VAN DER SCHAFT A. J., "Port-Hamiltonian modeling for control", *Annu. Rev. Control Robot. Autom. Syst.*, **3** (2020) 393.

[17] DUINDAM V., MACCHELLI A., STRAMIGIOLI S. and BRUYNINCKX H., *Modeling and Control of Complex Physical Systems: The Port-Hamiltonian Approach* (Springer-Verlag, Berlin/Heidelber, Germany) 2009.

[18] VAN DER SCHAFT A. J. and JELTSEMA D., *Port-Hamiltonian systems theory: An introductory overview* (Now Publishers Inc., USA) 2014.

[19] HAUSCHILD S.-A., MARHEINEKE N., MEHRMANN V., MOHRING J., BADLYAN A. M., REIN M. and SCHMIDT M., "Port-Hamiltonian modeling of district heating networks", in *Progress in Differential-Algebraic Equations II* (Springer, Cham, Switzerland) 2020, pp. 333–355.

[20] DOMSCHKE P., HILLER B., MEHRMANN V., MORANDIN R. and TISCHENDORF C., "Gas network modeling: An overview", preprint Collaborative Research Center TRR 154, 2021.

[21] VAN DER SCHAFT A. J., "Characterization and partial synthesis of the behavior of resistive circuits at their terminals", *Syst. Control Lett.*, **59** (2010) 423.

[22] SCHIFFER J., ZONETTI J., ORTEGA R., STANKOVIĆ A. M., SEIZI T. and RAISCH J., "A survey on modeling of microgrids–From fundamental physics to phasors and voltage sources", *Automatica*, **74** (2016) 135.

[23] MASCHKE B., ORTEGA R. and VAN DER SCHAFT A. J., "Energy-based Lyapunov functions for forced Hamiltonian systems with dissipation", *IEEE Trans. Autom. Control*, **45** (2000) 1498.

[24] SLOTINE J.-J., "Putting physics in control-the example of robotics", *IEEE Control Syst. Mag.*, **8** (1988) 12.

[25] ORTEGA R., VAN DER SCHAFT A. J., MAREELS I. and MASCHKE B., "Putting energy back in control", *IEEE Control Syst.*, **21** (2001) 18.

[26] TRIP S., CUCUZZELLA M., DE PERSIS C., VAN DER SCHAFT A. J. and FERRARA A., "Passivity-based design of sliding modes for optimal load frequency control", *IEEE Trans. Control Syst. Technol.*, **27** (2019) 1893.

[27] TRIP S., CUCUZZELLA M., DE PERSIS C., FERRARA A. and SCHERPEN J. M. A., "Robust load frequency control of nonlinear power networks", *Int. J. Control*, **93** (2020) 346.

[28] SILANI A., CUCUZZELLA M., SCHERPEN J. M. A. and YAZDANPANAH M. J., "Output regulation for load frequency control", *IEEE Trans. Control Syst. Technol.*, **30** (2022) 1130.

[29] JUSTO J., MWASILU F., LEE J. and JUNG J.-W., "AC-microgrids versus DC-microgrids with distributed energy resources: A review", *Renew. Sust. Energ. Rev.*, **24** (2013) 387.

[30] ZHAO J. and DÖRFLER F., "Distributed control and optimization in DC microgrids", *Automatica*, **61** (2015) 18.

[31] DE PERSIS C., WEITENBERG E. A. and DÖRFLER F., "A power consensus algorithm for DC microgrids", *Automatica*, **89** (2018) 364.

[32] GAO F., KANG R., CAO J. and YANG T., "Primary and secondary control in DC microgrids: a review", *J. Mod. Power Syst. Clean Energy*, **7** (2019) 227.

[33] CUCUZZELLA M., LAZZARI R., KAWANO Y., KOSARAJU K. C. and SCHERPEN J. M. A., "Robust Passivity-Based Control of Boost Converters in DC Microgrids", in *Proceedings of the 58th IEEE Conference on Decision Control (CDC) Nice, France, 2019* (IEEE) 2019, pp. 8435–8440.

[34] MACHADO J. E. and SCHIFFER J., "A passivity-inspired design of power-voltage droop controllers for dc microgrids with electrical network dynamics", in *Proceedings of the 59th IEEE Conference on Decision Control (CDC), Jeju, South Korea, 2020* (IEEE) 2020, pp. 3060–3065.

[35] KOSARAJU K. C., CUCUZZELLA M., SCHERPEN J. M. A. and PASUMARTHY R., "Differentiation and Passivity for Control of Brayton–Moser Systems", *IEEE Trans. Autom. Control*, **66** (2021) 1087.

[36] SILANI A., CUCUZZELLA M., SCHERPEN J. M. A. and YAZDANPANAH M. J., "Output Regulation for Voltage Control in DC networks with Time-Varying Loads", *IEEE Control Syst. Lett.*, **5** (2021) 797.

[37] CUCUZZELLA M., BOUMAN T., KOSARAJU K. C., SCHUITEMA G., LEMMEN N. H., JOHNSON ZAWADZKI S., FISCHIONE C., STEG L. and SCHERPEN J. M. A., "Distributed control of DC grids: integrating prosumers motives", *IEEE Trans. Power Syst.*, **37** (2021) 3299.

[38] SILANI A., CUCUZZELLA M., SCHERPEN J. M. A. and YAZDANPANAH M. J., "Robust output regulation for voltage control in dc networks with time-varying loads", *Automatica*, **135** (2022) 109997.

[39] CUCUZZELLA M., TRIP S., DE PERSIS C., CHENG X., FERRARA A. and VAN DER SCHAFT A. J., "A robust consensus algorithm for current sharing and voltage regulation in DC microgrids", *IEEE Trans. Control Syst. Technol.*, **27** (2018) 1583.

[40] TRIP S., HAN R., CUCUZZELLA M., CHENG X., SCHERPEN J. M. A. and GUERRERO J., "Distributed averaging control for voltage regulation and current sharing in dc microgrids: Modelling and experimental validation", *IFAC-PapersOnLine*, **51** (2018) 242.

[41] FERGUSON J., CUCUZZELLA M. and SCHERPEN J. M. A., "Exponential stability and local ISS for DC networks", *IEEE Control Syst. Lett.*, **5** (2020) 893.

[42] CUCUZZELLA M., TRIP S., FERRARA A. and SCHERPEN J. M. A., "Cooperative voltage control in AC microgrids", in *Proceedings of the 57th IEEE Conference on Decision Control (CDC) Miami, United States 2018* (IEEE) 2018, pp. 6723–6728.

[43] BOYD S. and VANDENBERGHE L., *Convex Optimization* (Cambridge University Press, United Kingdom) 2004.

[44] LOEFF K., JAFARIAN M. and SCHERPEN J. M. A., "Modeling of power distribution systems with solar generation: A case study", arXiv preprint, arXiv:1702.06113, 2017.

[45] FENG S., CUCUZZELLA M., BOUMAN T., STEG L. and SCHERPEN J. M. A., "An integrated human-physical framework for control of power grids", preprint, arxiv:2012.11208, 2020.

[46] GISELSSON P. and RANTZER A., "On feasibility, stability and performance in distributed model predictive control", *IEEE Trans. Autom. Control*, **59** (2013) 1031.

[47] NGUYEN D. B., SCHERPEN J. M. A. and BLIEK F., "Distributed optimal control of smart electricity grids with congestion management", *IEEE Trans. Autom. Sci. Eng.*, **14** (2017) 494.

[48] LARSEN G. K. H., VAN FOREEST N. D. and SCHERPEN J. M. A., "Power supply–demand balance in a smart grid: An information sharing model for a market mechanism", *Appl. Math. Model.*, **38** (2014) 3350.

[49] LARSEN G. K. H., VAN FOREEST N. D. and SCHERPEN J. M. A., "Distributed control of the power-supply balance", *IEEE Trans. Smart Grid*, **4** (2013) 828.

[50] Larsen G. K. H., van Foreest N. D. and Scherpen J. M. A., "Distributed MPC applied to a network of households with micro-CHP and heat storage", *IEEE Trans. Smart Grid*, **5** (2014) 2106.

[51] Alkano D. and Scherpen J. M. A., "Distributed supply coordination for power-to-gas facilities embedded in energy grids", *IEEE Trans. Smart Grid*, **9** (2016) 1012.

[52] Alkano D., Scherpen J. M. A. and Chorfi Y., "Asynchronous distributed control of biogas supply and multi-energy Demand", *IEEE Trans. Autom. Sci. Eng.*, **14** (2017) 558.

[53] Cenedese C., Fabiani F., Cucuzzella M., Scherpen J. M. A., Cao M. and Grammatico S., "Charging plug-in electric vehicles as a mixed-integer aggregative game", in *Proceedings of the 58th IEEE Conference on Decision Control (CDC), Nice, France, 2019* (IEEE) 2019, pp. 4904–4909.

[54] Dall'Anese E., Mancarella P. and Monti A., "Unlocking flexibility: Integrated optimization and control of multienergy systems", *IEEE Power Energy Mag.*, **15** (2017) 43.

[55] De Persis Claudio and Kallesoe Carsten Skovmose, "Pressure regulation in nonlinear hydraulic networks by positive and quantized controls", *IEEE Trans. Control Syst. Technol.*, **19** (2011) 1371.

[56] Scholten T., De Persis C. and Tesi P., "Modeling and control of heat networks with storage: the single-producer multiple-consumer case", *IEEE Trans. Control Syst. Technol.*, **25** (2016) 414.

[57] Strehle F., Vieth J., Pfeifer M. and Hohmann S., "Passivity-based stability analysis of hydraulic equilibria in 4th generation district heating networks", *IFAC-PapersOnLine*, **54** (2021) 261.

[58] Machado J. E., Cucuzzella M., Pronk N. and Scherpen J. M. A., "Adaptive control for flow and volume regulation in multi-producer district heating systems", *IEEE Control Syst. Lett.*, **6** (2021) 794.

[59] Machado J. E., Cucuzzella M. and Scherpen J. M. A., "Modeling and passivity properties of multi-producer district heating systems", *Automatica*, **142** (2022) 110397.

EPJ Web of Conferences **268**, 00017 (2022) https://doi.org/10.1051/epjconf/202226800017
LNES 2021

Conclusions — with personal reflections

Friedrich Wagner(*)

Max-Planck-Institute for Plasmaphysics - Garching, Greifswald, Germany

Summary. — Global warming necessitates a replacement of traditional technologies based on fossil fuels. The number of options is rather limited. Renewable energies will play a dominant role in this process. Whereas biomass and hydroelectricity are limited, wind and photovoltaic power (PV) can be upscaled to large capacities. The technical side of CO_2-free technologies has been covered in the bulk of these proceedings. In this paper, we concentrate on the main characteristics of an electricity system based mostly on wind and PV power and the consequences of integrating intermittent technologies. Most practical examples are taken from Germany. The question whether nuclear power can be considered as clean electricity source supplementing renewable energies will be discussed. Nuclear power will still play a role in 2050 and beyond in and outside of Europe.

1. – Introduction

Climate threats enforce a replacement of traditional energy producing and consuming technologies. The Earth power balance changes predominantly because of the excessive use of fossil fuels by human beings. Therefore, decarbonisation of the economy will become the global mission of this century. But we do not know how 11 bn people on Earth

(*) E-mail: fritz.wagner@ipp.mpg.de

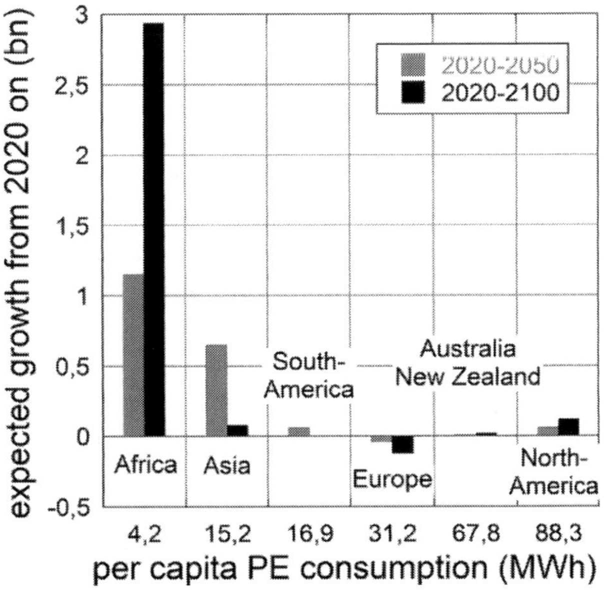

Fig. 1. – Expected population growth in different areas in two time periods —2020 to 2050 and 2020 to 2100 [1]— *vs. per capita* primary energy (PE) consumption of 2019 [2].

—expected by the United Nations for 2100— will organise themselves and co-exist. The primary problem is therefore the large population and its further growth. The energy problem is a direct corollary of overpopulation and therefore difficult to solve.

Figure 1 shows the expected growth of population in different areas of the world starting with 2020 *vs.* the present *per capita* primary energy consumption. Two cases are considered: in grey the expected growth up to 2050 and in black up to 2100. The continent whose population is expected to grow most is Africa, the one with the lowest *per capita* energy consumption. About 600 mill people in Africa do not have access to electricity now. As a contribution to fight global warming, energy saving cannot be expected from the African generations to come. The only serious options for Africa are clean energy supply technologies allowing an energy price, which moves fossil fuels from the market. The black continent needs urgently a green deal.

2. – Electricity as primary energy

There are only a few clean energy options available: to continue the use of fossil fuels on the basis of CCS technology (CCS stands for carbon capture and sequestration); fission with fast neutrons and breeder technology in the frame of the Generation-IV reactor concepts; nuclear fusion, and finally the different forms of renewable energies —biomass, geothermal power, hydro-electricity, maritime power (wave/tide/ocean currents), environmental heat in conjunction with heat pumps, onshore, offshore wind, and solar energies in the form of photovoltaic (PV) and solar thermal power([1]).

CCS is not supported in many countries and would not replace the need for transformation rather extend the time frame of it. Fission is often totally rejected even in countries with a long and successful history in science and technology and the demonstration of safe and efficient reactor operation for decades as it has been the case for Germany. Fusion will not contribute with a relevant share before the end of this century. Under these circumstances, only renewable energy forms remain. Therefore, their potential has to be discussed in detail. *E.g.*, the use of biomass is in conflict with nutrition and excessive production causes its own environmental problems. Hydro-, wave-, and geothermal power are limited to specific geographical conditions. Only wind and solar power can be scaled to large capacities and will, therefore, play a crucial role in the energy transformation process.

We conclude with the first major feature of the energy transition: electricity being produced largely by wind and PV power becomes the primary energy, serving all energy consuming sectors; chemical and thermal energies will be generated —not exclusively but mostly— from electricity.

3. – The epoch of the minerals

The second feature of the energy transition reflects the material needs of renewable energy systems (RES). The epoch of fossil fuels will be replaced by the one of the minerals. The consequence is that also the use of RES does not provide supply independence rather new dependences emerge. Figure 2 is taken from a very detailed recent report from IEA [3]. The critical minerals are listed here from copper to lithium to cobalt to the rare-earth minerals and others. Figure 2 shows the mineral use in the form of kg pro MW in case of power generation (fig. 2(a)) and demonstrates that renewable technologies need more of these minerals than conventional ones do. The case of transportation (fig. 2(b)) shows the demand in form of kg pro vehicle, again with a striking difference between electric and conventional cars. The large needs of copper (and aluminium) in electricity generation by wind and PV power are obvious because of low power density giving rise to distributed and complex networks. Zinc is needed to prevent corrosion as renewable energy technologies are located outside being exposed to the weather. Manganese is needed in steel production. The steel intensity (tons steel/MW [4]) of wind power is about a factor of four larger than that of nuclear power. Lithium, graphite and cobalt are used in Li-ion batteries, silicon in PV-panels. Rare-Earth minerals are needed for permanent magnets used in wind turbines (specifically in gearless systems) and electric cars.

The additional problem is that these minerals are available from a few countries only, less than in case of gas and oil, not to speak about abundant coal. The largest lithium

([1]) Technical issues are presented and discussed in the proceedings of the *Joint EPS-SIF International School on Energy - Energy Innovation and Integration for a Clean Environment* (`https://en.sif.it/courses/energy_school`) and in the proceedings of previous schools and will not be discussed in this paper.

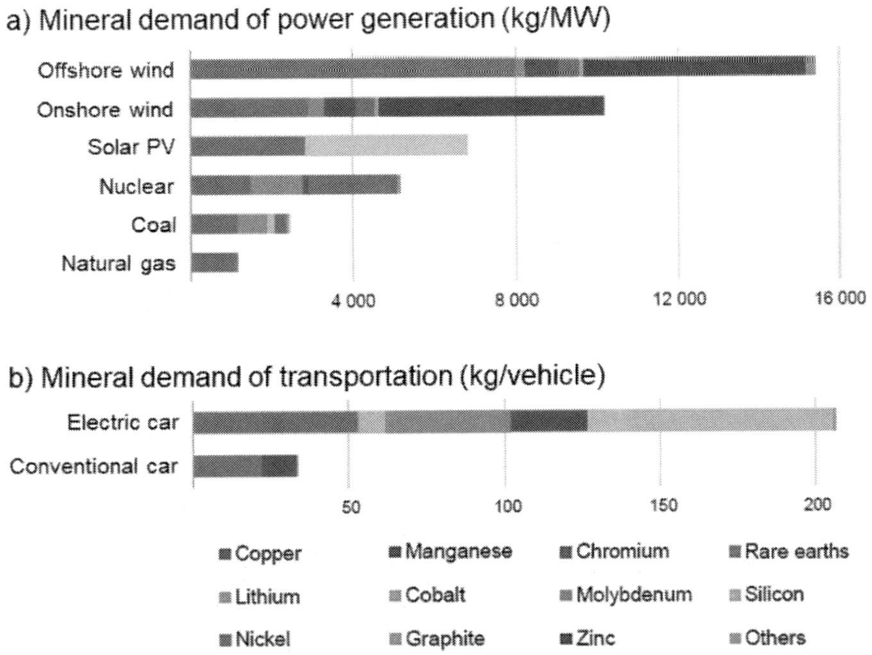

Fig. 2. – Mineral demand (a) of thermal and renewable power generation, and (b) of conventional and of electrical cars; the graph is reproduced from an IEA report [3], p. 26.

deposits are in Chile, followed by Australia, Argentina, China and USA [5]. Ganfeng lithium from China is the largest lithium miner in the world [6]. Tianqi Lithium of China controls nearly half of the world's production of lithium [3]. In addition, the processing technologies for these minerals are restricted to even less countries. Nearly 90% of the rare-earth minerals are said to be processed in China [7]. An additional problem is that the quality of mined minerals continuously degrades so that more energy and water (and money) is necessary for extracting a unit. On the other hand, mining of critical minerals has to be accelerated to meet the generally planned expansion of wind and PV power and of electric cars. *E.g.*, the demand of lithium for batteries is expected to grow by a factor of 40 between 2020 and 2040 [3]. Therefore, the energy transition needs to be backed by another transition in the way rare materials (and other commodities) are handled. Recycling will be a topic which will follow the process of decarbonisation and it is a wide field for research. Also, the notion of "sustainability" has to be expanded and applied to all technologies and used raw materials in a decarbonised economy. In a strict sense, only technologies whose realisation and operation are based on circular processes can be considered sustainable.

4. – Intermittent renewable energy systems

The work horses of the energy transition toward a comprehensive decarbonisation are wind and PV power. They are supplemented by sources with limited potential like

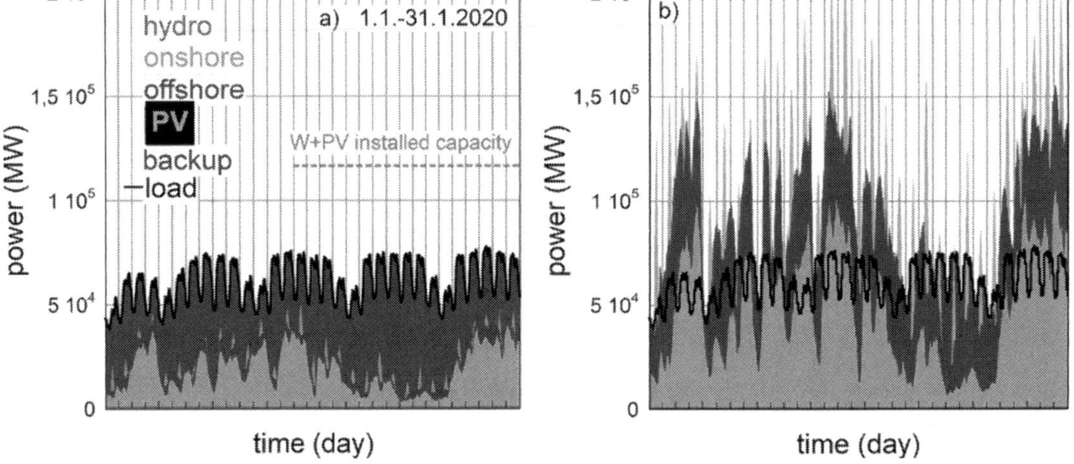

Fig. 3. – Power of renewable energy (RES) and thermal backup systems (a) for January 2020, and (b) up scaled to the production in 2045 according to the installations as suggested by "AGORA Energiewende" [9] with the following specification: onshore = 145 GW, offshore = 70 GW and photovoltaic power = 385 GW. The black curve represents the load of 2020 which has been left unchanged up to 2045 (alternative loads are discussed later in the text). For 2020, the installed intermittent renewable (iRE) power of 116 GW is indicated as dashed red line in fig. 3(a); for 2045, it is expected to be 600 GW [10].

hydro-electricity or energy forms partly needed for other sectors like biomass for aviation or heavy-duty mobility. The mix of renewable energies will vary strongly from country to country reflecting the prevailing natural conditions. Electricity from wind and PV installations will not only serve the electricity use proper but also contribute to meet the energy needs in other energy sectors like low temperature heat for buildings or high-temperature process heat for industry. Extended experience with the transformation exists already in case of electric cars introducing electricity into individual mobility. Because of the dominant role of electricity and the importance of wind and PV power in a future generation mix, it is necessary to discuss with some detail the characteristics of a supply system based predominantly on these two technologies.

4·1. *Consequences of intermittency* . – Wind and PV power have two characteristic features: variable power generation depending on the local weather conditions and low power density. First, we discuss the intermittent nature of wind and PV power [8]. In case of PV, there are short term —day/night— and seasonal variations. Figure 3(a) shows an example from January 2020 of Germany. Most examples in this paper are taken from Germany, which can be seen as a laboratory for the use of renewable energies in a still highly industrialised environment([2]). To a large extent, the German experience

([2]) Most data in this paper are obtained from the data banks of IEA (`https://www.iea.org/data-and-statistics`) and Entso-e (`https://www.entsoe.eu/data/`).

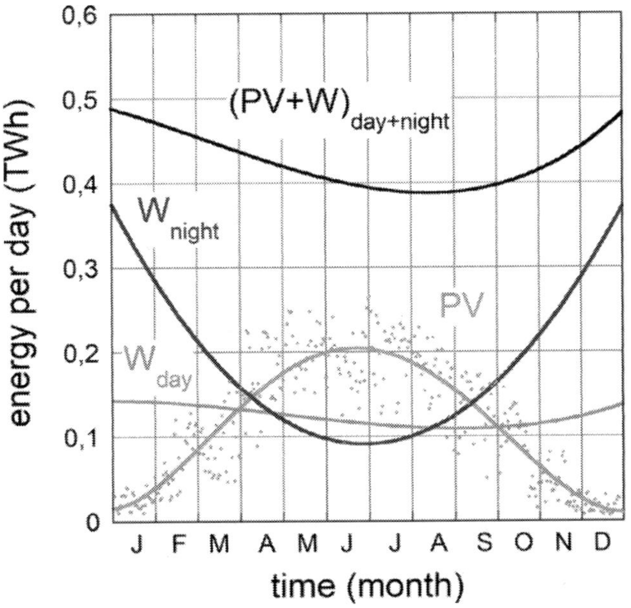

Fig. 4. – Daily energy generation by wind and PV power through the year 2018. For PV, the actual data points are shown along with a polynomial fit to guide the eye. For wind power and the sum of wind and PV, only the fits are given to simplify the diagram. For wind, day and night generations are separately shown. Definition: a day starts or ends when PV has reached 5% of peak power of this day (see main text).

can be transferred to other EU countries. The black curve in fig. 3(a) represents the load and it shows the day-to-day variation and the drop in consumption during the weekends. Plotted are also wind and PV power, with PV represented by the small orange needles sitting on top of blue onshore and offshore wind traces. PV is not very effective in winter. Wind and PV power capacities add up to 116 GW (end of 2020), shown as horizontal dotted line in fig. 3(a) being much above the maximal load of $\sim$ 80 GW. In spite of the high installed capacity, wind and PV power cover only a fraction of the demand in January 2020. The gap is filled by backup power (brown) at present mostly by gas, coal and nuclear power.

In case of intermittent generation, there is electricity with two qualities: from wind and PV one has to take what one gets whereas the controlled backup electricity is dispatchable power which is added to exactly meet the load for every moment. The intermittent production necessitates a backup system, which —at a later stage— can be replaced by storage, producing clean secondary electricity. Storage will be discussed in sect. **5**˙3.

Figure 4 shows the daily electricity from wind and PV generation through the year 2018 highlighting the seasonal variation of renewable energy production. In case of PV, the daily energy values are plotted together with a polynomial fit to the data to guide the eye. In the other cases, only the fits are shown for clarity. The main purpose of this diagram is to demonstrate the availability of wind power during and outside PV production. Therefore, the night is defined here when PV power is below 5% of the peak PV power of that day. The drop of PV energy toward the winter months is, of course, the consequence of lower radiation levels and shorter days (and possible snow coverage of PV panels).

PV electricity in January and December 2018 is about 11% of the one in July and August. In these periods, electricity supply from iRE is dominated by wind power. Average wind generation during the day is rather constant throughout the year (with the definition of day and night as used here) whereas wind during the night shows a strong increase toward the winter months. In spite of strong PV generation in summer, the total iRE production maximises in winter (black curve in fig. 4). This seasonal behaviour agrees with the seasonal variation of the demand in most European countries at least as long as global warming does not enforce widespread use of air conditioning in summer. Average electricity generation is larger during the day in summer and during the night in winter. It can be expected that the resulting electricity price variation will have a strong impact on user response (*e.g.*, loading car batteries in summer favourably during the day and in winter during the night).

In the transition period of the energy transition, the continuously operating RES —hydropower and biomass— are available along with growing PV and wind power supplemented by a backup system. It is generally expected that backup will finally be replaced by storage. The grid has to accommodate systems delivering several 100 GW power, feeding in at different voltage levels, being partly decentralised like onshore wind and PV power but possibly also with strongly centralised components like offshore wind. The save distribution of power from sources to sinks will be a major technical challenge. The complexity of such a supply system is further increased because PV power at a level of several 100 GW (*e.g.*, in case of Germany) leads to dramatic dynamics during sunrise and a few hours later during sunset and has the additional feature to basically drop out in the winter months. The dominant use of iRE necessitates therefore three critical components: wind and PV installations, a powerful and "smart" grid providing all necessary system services and a backup/storage system to generate secondary electricity. The visions of a distributed power system with prosumers adopting simultaneously the role of sources and sinks and which are seen as major market participants seems to be an idealistic view of the future supply reality.

4˙2. *Low power density of iRE and national generation potentials.* – In this section, we will address the issue of low power density of iRE which will lead us to address national installation potentials. The identification of the actual potential for wind and PV installations is not a scientific rather a political issue. Non-scientific considerations play a decisive role like the conflict between wind farms and the protection of nature and habitats or the one between PV parks and agriculture. For the purpose of this paper, we select the proposed capacities AGORA has defined for Germany for 2045, when Germany is supposed to reach the net-zero CO_2 emission status. "AGORA Energiewende" [9] is a so-called think tank dealing mostly with the "Energiewende" of Germany. The AGORA paper from 2021 [10] is one of the most recent publications on this topic. AGORA suggests 145 GW onshore, 70 GW offshore, and 385 GW PV power, in total 600 GW for Germany. Up to now, the maximal load is about 80 GW; at the end of 2021, 122 GW wind and PV power have been installed.

In fig. 3(b), the January 2020 data shown in fig. 3(a) are extrapolated to 2045 on the basis of AGORA's installation figures. The reference load profile assumed in this graph is the one of 2020, which is unrealistic for 2045 but will not change the conclusions drawn in his paper. Even at 600 GW, there is the need of backup at substantial power. In spite of 385 GW installed power, the PV contribution is still very small in winter. In addition, substantial surplus is produced, which will be conceptionally used to fill storage and replace the backup and/or to produce, *e.g.*, hydrogen to serve other energy sectors. Therefore, the energy transition falls into two periods: 1) the build-up period where backup is needed —preferably in a technology with low or no CO_2 emissions([3])— and 2) the final state of net zero-CO_2 emissions using clean surplus and secondary electricity from clean storage technologies.

4·3. *Full-load-hours and capacity factors.* – On the basis of well-defined national potentials of wind and PV powers, the annual electricity generation E can be calculated using full-load-hour (flh) or capacity factor (C) values:

$$\begin{aligned} E &= flh^{\mathrm{W_{on}}} \times P_{\mathrm{W_{on}}} + flh^{\mathrm{W_{off}}} \times P_{\mathrm{W_{off}}} + flh^{\mathrm{PV}} \times P_{\mathrm{PV}} \\ &= 8760 \times \left(C^{\mathrm{W_{on}}} \times P_{\mathrm{W_{on}}} + C^{\mathrm{W_{off}}} \times P_{\mathrm{W_{off}}} + C^{\mathrm{PV}} \times P_{\mathrm{PV}}\right). \end{aligned}$$

flh and C depend on the weather conditions during a year and on the technical status of the employed generation technologies. This is of specific importance for new wind turbines, which reach the market with continuously increasing power both for onshore and offshore applications.

In the prediction of future iRE electricity generation, flh and C play an important role, with the complexity to anticipate the technological progress. Optimistic forecasts often ignore that specifically for wind generation the high-yield locations are already occupied and less favourable sites have to be selected in the future, that larger rotor areas are needed to obtain larger capacity factors [11] enforcing larger distances between wind turbines and that the interference of wind mills in extended wind farms cause flh and C to drop [12]. A natural power limitation for onshore wind is given by the logistics to transport the large rotor blades to the desired assembly sites.

Figure 5 compares the flh-values of the three iRE technologies and the general growth and annual variation of flh-values for wind and PV power over the years. flh of PV averaged over several years is about 55% of that of onshore wind, which itself is 52% of that of offshore wind. The lines in fig. 5 are linear fits to the data. The numbers on the

([3]) The concept in Germany is to use gas as backup. The recent steep price rise of gas and the supply security of Russian gas as a consequence of the Russian-Ukrainian war jeopardize this strategy.

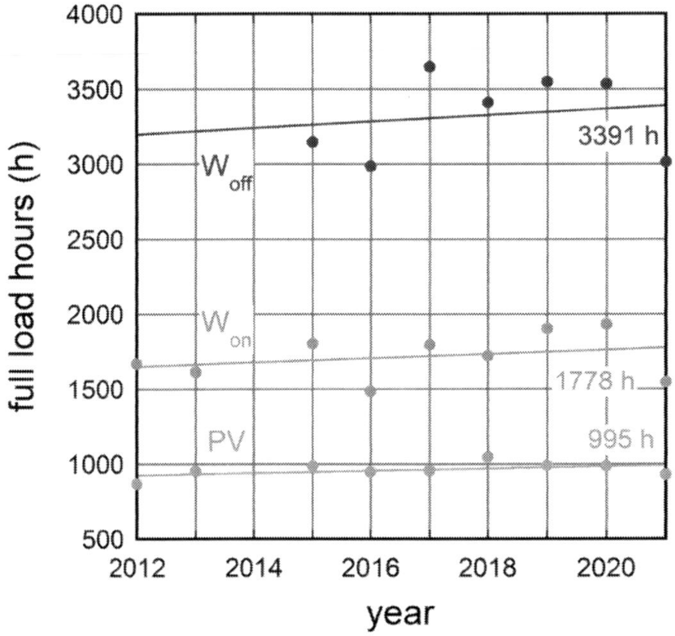

Fig. 5. – Development of full-load hours from 2012 (in case of offshore wind, W_{off}, from 2015) to 2021 for the three iRE technologies, onshore, W_{on}, offshore wind, W_{off}, and photovoltaic power, PV. The lines are linear fits to the data. The numbers on the right side represent the *flh*-values of 2021 as obtained from the fit-lines.

right side represent the *flh*-values of 2021 as obtained from the fit-lines. For all three technologies, *flh* slightly increases with time. *E.g.*, for onshore wind, *flh* increases by 14 hours per year in the average. The trend is small compared to annual variations *e.g.*, from 2016 to 2017, or 2020 to 2021. An increase of *flh*-values could be expected because the average turbine power of all onshore wind installations in Germany increased from 2012 to 2021 from 1.3 to 1.9 MW. Nevertheless, there is no basis for extreme expectations by selecting overly optimistic *flh*-values in simulation studies as one can often find in the literature.

5. – Simulation studies

5·1. *Electricity generation.* – Figure 6 shows the results of model studies on the basis of 2019 generation and consumption data. Basis are load, wind and PV power data tabulated with 15 min time resolution [13]. In the simulations, electricity generation is exclusively by hydro power at an unchanged level and by iRE —wind and PV power— upscaled to replace fossil and nuclear generation. Biomass is not considered though today typically 45 TWh electricity are generated by biogas. We assume that biogas is used for more critical applications in aviation or heavy-duty transportation.

In the first step of the simulations, the annual load L_a agrees with the present net electricity consumption of Germany: $L_a = 520$ TWh. The temporal course of the load profile

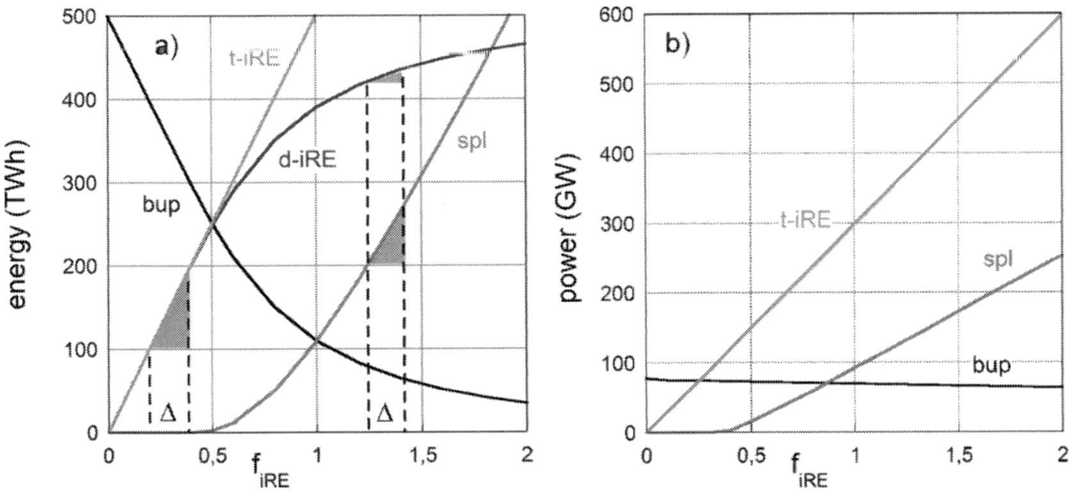

Fig. 6. – (a) Energy generation *vs.* the normalised intermittent iRE share, $f_{\mathrm{iRE}} = E_{\mathrm{W+PV}}/T_{\mathrm{a}}$, for the conditions of 2019. This relation is shown for the total iRE generation, (t-iRE), the directly usable iRE share, (d-iRE), the surplus generation (spl), and the backup energy (bup). The triangles indicate how an increment Δ in wind+PV generation leads to an increase in d-iRE in the linear range of this curve and how d-iRE and spl share the increase in the saturation range. (b) Corresponding powers.

$L(t)$ during the year is maintained. The impact of alternative loads are briefly mentioned further below. 20 TWh are provided by hydro-electricity and the rest, 500 TWh, by wind and PV defining the annual iRE target $T_{\mathrm{a}} = L_{\mathrm{a}} - 20\,\mathrm{TWh} = 500\,\mathrm{TWh}$. In the model, the ratio of PV to wind energy is selected according to the "optimal mix" principle [14], offshore wind is arbitrarily set to 1/3 of the total wind generation. The energy and power shares of the optimal mix are typically: energy: $E_{\mathrm{PV}} \sim 22\%$, $E_{\mathrm{W_{on}}} \sim 52\%$, $E_{\mathrm{W_{off}}} \sim 26\%$; power: $P_{\mathrm{PV}} \sim 37\%$, $P_{\mathrm{W_{on}}} \sim 51\%$, $P_{\mathrm{W_{off}}} \sim 12\%$.

In the second step, the iRE electricity generation, $E_{\mathrm{W+PV}}$, is varied. Variable is $f_{\mathrm{iRE}} = E_{\mathrm{W+PV}}/T_{\mathrm{a}}$. In fig. 6. f_{iRE} is varied between $0 \leq f_{\mathrm{iRE}} \leq 2$ implying an annually generated iRE electricity varying from 0 to 1000 TWh. $f_{\mathrm{iRE}} = 1$ is denoted as "100% case" because wind and PV produce just the target T_{a} (= 500 TWh).

Plotted in fig. 6(a) are the *annually* generated iRE electricity (t-iRE), the directly used one (d-iRE), surplus (spl), and backup (bup) energies and in fig. 6(b) the respective powers. The integral annual values of fig. 6(a) are deduced from time-dependent calculations. Directly used is the iRE power ≤ the momentary target load $T(t)$; backup supplements iRE in cases where t-iRE$(t) < T(t)$: bup$(t) = T(t) -$ d-iRE(t); in case of excess-generation, surplus is the power which exceeds the momentary target load: spl$(t) =$ t-iRE$(t) - T(t)$.

As shown in fig. 6(a), directly used electricity (d-iRE) increases initially linearly with f_{iRE}. At $f_{\mathrm{iRE}} = 0$, backup starts at 500 TWh, the annual target T_{a}. Increasing d-iRE causes the backup energy, bup, to decrease in a symmetric form: d-iRE + bup = T_{a}. In this energy range, no surplus is generated because even the peaks of the iRE production fall below the load and can directly be used. This changes at an iRE penetration of about 40%. From there on, iRE-surplus is generated. Beyond the 40% share, d-iRE and bup start to level off whereas the further growth of iRE power progressively generates surplus. The "100% case" is reached when iRE production t-iRE = target load T_a. In this case, surplus is equal to backup energy. From the 500 TWh generated, 390.3 TWh are directly used and 109.7 TWh appear as surplus necessitating the same amount as backup energy: In the 100% case, when iRES produces just as much as needed, under 2019 conditions 22% appear as surplus and have to be replaced by controllable residual power to meet the load. This is the reason why the promises of a 100% supply by renewable energies have to be scrutinized with some detail.

Beyond the 100% case, overproduction starts. The non-linear relation of d-iRE in the range of overproduction is an important feature of intermittent supply. The grey triangles in fig. 6(a) represent the response of d-iRE in case of an increment Δ in f_{iRE}. In the linear range, the increase in iRE energy (and power) leads to directly usable energy. In the saturation range, most of the increase is in the form of surplus and only a decreasing fraction can directly be used.

Surplus electricity used after transformation, storage and anew generation as secondary electricity will be in a different cost category as backup energy is in the period of iRE capacity build-up. In the saturation zone of d-iRE the question of supplementing or substituting renewable energies by alternative clean technologies like nuclear power arises. The cost assessment between renewable and nuclear technologies has actually to be done by comparing, *e.g.* nuclear electricity costs with those of secondary and less with the costs of primary iRES generation. The cheapest way is, of course, to adjust the load to generation (responsive load) and to reduce surplus production by demand adaptation. This important aspect represents the fourth major feature of the energy transition —the replacement of the traditional supply scheme of "generation follows the load" by "the load has to be adapted to generation". It can, however, be doubted that such a scheme is viable at large scales facing the strong day-to-day generation variability preventing reliable day-ahead planning. The first field tests of demand adaptation yielded disappointing results in Germany [15].

Figure 6(b) plots the power values for the three parameters of relevance, the total iRE power (t-iRE), the backup and the surplus powers. iRE power and energy are linearly related. The model is set up linearly and possible nonlinearities introduced, *e.g.*, by transmission losses or peak power catenation are not considered. Two further features of intermittent power supply are of importance: 1) the power level of surplus power is high, much above the present peak load of ~ 80 GW. It will be necessary to catenate the power peaks in order to simplify and secure grid operation and limit the grid operational costs. Nevertheless, the remaining power level defines the consumer technologies like electrolysers for hydrogen generation and thermal storage systems. 2) the backup power is only moderately reduced though the backup generation drops significantly in the range of overproduction (see fig. 6(a)). flh of the backup system falls to 540 h for $f_{\mathrm{iRE}} = 2$ raising the question of operational economy. On the other hand, the backup system is needed to meet the demand and the economic concerns could be solved introducing a

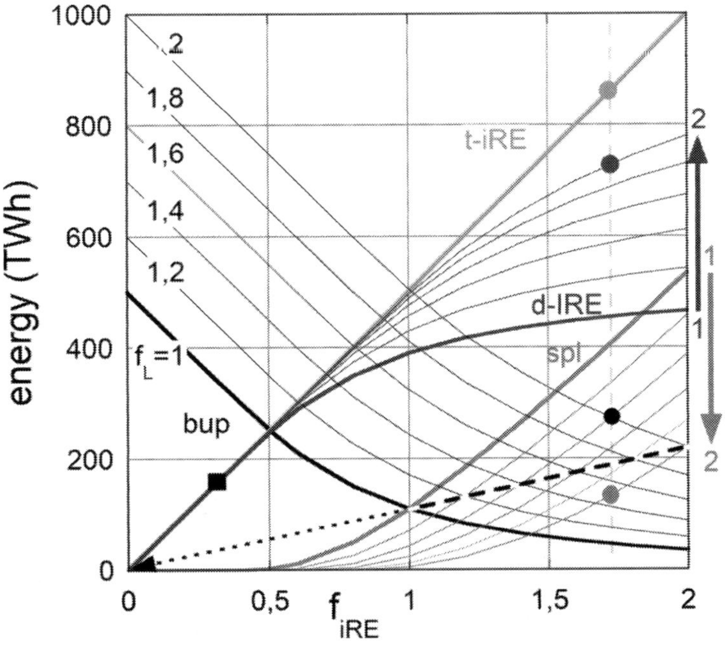

Fig. 7. – This diagram is an extension of fig. 6(a) when additionally, the consumption is varied. The consumption is parametrised by f_L = annual consumption/target load. The fat curves are reproduced from fig. 6 with $f_L = 1$. The thin lines represent the variation of bup, t-iRE, d-iRE and spl when f_L is varied from 1 to 2 (see arrows), whereas each set starts with the fat curves of $f_L = 1$. The dashed linear line denotes the special cases where spl = bup. They are dubbed 100% cases because spl would conceptually allow replacing bup by storage if storage losses were ignored. The square denotes the German situation in 2018; the dots the situation of Germany in 2045 expected from the AGORA installed iRES capacity [10] and processed on the basis of 2019 data.

so-called capacity market: In this case, backup operation is not funded for generating electricity rather for the potential to generate it.

The backup power P_{bup} drops by 18% at an overproduction factor of 2 (1000 TWh iRE generation). This reduction is often seen as evidence that wind, unlike PV power, contributes with a generation minimum P_{min}, which is always > 0 and which increases with total installed power. This is not truly the case [16]. For 2012 to 2020 (2016–2020), the average P_{min} of W_{on} (W_{off}) was 148 MW (0 MW); the annual increase of P_{min} of onshore wind was about 9 MW whereas the installations grew by 3.4 GW per year. The reduction of the P_{bup} peak is rather caused by PV power whose peak closely coincides with the load peak and moves the maximal backup demand toward the wings of the PV profile where consumption is reduced both in summer and winter.

5·2. *Electricity consumption.* – Up to now, we used the model to portray the conditions for an RES electricity generation of 520 TWh, 500 TWh from iRE and 20 TWh from hydro-electricity. Figure 7 shows the model results when we next allow the reference load to change between 500 TWh and 1000 TWh. In fig. 7 the demand is represented by

f_L = annual consumption/annual target being varied between $f_L = 1$ and 2. The results with $f_L = 1$ shown in fig. 6(a) are reproduced and given as thick lines in fig. 7. Directly used iRE energy (d-iRE) increases with f_L. This goes at the expense of surplus energy generation, which decreases with increasing f_L. Also, backup needs increase with f_L, representing the overall increase of demand. The black square denotes the end of 2021 situation of Germany with $f_{iRE} = 0{,}33$; 56 GW Won, 7.8 GW Woff and 56 GW PV having generated 165 TWh electricity. We have already mentioned (fig. 6(a)) that at $f_{iRE} = 1$ (100% case) surplus and backup energy curves cross. Disregarding losses, surplus could —in this case— directly replace backup. The dashed line denotes the crossing of these two parameters for $1 \leq f_L \leq 2$. Under optimal mix conditions with $E_{bup} = E_{spl}$, the two quantities increase linearly with f_{iRE}. The situation changes as soon as we introduce storage with transformation losses.

5·3. *Storage.* – Storage will eventually replace backup. Storage may fulfil two tasks: providing power and energy. Power for grid control will most probably be done by batteries; seasonal storage is in the TWh capacity range. In this case, the energy carrier has to be changed to one allowing cheap storage and this is not in the form of electricity rather of hydrogen as initial step. The transformation from electricity to hydrogen leads to energy losses. In addition, hydrogen is not simple to store. The lower calorific value of hydrogen is 3 kWh/Nm^3 compared to 10 kWh/Nm^3 of methane. Correspondingly larger has to be the storage volume. If required, hydrogen can be further transformed into ammoniac (*e.g.*, Haber-Bosch process) or carbon hydrates (*e.g.*, Fischer-Tropsch process employing additionally CO) but at the expense of further losses.

The basic scheme of producing secondary electricity is shown in fig. 8 [17]. Surplus represents primary electricity producing hydrogen via electrolysis, which is stored and used, *e.g.* in fuel cells to generate secondary electricity. The numbers in fig. 8 exemplify a case with surplus generation of 1.5 TWh, representing the average daily demand of electricity in Germany. Heat losses and separate electricity demand of the compressor are shown in the diagram. The water supply for the electrolyser of 243000 m^3 compares with the daily consumption of the city of Munich. Because of the system losses, 1.5 TWh primary electricity yield 0.45 TWh secondary electricity. The system efficiency is 30%. We conclude that the price difference between primary and secondary electricity will at least be a factor of three.

In the following, the basic relations which govern storage will be discussed. In the modelling employed here, the storage is split up into a short-term (day) storage (DSt, with a capacity of 160 GWh) and a seasonal storage (SSt) with a capacity to be calculated. DSt is foreseen for grid services (not specified any further)([4]). Input for storage calculations are wind and PV power values and the load profile of a year, here of 2019. The storage is filled by surplus power which depends on f_{iRE} and f_L (see fig. 7). A storage fulfils the conditions that the "filling level" at the end of the year matches the one at the

([4]) Details of the storage calculations will be given in another publication.

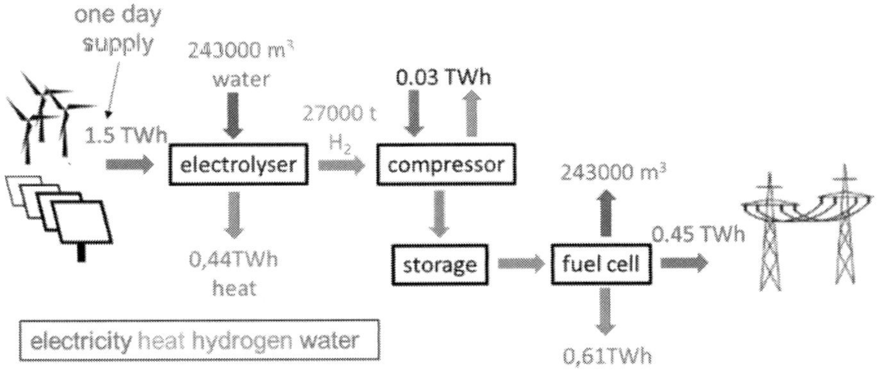

Fig. 8. – Schematic of a system to generate secondary electricity from primary iRE power using an electrolyser to produce hydrogen and a fuel cell to generate secondary electricity. The numbers exemplify the situation of one day of German electricity consumption of $\sim$ 1.5 TWh. Heat generation and water consumption are also given. The schematic follows [17].

beginning (periodic boundary conditions). In the model, storage has to be fully emptied at least once a year. Then the peak storage level defines the capacity. Though the basic layout of the problem is linear, the storage filling shows a nonlinear behaviour. The work by the combined storage system should agree with the backup needs as presented in sect. **5**˙1 (see also fig. 6(a)). Transformation losses are proportional to the storage work. For calculating losses, we assume that DSt could be based on batteries with an efficiency of 85%; in case of seasonal storage, we assume transformation to hydrogen with an electrolyser *plant* efficiency of 70% followed by a fuel cell with 50% efficiency. In case the system produces surplus, after the load has been met, it is considered chemically stored energy in the form of hydrogen.

Figure 9 shows the combined storage capacity and storage work with the original backup energies to compare with as black dots and the hydrogen production from remaining surplus([5]). $f_{\mathrm{L}} = 1.31$ in the case depicted in fig. 9, which corresponds to the political expectation in Germany for 2030([6]). Storage capacity and work increase with f_{iRE} and at $f_{\mathrm{iRE}} = 1.625$ a maximum of 60 TWh capacity and 100 TWh of work is reached. Beyond this f_{iRE} value, both parameters decrease. In the rising branch, storage is limited by the availability of surplus; additional backup energy is still needed to meet the demand (difference to the dots in fig. 9). At the capacity/work maxima, processed surplus (= secondary electricity) just meets the demand and the storage work fully replaces backup. Beyond the capacity/work maxima, additional surplus is generated which can be turned into hydrogen as shown in fig. 9.

The storage analysis starts at $f_{\mathrm{iRE}} = 1$, though surplus is generated starting with $f_{\mathrm{iRE}} \sim 0{,}4$. For $f_{\mathrm{iRE}} < 1$, large storage capacities are needed to contribute with storage

([5]) 4.5 kWh needed to generate 1 Nm^3 H_2.
([6]) At the end of 2020.

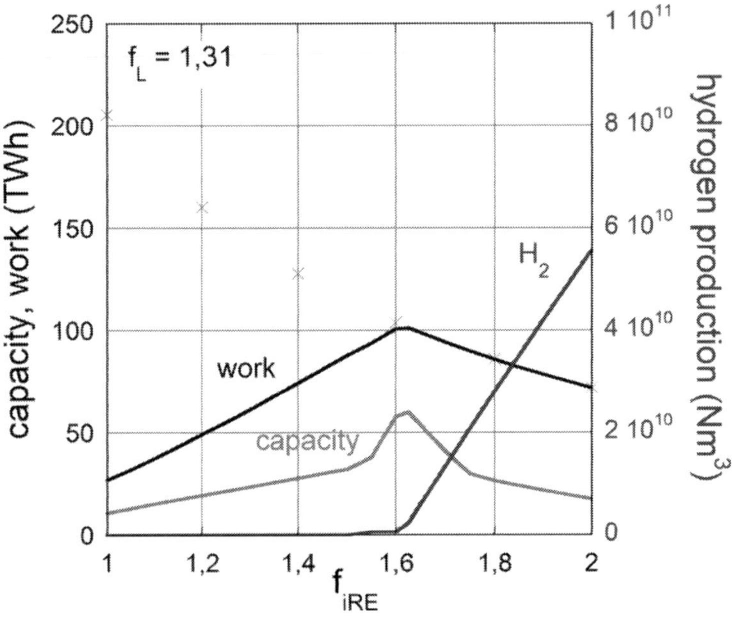

Fig. 9. – For $f_L = 1.31$, corresponding to the consumption of 650 TWh as expected for Germany for 2030 [18], storage capacity, storage work and possible hydrogen generation are plotted *versus* the normalised intermittent iRE share f_{iRE}. The symbols ($\times$) represent backup energy in case of no storage.

work, which, on the other hand, is by far not able to replace the backup needs. This is the reason why storage enters the electricity transition at an advanced transition stage *e.g.* at $f_{iRE} \sim 1$. With increasing overproduction, the seasonal storage capacity decreases. At $f_L = f_{iRE} = 1$ storage capacity is 18 TWh based on 2019 conditions. The capacity drops down to 0.3 TWh in case of $f_L = 2$ and rises to 36 TWh, when f_{iRE} is allowed to increase to 2. The final electricity generation will be in the range $f_{iRE} > 1$ as the national potential has to be fully exploited to serve all energy consuming sectors. It is questionable whether under these conditions storage exclusively dedicated to secondary electricity generation is viable. Such storage systems must be able to operate under high power levels but, on the other hand, are limited to the production of low quantities of secondary electricity.

6. – The German situation: an example in numbers

Up to now, we have discussed the main characteristics of electricity generation mostly by wind and PV power using simulations based on the projection of historical data. For this purpose, we have employed model cases like the 100% case. Now, we will try to analyse the German energy and electricity demand and supply situation expected for 2045 in more detail and specifically try to provide actual numbers for major parameters. 2045 is the year in which, according to German politics, the country should be net-CO_2-emission free. Reference points for the assessment of the 2045 German energy and electricity situation is an end electricity demand of about 520 TWh (an average value over the last decade) and an end energy need of about 2600 TWh.

As it is pointless to form one owns opinion on the actual iRE potential of Germany, we resort here to the figures provided by AGORA: $W_{on} = 145$ GW, $W_{off} = 70$ GW and PV = 385 GW [10]. The *average* energy generation from 2012 to 2020 calculated from the actual iRE electricity generation and scaled from the historical installations to those suggested by AGORA is 850 TWh per year. The highest value, 908 TWh, is from 2020, the lowest one, 723 TWh, from 2013. A year-to-year variation of 22% in iRE energy generation has to be considered. The year 2019, which we will use for a more in-depth analysis in the following, yielded 860 TWh. The comparison of 2021, a low-wind year, with 2020 exemplifies the high level of variability and demonstrates painfully the dependence of the electricity generation by iRE on unpredictable weather conditions. This aspect also sets a natural limit to the relevance of overly precise predictions.

For calculating the future electricity generation on the data basis of a specific year the full load hour figures, flh, of this specific year is used having in mind that further technical progress will partly be compensated by progressively less favourable generation sites. Nevertheless, the projected energy values can be expected to be lower limits. Figure 5 showed, however, that in the period 2012 to 2021, flh for the three iRE technologies strongly changed from year to year but shows only a weak upward trend in this period.

The annual iRE electricity generation has to be compared with the electricity demand of a completely de-carbonised economy. The future demand is, however, also difficult to predict. Maybe the basic electricity use might decrease thanks to the demographic factor, behaviour changes and the availability of low-consumption technologies; but industry 4.0, digitisation and smart systems will increase the demand. We stay with the present average value of 520 TWh for the net electricity consumption proper (ignoring losses) but new electricity consumers are mobility ($\sim$ 100 TWh), the use of heat pumps ($\sim$ 80 TWh), and the transformation of steel industry (40–80 TWh). The chemical industry association alone claims 628 TWh for a complete decarbonisation of its activities [19]. These figures would add up to more than 1400 TWh —much more than what can nationally be produced by RES.

AGORA concludes that Germany will need 1017 TWh electricity in 2045. Also, this lower amount cannot be provided by RES. In addition to the average iRE production of $\sim$ 850 TWh, 20 TWh are generated by hydropower and possibly about 50 TWh can be added using biogas. This contribution can be questioned, however, because biomass may have to serve other energy sectors as mentioned above.

We consider two cases, the one with maximal hydrogen production from surplus electricity and the second one of electricity autonomy. How much hydrogen can Germany produce with its own means? The AGORA installations allow the generation of 860 TWh electricity under 2019 starting conditions. 705 TWh can be used directly([7]); 155 TWh are

([7]) This value is obtained using the present load profile. In case of an alternative —a load profile surplus, which can be turned into 34 bn Nm^3 of H_2, about twice the present hydrogen use of Germany's industry. In this case, 292 TWh electricity have to be imported or provided by a separate backup system in order to meet the AGORA electricity target. These results are marked as dots in fig. 7.

In the next step, we analyse the electricity autonomous situation of Germany in 2045. In this case, 550 TWh can be directly used. This is less than in the "hydrogen production case" because the additional losses in the fuel cell have to be accounted for. The rest is handled via a storage with a total capacity of 79 TWh. DSt contributes with 30 TWh and SSt with 92 TWh secondary electricity, in the sum 672 TWh. In this scheme, Germany can generate slightly more than the present *total* electricity generation (about 650 TWh). Considering 20 TWh of hydro-electricity, 325 TWh electricity have to be imported to meet the AGORA demand target. The transformation losses add up to 173 TWh, surpassing the storage work.

DSt contributes with 30 TWh at a storage capacity of 0.16 TWh, corresponding to 190 annual full-equivalent cycles, SSt with 92 TWh at a capacity of nominally 79 TWh. The storage capacity of 79 TWh depends on the peak power to be processed by SSt and can be reduced by catenating power peaks. Unlike to DSt, a dedicated seasonal storage for electricity seems to be unrealistic specifically when power specification, actual work, the number of full equivalent cycles, and transformation losses (168 TWh of SSt) are considered. More electricity can be directly used and less has to be imported if one abstains from the autonomous solution. In addition to short-term storage mostly for grid services (with 0.16 TWh, only about 2.5 hours of present electricity needs can be bridged) the results suggest to better generate hydrogen from surplus. Depending on the actual economic situation, secondary electricity can be imported or generated from the stock of nationally generated and imported hydrogen.

Regarding the net energy situation of Germany in 2045, in addition to 850 TWh iRE and 20 TWh hydro-electricity, Germany can generate nationally about 300 TWh biomass and about 240 TWh environmental heat in conjunction with heat pumps [20] and $\sim$ 80 TWh solar thermal heat [21]. The energetically used share of biomass can hardly be increased. The present conditions lead already to a loss of bio-diversity and a strong reduction of insect and of bird populations. In total, the national energy production is in the range of 1500 TWh, about 1100 TWh short of the present average net energy consumption. Up to now, Germany's end energy consumption has slightly increased from 1990 to 2019. Savings of 1100 TWh do not seem to be possible on an evolutionary path.

All these numbers are indicative only and can be questioned in detail. Nevertheless, it is clear that Germany will remain an energy importer but, unlike in the past, not for low-level still abundant fossil fuels rather for electricity and hydrogen generated by renewable technologies.

---

constructed from the expected price variation— 773 TWh can be directly used and 224 TWh have to be imported (details to be published elsewhere).

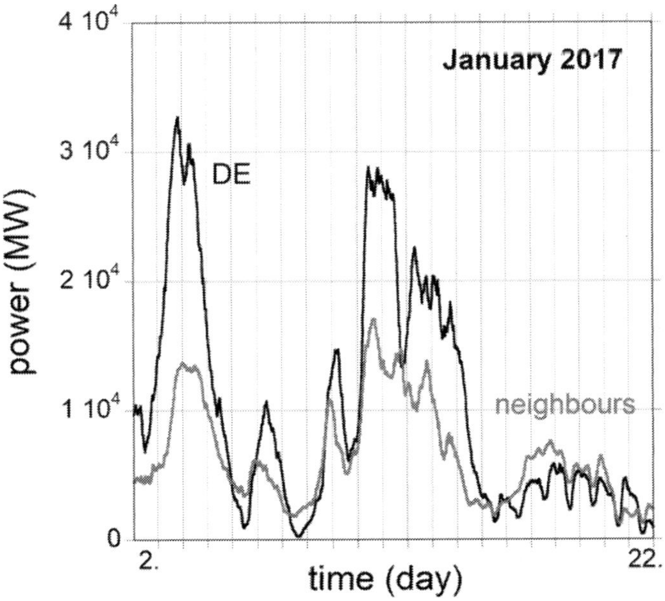

Fig. 10. – Onshore wind power generation of Germany and, in the sum, of its 9 neighbours for January 2017.

7. – Electricity import from Germany's neighbours

With the AGORA electricity target of 1017 TWh, Germany has to annually import electricity of about 300 TWh. Because of transportation losses, the import should be from the closest neighbours. This implies a strong increase in EU export from the present level of typically 50 TWh of France and also exported by Germany in the recent years. In the future, it has to be renewable and possibly nuclear electricity to be exported. The export potential of RE electricity will depend on the weather pattern over Europe, which has typically an extension covering central and northern Europe. The consequence is that its neighbours also have wind and sun, when Germany enjoys these conditions and vice versa. Figure 10 compares wind power of Germany for three weeks in January 2017 with the one of its 9 neighbours. The comparison shows strong correlation [13]. The "lull-year" 2021 gave rise to a 13% lower onshore wind harvest in Germany and in the average to an 8% lower one in Germany's neighbour states.

The analysis of wind power in Europe reveals two uncorrelated zones, a northern one and a southern one. The correlation matrix of fig. 11(left) shows high correlation between wind in Greece, Portugal, Spain, Romania and Italy and separately between the central and northern countries. Apart from autocorrelation, the Spaerman correlation coefficients vary between 0.8 and 0.25 from reddish to yellow. This splitting became also evident in the comparison of the onshore wind generation of 2020 and 2021. The northern countries experienced a reduction in W_{on} generation by -7% in the average whereas the obviously decoupled southern countries enjoyed a gain by $+8\%$.

The consequence of the European weather pattern is that countries at larger distances

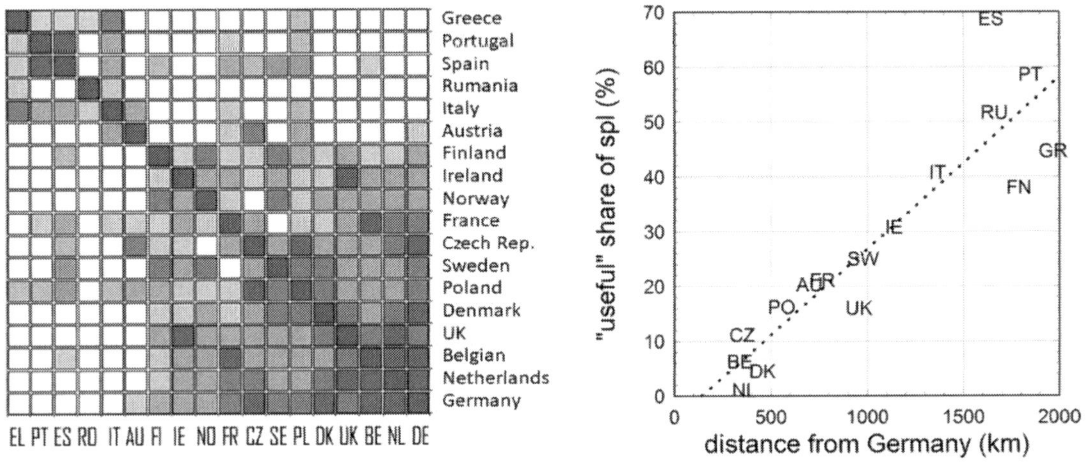

Fig. 11. – Left: Spaerman correlation matrix of onshore wind energy between the countries listed (2016). Wind generation is correlated in two well separated areas. Apart from autocorrelation, the correlation coefficients vary from reddish (0.8) to yellow (0.25). Only a small correlation exists between the southern and northern countries. Right: the "useful" share of a country's surplus is plotted against the distance of the wind park core of this country from Germany's wind park core (see text). "Useful" is defined as surplus energy of a country at a moment when Germany actually needs backup.

from Germany produce a higher share of "useful" surplus, defined as being available when Germany is in need (of backup power). Figure 11(right) shows the results of this analysis for December 2016 data. Plotted is the national share of normalised surplus summed up during periods of backup needs of Germany against the distance from Germany. A national wind park centre was defined from the averaged latitude and longitude readings of wind parks [22] whereas the individual wind park contribution was weighted by the installed power. The separation of these centres to the one of Germany yielded the abscissa in fig. 11(right).

The import of electricity requires long high-voltage transmission lines and powerful interconnectors at the borders (presently 18.5 GW for Germany). Transmission lines are unpopular as Germany itself has experienced in its efforts to build electricity "highways" from Thuringia to the neighbour state Bavaria. Dismissive public positions can be expected specifically when transfer lines do not serve the national supply needs.

As the German electricity import needs go beyond any precedent, electricity import at the necessary level will hardly be possible when all neighbours of Germany also use mostly RES. The consequence is that Germany has predominantly to import, *e.g.*, hydrogen to generate secondary electricity to fill the gap. 300 TWh secondary electricity correspond to about 900 TWh primary RES electricity generated by exporting countries. This is comparable to the national potential of Germany as we have seen above. To meet the future electricity needs of Germany, a similar build-up of iRE installations is necessary together with the necessary hydrogen/ammoniac transformation technology by future trading partners. It is not clear which countries qualify for hydrogen export at these levels and be ready along the foreseen time scales. The qualification is based on sufficient sun, wind and ultra-clean water availability for electrolysis and large port capabilities for hydrogen export. 100 TWh energy correspond to 3 mill tons of hydrogen based on the lower calorific value. Transported as liquid, about 93 ships of the present specification [23] have to arrive in Germany every day. The transport in the form of ammoniac implies gravimetrically predominantly the transport of nitrogen.

8. – Nuclear energy

During writing of this paper, the European Commission has published its draft paper on the EU Taxonomy Regulations [27] stating that nuclear energy can contribute to decarbonisation. The comparison of CO_2 emissions for power generation of France and Germany supports the position of the Commission. Figure 12(a) shows the historical development of the electricity mix of the two neighbour countries starting from 1970 [28]. France developed nuclear power, Germany burnt predominantly coal and is for years the world's largest user of lignite, the "dirtiest" form of coal. This is reflected by the CO_2 emission as compared in fig. 12(b). For electricity generation France emitted about 300 mill tons per year of CO_2 less than Germany over 4 decades. This is a substantial service to the environment. The red squares in fig. 12(b) show the decrease of CO_2 emission in case Germany had kept its nuclear power arsenal and moved out of lignite in the steps nuclear energy was removed from the grid. Such an act would have helped the environment.

As Germany stopped three nuclear power stations at the end of 2021, 103 nuclear power reactors operate in 13 EU countries producing 1/4 of the electricity. Thus 70% of the EU population is served by nuclear power. The future of nuclear energy is not spelled out in detail. Some countries plan to terminate nuclear power (Belgian, Germany, Spain), others to prolong it to 60 years (Hungary, Sweden, Switzerland). Finally, there is a group of countries which builds new reactors or has plans to do so (Belarus, Czech Rep., Finland, France, Netherlands, Hungary, Poland, UK but also Belgium([8])). It can be expected that these plans will not be realised in all cases but it can be expected that nuclear power will be in operation in Europe up to 2050 and beyond. A critical issue specifically for France is, however, the aging of the pioneering reactors of the 70ies.

The discussion on the EU Taxonomy proposal is around the term "sustainability". This is a very complex criteria also for RES facing the material and mineral demands as presented in sect. **3** and the little hope that critical materials can efficiently be recycled in near future. A rather obvious criteria would be whether there is a need for clean energy sources in addition to RES. The analysis of the Germany strategy shows that such a need indeed exists. If France uses is potential for RES, which seems to be higher than the one of Germany, it still can add short of 400 TWh of nuclear electricity, which is

([8]) There are recent plans to consider small modular reactors (SMR) for Belgium.

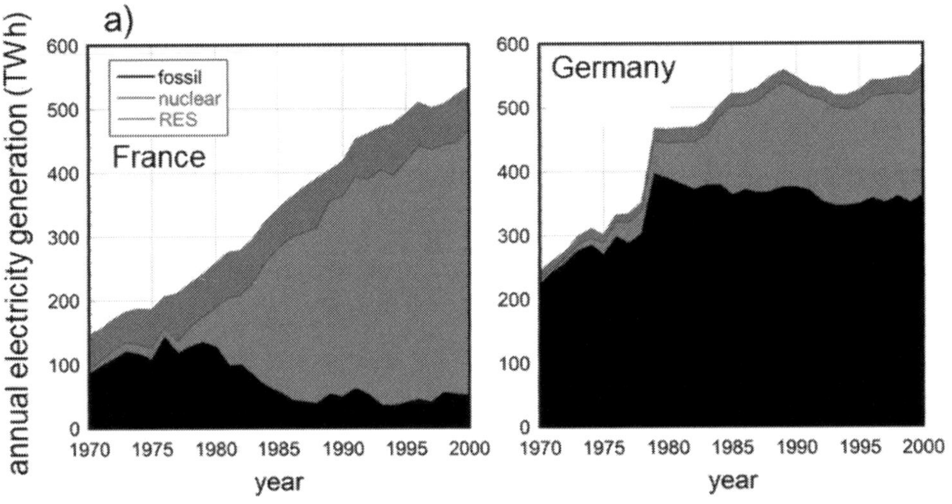

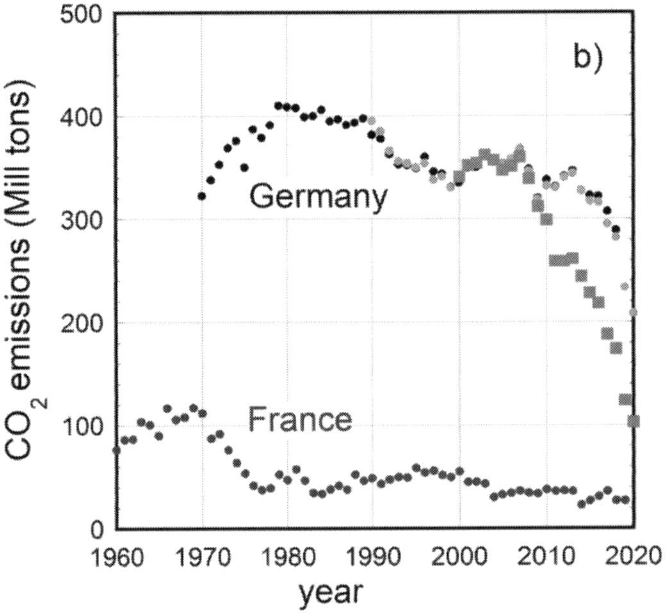

Fig. 12. – (a) Annual electricity generation of France and Germany from 1970 to 2000 split up into fossil, nuclear and RES shares. (b) CO_2 emission from electricity generation of France and Germany starting from 1960 or 1970, respectively. Data sources: black and blue: EDGAR-EU [24]; the grey dots represent emission data averaged from UBA [25] and STATISTA [26] compilations. The red squares represent Germany's CO_2 emissions in case nuclear energy had remained at the year 2000-level and lignite had been reduced in the following years in the steps nuclear energy was removed from the grid.

a tremendous asset facing energy shortages or the environmental damages of low-power density technologies operated in densely populated countries.

The nuclear accident of Chernobyl was caused by bad technology (positive void coefficient) and an irresponsible experiment carried at with the Soviet RBMK reactor type. The Fukushima accident [29] was so dramatic because the protection wall against Tsunamis was not high enough ignoring historical experience. The safety installations (rupture discs, use of hydrogen recombiners) did not meet modern standards; backup power supplies were not correctly placed. These deficiencies were obviously only possible because regulation bodies did not function properly. The official National Diet report on the Fukushima accident states [30]: *The incestuous relationships that existed between regulators and business entities must not be allowed to develop again.* All these problems can be avoided by proven technology and independent and transparent regulations.

Nuclear waste storage is for sure a severe problem with which the above-mentioned European nuclear countries have to deal with. There is no way out and a best effort solution has to be found by each nuclear country. Finland will be first to open its waste repository in $\sim$ 2025 in Onkalo. This repository will finally be sealed. It might be advisable to leave repositories open and await the results from MYHRRA transmutation reactor [31] and Gen-IV reactor concepts [32] which are expected to shorten the decay time of nuclear waste. In this case, the public position to waste handling and disposal could change over the next years.

The operation of nuclear power stations in load-following mode has been demonstrated by many power reactors with the conclusion of a "*full complementarity between nuclear and renewable energies*" [33]. It seems to be specifically possible to cope with the daily rise and fall of PV power which would lead to two output variations a day, which seem to be manageable.

A frequent objection against the use of nuclear energy is the high capital costs and the resulting high electricity costs. As argued in sect. **5**˙1, in case of iRE, the electricity costs of nuclear power has to be compared not with those of primary electricity generation rather with those of the inescapable production of secondary electricity.

9. – Final remarks

We come back to the beginning of this paper. Because of the few options for a clean electricity and energy supply, the strategies seem to be clear —to maximally use the available natural options like hydro-electricity, biomass and geothermal power. In many cases, the dominant supply must come from wind and solar power. The employment of CCS may be required to handle unavoidable industrial CO_2 emissions, *e.g.*, in the production of cement [34]. The use of nuclear power is a political and societal issue. Germany observes strictly a schedule of terminating secured power replacing the missing power by a strongly delayed rollout plan for iRE and the necessary grid expansion. With such an approach, Germany violates a proven electricity supply principle of Europe that each country secures first its own demand. With predominantly iRE, electricity supply safe-guarding is not possible. Being the European country with the largest electricity consumption, the German electricity strategy has strong consequences on her neighbours and should be of concern. The European-wide grid separation from Jan. 8th, 2021 is a good example of what could happen more frequently and more severely in the future [35].

We have seen that a focus on iRE demands the use of a backup system, which uses preferentially gas as long as hydrogen is not available in larger quantities. Burning gas has lower CO_2 emissions as consequence but CH_4 leakage on the way from the gas fields in Siberia to the user in Europe is specifically detrimental as methane has a much larger global warming potential than CO_2 has (about a factor 30, averaged over 100 years). For $f_{\mathrm{iRE}} = 1$, the backup capacity is 70 GW, 10% less than peak power. 7,5 GW could be covered by hydro and biogas electricity. According to [36] 82 gas power stations with, in the sum, 22.5 GW electrical power are operated in Germany in 2020. To meet the backup needs, additional 40–50 GW have to be realised, which correspond to 150 gas power stations of the 300 $\mathrm{MW_{el}}$ class (corresponding to the average power of the presently installed units). Up to 2045, every two months a new unit has to be connected to the grid. 2/3 of the present gas power stations are of GuD type producing electricity and heat. It can be questioned whether their specific dedication allows them to provide electricity services only. The consequences of the Russian-Ukrainian war on international gas trade could jeopardize the strategy to supplement the expansion of iRES by a backup system based on burning gas.

The major installation thrust has to be the expansion of wind and PV power. To reach the AGORA 2045 development goals for the three iRE technologies, the annual additions of onshore, offshore and PV power have to be 3.8, 2.6 and 13.6 GW. The average expansion rates from 2010 to 2021(9) were 2.8, 1.0, and 4.1 GW per year; the peak values were 5.3 (2017), 2.3 (2015), and 7.6 (2012) GW. For all cases, the expansion rate has to be doubled compared to the average historical values and kept at this level for more than two decades. Only in case of onshore wind, the future expansion rate has already been demonstrated. One has to consider, however, that, in addition, end-of-life systems have to be replaced. Assuming a lifetime of 25 years, the maximal addition rates for W_{on}, W_{off} and PV rise to 9.1, 4.9 and 21 GW per year. It is questionable whether such goals can be met. In 2021, about 30 GW of onshore wind have been installed worldwide outside of China.

For Germany, this ambitious expansion of wind and PV power necessitates additionally thousands of km of high-voltage lines and 100 thousands of middle- and low-voltage lines either newly built or upgraded. This development has already started years ago whereas the intermediate targets have rarely been reached.

A large part of primary energy is used to heat the housing stock. In Germany 2020, 50% of it is heated by gas, 25% by oil and only 2.6% by heat pumps [37]. It is another tremendous task to replace oil and gas heating specifically in older buildings with their additional difficulties to realise proper thermal insulation.

The purpose of these examples is to highlight the tremendous task of the energy transformation for a still strongly industrialised country, which should happen within a short time frame in spite of rather unfavourable boundary conditions —the demographic factor, rising material and energy costs, lack of specialists and craftsmen.

(9) In case of offshore wind from 2016 to 2021.

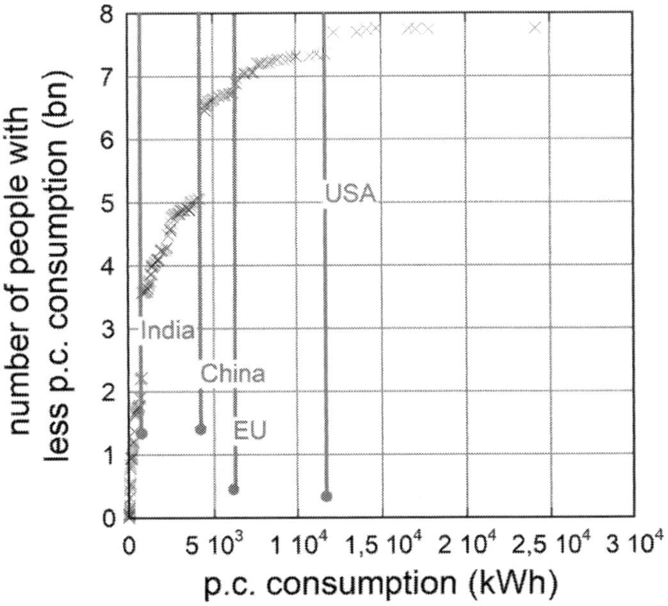

Fig. 13. – Plotted is the number of people with less *per capita* (p.c.) electricity consumption than the respective number of it at the abscissa. The individual data of India, China, and the USA are given as red dots. The vertical lines mark the gaps in the ordinate when the population of the respective country is added. For the European Union, there is no gap because in this case the individual countries are plotted.

But the electricity supply and consumption conditions of Germany and the EU are not representative with 6440 kWh/p.c. or 6156 kWh/p.c., respectively. Figure 13 addresses the accessibility to electricity and shows the number of people without access to a specified per capita (p.c.) electricity consumption (abscissa). Taking the example of

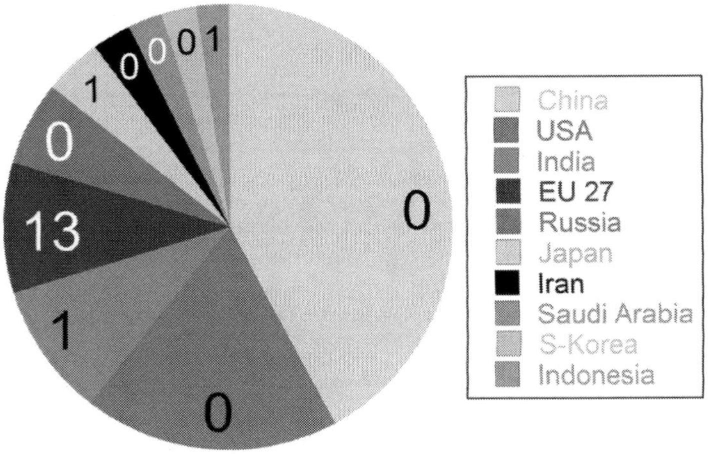

Fig. 14. – Pie chart of the 2020 CO_2 emission of the primary 10 emitters [39] and the number of United Nations COP conferences organised by the respective countries [40].

the EU (red dot), close to 7 bn people have to comply with less p.c. electricity use. The p.c. electricity consumption for India, China, and USA is given as red dots. The vertical lines mark the gaps in the curve where their population enters the graph. A gap is not shown for the EU because the EU members are plotted individually.

This diagram plots the electricity consumption because electricity will be the future primary energy. Its technical development status is crucial for the energy transformation up to 2050. The graph demonstrates that today only a low level of electricity supply exists for, by far, the majority of the world population. This is the situation though today the energy supply is by established technologies and up to now cheap and available fossil fuels. It is difficult to imagine that the countries with supply limitations, *e.g.*, the African generations to come (see fig. 1) will be able to master their energy transition to the use of clean electricity, which is anyway less of a transition rather a system set-up.

Considering the political interests and agendas of the five main emitters —China, USA, Europe, India, and Russia— one may also worry in their case about a sustainable effort to go jointly through the transition up to 2050. Figure 14 gives the number of COP conferences organised by different countries on the background of their CO_2 emission share [38]. The COP conference is the conference of the parties, the UN climate conference. China and the US, the largest CO_2 emitters, have not yet organised a COP conference. India has organised one, Russia none. Europe has organised 13, the last one in Glasgow in 2021. The political interest on a concerted approach to fight climate change seems to strongly differ and the EU and Europe's governments have to be careful that the climate change issue will not become their exclusive project.

Because of the obvious risk of failure, it is mandatory to take additionally appropriate measures of protection against the consequences of global warming. Politics may not exclusively concentrate on prevention but has to include adaptation measures. Unlike the global efforts, adaptation measures promise immediate and traceable effects. This is a wide field which should receive more attention whereas science will also play a crucial role therein.

* * *

Many thanks to Th. Franke and M. Wanner for critically reading the manuscript.

REFERENCES

[1] https://population.un.org/wpp/Download/Standard/Population/; Medium variant.
[2] https://ourworldindata.org/grapher/per-capita-energy-stacked?country=USA~GBR ~OWID_WRL~CHN~IND~FRA~DEU~SWE~ZAF~JPN~BRA.
[3] IEA, The Role of Critical Minerals in Clean Energy Transitions: Flagship report 2021, https://www.iea.org/reports/the-role-of-critical-minerals-in-clean-energy-transitions. All rights reserved; as modified by F. Wagner.
[4] https://www.sif.it/static/SIF/resources/public/files/va2019/vidal1.pdf.
[5] https://de.statista.com/statistik/daten/studie/159933/umfrage/laender-mit-den-groessten-lithiumreserven-weltweit/.
[6] https://miningglobal.com/top10/5-largest-lithium-mining-companies-world.

[7] https://www.wsj.com/articles/china-set-to-create-new-state-owned-rare-earths-giant-11638545586.
[8] Wagner Friedrich, "Electricity by intermittent sources: An analysis based on the German situation 2012", *Eur. Phys. J. Plus*, **129** (2014) 20.
[9] https://www.agora-energiewende.de/.
[10] https://www.agora-energiewende.de/veroeffentlichungen/klimaneutrales-deutschland-2045/.
[11] Wagner Hermann-Josef, "Introduction to wind energy systems", *EPJ Web of Conferences*, **246** (2020) 00004.
[12] *Making the Most of Offshore Wind - Re-Evaluating the Potential of Offshore Wind in the German North Sea*, https://static.agoraenergiewende.de/fileadmin/Projekte/2019/Offshore_Potentials/176_A-EW_A-VW_Offshore-Potentials_Publication_WEB.pdf.
[13] https://www.google.com/search?client=firefox-b-d&q=entsoe+data+transparency.
[14] Wagner F., "Electricity generation by intermittent sources", *EPJ Web of Conferences*, **98** (2015) 04006.
[15] https://www.dena.de/newsroom/publikationsdetailansicht/pub/studie-roadmap-demand-side-management/.
[16] Linnemann Th. and Vallana G. S., "Wind energy in Germany and Europe. Status, potentials and challenges for baseload application. Part1: Developments in Germany since 2010", *VGB PowerTech*, **97** (2017) 70; "Wind Energy in Germany and Europe: Status, potentials and challenges for baseload application Part 2: European Situation in 2017", *VGB PowerTech*, **3** (2019) 64.
[17] Wanner M., "Transformation of electrical energy into hydrogen and its storage", *Eur. Phys. J. Plus.*, **136** (2021) 593.
[18] https://www.bmwi.de/Redaktion/DE/Pressemitteilungen/2021/11/20211116-bundeswirtschaftsministerium-legt-ausfuehrliche-analyse-des-stromverbrauchs-2030-vor.html.
[19] https://www.vci.de/langfassungen/langfassungen-pdf/vci-study-greenhouse-gas-neutrality-in-the-german-chemical-industry.pdf.
[20] 2020: 16 TWh, https://www.umweltbundesamt.de/themen/klima-energie/erneuerbareenergien/umgebungswaerme-waermepumpen#undefined.
[21] 2020: about 8 TWh, 2017: https://blog.paradigma.de/solarthermie-in-deutschland-aktuelle-zahlen-und-fakten-stand-2017/; extrapolation to 2020: https://de.statista.com/statistik/daten/studie/20000/umfrage/anzahl-der-solarwaerme-anlagen-in-deutschland-seit-1999/.
[22] https://www.thewindpower.net/.
[23] Suiso Frontier, Japan.
[24] https://edgar.jrc.ec.europa.eu/.
[25] https://www.umweltbundesamt.de/daten/klima/treibhausgas-emissionen-in-deutschland#emissionsentwicklung.
[26] https://de.statista.com/statistik/daten/studie/2275/umfrage/hoehe-der-co2-emissionen-in-deutschland-seit-1990/.
[27] https://www.euractiv.com/wp-content/uploads/sites/2/2022/01/draft-CDA-31-12-2021.pdf.
[28] Wagner Friedrich, "CO 2 emissions of nuclear power and renewable energies: a statistical analysis of European and global data", *Eur. Phys. J. Plus*, **136** (2021) 562.
[29] Muraoka K., "Energy situations in Japan before and after the Fukushima nuclear accident", *EPJ Web of Conferences*, **98** (2015) 05003.

[30] https://www.nirs.org/wp-content/uploads/fukushima/naiic_report.pdf.
[31] Abderrahim H. A., "Realization of a new large research infrastructure in Belgium: MYRRHA contribution for closing the nuclear fuel cycle making nuclear energy sustainable", *EPJ Web of Conferences*, **246** (2020) 00012.
[32] Caciuffo R., Fazio C. and Guet C., "Generation-IV nuclear reactor systems", *EPJ Web of Conferences*, **246** (2020) 00011.
[33] https://hal-edf.archives-ouvertes.fr/hal-01977209/document.
[34] https://www.heidelbergcement.com/de/pi-02-06-2021.
[35] https://www.entsoe.eu/news/2021/01/15/system-separation-in-the-continental-europe-synchronous-area-on-8-january-2021-update/.
[36] https://www.umweltbundesamt.de/dokument/datenbank-kraftwerke-in-deutschland.
[37] https://www.bdew.de/service/daten-und-grafiken/beheizungsstruktur-wohnungsbestand-deutschland/.
[38] https://edgar.jrc.ec.europa.eu/overview.php?v=booklet2019.
[39] https://ourworldindata.org/greenhouse-gas-emissions.
[40] https://www.downtoearth.org.in/climate-change/coplist.

AUTHOR INDEX